Innovation in Wind Turbine Design

Innovation in Wind Turbine Design

Second Edition

Peter Jamieson
Strathclyde University, UK

Registered Offices
John Wiley & Sons, Inc., 111 River Street, Hoboken, NJ 07030, USA
John Wiley & Sons Ltd, The Atrium, Southern Gate, Chichester, West Sussex, PO19 8SQ, UK

Editorial Office
The Atrium, Southern Gate, Chichester, West Sussex, PO19 8SQ, UK

For details of our global editorial offices, customer services, and more information about Wiley products visit us at www.wiley.com.

Wiley also publishes its books in a variety of electronic formats and by print-on-demand. Some content that appears in standard print versions of this book may not be available in other formats.

Library of Congress Cataloging-in-Publication Data applied for

Hardback ISBN- 9781119137900

Cover design by Wiley
Cover image: © Mischa Keijser/Gettyimages

Set in 10/12pt WarnockPro by SPi Global, Chennai, India

To Adele and Rose

Contents

Foreword

Those of us who have been active in the wind energy industry for the past few decades have been lucky. We have been involved in an industry that is technically fascinating, commercially exciting and thoroughly worthwhile. We have seen turbines increase in diameter from 10 to 120 m and in power from 10 to 10 000 kW – what a fantastic journey!

The size of the turbines is the most obvious characteristic because it can be so clearly seen – wind turbines are now by far the biggest rotating machines in the world. Less visible is the ingenuity of the designs. Looking back a couple of decades, there were many 'whacky' ideas that were seriously contemplated and even offered commercially and some of those whacky ideas have become conventional. Superficially, the latest generation of turbines may all look the same, but underneath the nacelle and inside the blades there are many fascinating differences. For a long time, the mantra of the wind turbine industry has been 'bigger and bigger', but now it has moved to 'better and better' and this change marks a change in the areas of innovation.

Peter Jamieson is one of the clearest thinkers in the industry and I am delighted and honoured to have worked with him for almost 20 years. He is a real blue sky thinker unimpeded by convention and driven by a strong sense of rigour. Innovation in wind turbine design is what Peter has been doing for the past 30 years and it is about time he wrote a book about it. I fully supported Peter's idea that he should put his professional thoughts on record and now he has done so.

Anyone interested in the technical aspects of both the past developments and the exciting future of wind turbines should read this book carefully and be inspired. This is no arid technical text or history – this is real intellectual capital and, of course, innovation.

Andrew Garrad

Preface

This book is about innovation in wind turbine design – more specifically about the evaluation of innovation – assessing whether a new concept or system will lead to improved design-enhancing performance or reducing cost. In the course of a working life in wind energy that began in 1980, the author's work has increasingly been, at the request of commercial clients and sometimes public authorities, to evaluate innovative systems providing reports which may or may not encourage further investment or development. In some cases, the clients are private inventors with a cherished idea. Other cases include small companies strategically developing innovative technologies, major industrial companies looking for an entry to the wind turbine market or major established wind turbine companies looking to their next-generation technology.

There is substantial conservatism in the wind industry as in most others and largely for the best possible reasons. Products need to be thoroughly proved and sound, whereas change is generally risky and expensive even when there is significant promise of future benefit. To some extent, change has been enforced by the demands year by year for larger wind turbines and components. There has been convergence in the preferred mainstream design routes but, as new players and new nations enter the wind business, there is also a proliferation of wind technology ideas and demand for new designs. The expansion of wind energy worldwide has such impetus that this book could be filled with nothing but a catalogue of different innovative designs and components.

It may initially seem strange that as much of the book is devoted to technology background as to discussion of specific innovative concepts. However, innovation is not a matter of generating whacky concepts as an entertainment for bored engineers. The core justification for innovation is that it improves technology, solving problems rather than creating them. To achieve that, it is crucial that the underlying requirements of the technology are well understood and that innovation is directed in areas where it will produce most reward. Hence is the emphasis on general technology background. Within that background some long-established theory is revisited (actuator disc and blade element momentum) but with some new equations developed.

Among much else, this second edition contains predictable updates regarding new larger turbines and new systems plus expanded sections on the ever-growing offshore applications and the developing interest in airborne wind power. There is also new content relating to presentation of basic theory, a fuller evaluation of many issues concerning ducted rotors, various new top-level analyses of the low-induction rotor concept, flow relativity (relating to driving rotors through still air as a means of performance measurement) and kite performance, for example.

Innovative ideas by definition break the mould. They often require new analytical tools or new developments of existing ones and, in general, fresh thinking. They do not lend themselves to a systematised, routine approach in evaluation. Evaluating innovation is an active process like design itself, always in evolution with no final methodology. On the other hand, there are basic principles and some degree of structure can be introduced to the evaluation process.

In tackling these issues, a gap was apparent – between broad concepts and detailed design. This is territory where brainstorming and then parametric analyses are needed, when pure judgement is too limited but when heavyweight calculation is time consuming, expensive and cannot be focused on with any certainty in the right direction. This why 'detailed design' is not much addressed. It is the subject of another book. This one concerns building bridges and developing tools to evaluate innovative concepts to the point where investment in detailed design can be justified. Innovation in wind energy expresses the idealism of the designer to further a sustainable technology that is kind to the planet.

Peter Jamieson

Acknowledgement

My professional life in wind technology began in 1980 in the employment of James Howden and Company of Glasgow and I very much appreciate many colleagues who shared these early days of discovery. Howden regrettably withdrew from turbine manufacture in 1988, but by then my addiction to wind was beyond remedy.

In those days I much admired a growing wind energy consultancy, Garrad Hassan and Partners. I was delighted to join them in 1991 and, as it happened, founded their Scottish office. I felt that it would be great to have a working environment among such talented people and that I would have a continuing challenge to be worthy of them.

In particular, I would very much like to thank Andrew Garrad and Dave Quarton for encouragement, practical support and great tolerance over 4 years in the preparation of the first edition. At the end of 2013 I retired from Garrad Hassan, by then part of DNV GL. Commencing in 2009, I was employed part time in the Centre for Doctoral Training in Wind Energy in the University of Strathclyde and enjoy working with great teams of staff and students who, now numbering over 40, are studying wind and marine topics at PhD level. I am much indebted to Bill Leithead, director of the centre, especially for many valuable brainstorming sessions on wind technology over the years.

I have to say special thanks to the late Woody Stoddard, who was an inspiring friend and enormously supportive, especially considering the few times we met.

Considering the very many times I have imposed on his good nature, I have equally to thank Mike Graham for his freely given help in so many projects and as an excellent, unofficial aerodynamics tutor. Much thanks also to Henrik Stiesdal, who, as an extremely busy man at the technical helm of a large wind turbine manufacturing company, found time to contribute a chapter to this book.

My warm thanks also go to very many other work colleagues and associates who, knowingly or otherwise, have made valuable contributions to this book. Among them are:

Albert Su, Alena Bach, Alexander Ovchinnikov, Andrew Latham, Anne Telfer, Ben Hendriks, Bob Thresher, Carlos Simao Ferreira, Chai Toren, Charles Gamble, Chris Hornzee-Jones, Chris Kirby, Christine Sams, David Banks, David Milborrow, David Sharpe, Ed Spooner, Emil Moroz, Ervin Bossanyi, Fabio Spinato, Fatma Murray, Iain Dinwoodie, Jan Rens, Geir Moe, Georg Böhmeke, Gerard van Bussel, Herman Snel, Irina Dyukova, Jamie Taylor, Jega Jegatheeson, Jim Platts, John Armstrong, Kamila Nieradzinska, Kerri Hart, Leong Teh, Lindert Blonk, Lois Connell, Lutz Witthohn, Magnus Kristbergsson, Marcia Heronemus, Mark Hancock, Martin Hansen, Masaaki Shibata, Mauro Villanueva-Monzón, Mike Anderson, Mike Smith, Nathalie Rousseau,

Nick Jenkins, Nils Gislason, Patrick Rainey, Paul Gardner, Paul Gipe, Paul Newton, Paul Veers, Peter Dalhoff, Peter Musgrove, Peter Stuart, Rob Rawlinson-Smith, Roger Haines, Roland Schmehl, Roland Stoer, Ross Walker, Ross Wilson, Ruud van Rooij, Sandy Butterfield, Seamus Garvey, Stephen Salter, Steve Gilkes, Stuart Calverley, Tim Camp, Takis Chaviaropoulos, Theo Holtom, Tomas Blodau, Trevor Nash, Uli Goeltenbott, Unsal Hassan, Uwe Paulsen, Varan Sureshan, Vidar Holmöy, Win Rampen, Wouter Haans, Yuji Ohya.

Peter Jamieson

Introduction

0.1 Why Innovation?

Fuel crises, concerns about global environmental threats and the urgent needs for energy in expanding new economies of the former third world have all contributed to an ever-increasing growth of renewable energy technologies. Presently, wind energy is the most mature and cost-effective of these.

While other more diverse applications are discussed, this book keeps the main focus on wind energy converters that produce electricity. This is primarily because the greater part of the author's experience is with such systems. However, in a more objective defence of that stance, it may be observed that by far the largest impact of wind technology on the world's energy supply presently comes from systems generating into electrical networks.

Innovation is about new ideas, and some quite unusual designs are evaluated in this book. Why give attention to such designs which may not be in the mainstream? Exploring alternative concepts not only deepens understanding of why the mainstream options are preferred but also suggests where they should be challenged by alternatives that have significant promise. In any case, ideas are grist to the mill of technological progress and those which fail in one embodiment may well later be adapted and successfully reincarnated.

As is discussed shortly, the generation of power from the wind presents unique challenges. Unlike cars and houses, for example, energy is a commodity which has utilitarian value only. No one prefers a particular petrol because it has a nicer colour. The wealthy may indulge in gold or gold-plated bathroom taps, but no one can purchase gold-plated electricity. Energy must meet generally stringent specifications of quality in order to be useful (voltage and frequency levels particularly in the case of electricity). Once it does, the main requirement is that it is dependably available and as cheap as possible.

The end purpose of innovation in wind turbine design is to improve the technology. Usually, this means reducing the cost of energy and this is the general basis of evaluating innovation in this book. However, even this simply stated goal is not always the final criterion. In some instances, for example, the objective is to maximise energy return from an available area of land. Sometimes capital cost has a predominant influence. The bottom line is that any technology must be tailored accurately to an engineering design specification that may include environmental, market, cost and performance issues.

The detailed design of a wind turbine system is not a minor or inexpensive task. By the time an innovative design is the subject of a detailed design study, although it may yet

Innovation in Wind Turbine Design, Second Edition. Peter Jamieson.
© 2018 John Wiley & Sons Ltd. Published 2018 by John Wiley & Sons Ltd.

be some way from market, it has already received significant investment and has passed preliminary tests as to the potential worth of the new concepts.

Thus, there is an intermediate stage between first exposure of a concept up to the stage of securing investment in a prototype when the concepts are examined and various levels of design are undertaken. Usually, a search for fatal flaws or obvious major shortcomings is the first stage. The design may be feasible but will have much more engineering content than its competitors and it is therefore unlikely to be cost-effective. More typically, there is no clear initial basis for rejecting the new concepts and a second level of appraisal is required. A systematic method is needed to review qualitatively, and where possible quantitatively, how the design compares to existing technology and for what reason(s) it may have merit. At this stage, detailed, expensive and time-consuming analyses are precluded, but there is a great need for parametric evaluations and simplified analyses that can shed light on the potential of the new concepts.

This book is very much about these preliminary evaluation stages, how simple insight-ful methods can provide guidance at a point where the value of the innovation is too uncertain to justify immediate substantial investment or detailed design.

0.2   The Challenge of Wind

According to Murray [1], the earliest written reference to windmills is of the fifth century BC. Windmills (although probably only then existing as children's toys) are listed, among other things, as something a devout Buddhist would have nothing to do with! The aerodynamic rotor concept is evidently ancient.

To generate electricity (by no means the only use for a wind turbine but certainly a major one under present consideration), requires the connection of such a rotor to an electric generator. Electric motor/generator technology began in Faraday's discoveries in the mid-nineteenth century. About 70 years ago and preceding the modern wind industry, the average household in the United States contained about 40 electric motors. The electric motor/generator is therefore not ancient but has been in mass production for a long time in recent history. What then is difficult about the marriage of rotor and generator into successful and economic power generating systems? The challenge of modern wind technology lies in two areas, the specification of an electricity-generating wind turbine and the variability of the wind.

0.3   The Specification of a Modern Wind Turbine

Traditional 'Dutch' windmills (Figure 0.1) have proliferated to the extent of 100 000 over Europe in their heyday. Some have survived 400–600 years, the oldest still operating in the United Kingdom being the post mill at Outwood, Surrey built in 1628. A short account of the history of early traditional wind technology in Eggleston and Stoddard [2] shows that they exhibit considerable practical engineering skill and empirical aerodynamic knowledge in their design and interesting innovations such as variable solidity blades (spilling the air through slats that can open or close) that have not surfaced in modern wind turbine design. However, these machines were always attended, were controlled manually for the most part, were integrated parts of the community and

Figure 0.1 Jill post mill at Clayton Sussex. Reproduced with permission of Paul Barber.

were designed for frequent replacement of certain components, and efficiency was of little consequence.

In contrast, to generate electricity cost-effectively is the specification of a modern power-generating wind turbine. To meet economic targets, it is unthinkable for the wind turbine to be permanently attended, and unacceptable for it to be much maintained. Yet, each unit is a self-contained mini-power station, requiring to output electricity of standard frequency and voltage into a grid system. Cost-effectiveness is overriding, but the efficiency of individual units cannot be sacrificed lightly. Energy is a prime value; whereas the lifetime costs comprise many components, each one of which has a lesser impact on cost of energy. Also, the total land area requirements per unit output will increase as efficiency drops.

It should be clear that wind technology embraces what is loosely called 'high-tech' and 'low-tech' engineering. The microprocessor plays a vital role in achieving self-monitoring unmanned installations. There is in fact nothing particularly simple about any kind of system for generating quality electricity. Diesel generators are familiar but not simple, and have a long history of development.

Thus, it is by no means enough to build something 'simple and rugged' that will survive any storm. Instead, the wind turbine must be value engineered very carefully to generate cheap electricity with adequate reliability. This is the first reason why the technology is challenging.

0.4   The Variability of the Wind

The greatest gust on record was on 12 April 1934 at the peak of Mount Washington in the Northern Appalachians [3]. 'On record' is a revealing phrase as anemometers have usually failed in the most extreme conditions. At 103 m/s, a person exposing 0.5 m^2 of frontal area would have experienced a force equivalent to about 1/3 of a tonne weight. In terms of annual mean wind speed, the windiest place in the world [3] is on the edge of Antarctica, on a mountain margin of East Adelie land. At 18 m/s annual mean wind speed, the available wind energy is about 200 times that of a typical European wind site. These are of course extreme examples and there are no plans to erect wind turbines on either site.

Nevertheless, it underlines that there is enormous variation in wind conditions. This applies both on a worldwide basis but also in very local terms. In the rolling hills of the Altamont Pass area of California, where many wind farms were sited in the 1980s, there are large differences in wind resource (100%, say, in energy terms) between locations no more than a few hundred metres apart. Wind turbines are situated right at the bottom of the earth's boundary layer. Their aerofoils generally travel much more slowly than aircraft or helicopter rotors, and the effect of wind turbulence is much more consequential for design. The crux of this is that it is hard to refine a design for such potentially variable conditions, and yet uneconomic to design a wind turbine fit to survive anywhere. Standardisation is much desired to cheapen production, but is in conflict with best economics at specific local sites. Designs often need to have adaptive features to accommodate larger rotors, uprated generators or additional structural reinforcement as necessary.

Anemometry studies, both to determine suitable sites and for the micro-siting of machines within a chosen area, are not academic exercises. Because of the sensitivity of wind energy to wind speed and wind speed to short- and longer term climatic patterns, the developer who is casual about wind resource estimates is playing a game of roulette on profit margins. Thus, the variability of the wind is the second major reason wind turbine design is challenging.

0.5   Early Electricity-Generating Wind Turbines

Rather presciently, from a twenty-first century viewpoint, Sir William Thomson (later to become Lord Kelvin) suggested in his address to the British Association meeting in York in 1881 that, as fossil fuel resources were consumed and become more expensive, wind power might be used to generate electricity. Professor James Blyth, of Anderson's college, Glasgow (later to become Strathclyde University) was thus inspired to build and test in 1887 the first windmill to be used for electricity generation. Power stored in a battery was used to light up to 10, 8 candlepower, 25 V, incandescent lamps in Blyth's cottage.

Figure 0.2 Blyth windmill commercial prototype. Reproduced with permission of the Andersonian Library, University of Strathclyde.

There is no photographic record of Blyth's first electricity-generating wind turbine of 1887, a horizontal-axis, American-style multi-bladed rotor. A little later in 1891, he developed a vertical-axis wind turbine which provided lighting for his holiday home at Marykirk, a small village in Scotland about 45 km from the city of Dundee. The diameter was about 10 m and the 'blades', 8 semi-cylindrical boxes, as Blyth called them, are each about 1.8 m wide and 1.8 m high. Soon afterwards he had this design engineered more professionally (Figure 0.2) and sold a small number of these, the first 'commercial' electricity generating wind turbines in the world.

Blyth was succeeded by the American, Charles Brush who in 1888 used a large multi-bladed windmill to illuminate his Cleveland mansion. The rotor had 144 blades and a diameter of 17 m and with the tower weighed about 36 tonnes. At full load, the dynamo would then turn at 500 rpm and give an output of 12 kW. The early stand-alone electricity-generating windmills had significant problems with highly variable input, affecting the reliability of the accumulators. The Dane, Poul la Cour built his first windmill at Askov (in Jutland, about 40 km east of Esbjerg) in 1891 with a diameter of 11.6 m and four sails each 2 m wide. In 1891, la Cour invented the 'kratostata' to smooth out the power fluctuations that result from the turbulence in the wind. This was a mechanical device allowing some slip in a belt transmission alleviating rapid changes in load from turbulent wind variations, and it appreciably helped the problems with batteries. However, at the end of the nineteenth century, there was neither sufficient technology development nor a suitable market context for wind-generated electricity to progress further and become cost-effective.

0.6 Commercial Wind Technology

In the twentieth century, wind technology headed towards mainstream power generation beyond the water pumping and milling applications that had been exploited for several thousand years. The Gedser wind turbine is often credited as the seminal design of the modern wind industry. With assistance from Marshall Plan post war funding, a 200 kW, 24 m diameter, three-bladed wind turbine was installed during 1956–1957 on the island of Gedser in the south east of Denmark. This machine operated from 1958 to 1967 with about 20% capacity factor.[1]

In the early 1960s, Professor Ulrich Hütter developed high tip speed designs, which had a significant influence on wind turbine research in Germany and the United States.

In the early 1980s, many issues of rotor blade technology were investigated. Steel rotors were tried but rejected as too heavy, and aluminium as too uncertain in the context of fatigue endurance. Wood was a logical natural material designed by evolution for high fatigue bending strength-to-weight ratio. The problem of moisture stabilisation of wood was resolved in the wood-epoxy system developed by Gougeon Brothers in the United States. This system has since been employed in a number of small and large wind turbines (e.g. the former NEG-Micon NM82). Wood-epoxy blade technology was much further developed in the United Kingdom, latterly by Taywood Aerolaminates who were assimilated by NEG-Micon and then in turn by Vestas. The blade manufacturing industry was, however, dominated by fibreglass polyester construction which evolved from a boat building background, became thoroughly consolidated in Denmark in the 1980s and has since evolved into more sophisticated glass composite technologies using higher quality fibres (sometimes with carbon reinforcement), and more advanced manufacturing methods such as resin infusion.

During the 1980s, some megawatt-scale prototypes had appeared and this history is well documented by Spera [4] and Hau [5]. In general, these wind turbines had short lives and, in some cases, fatal flaws in design or manufacture. Valuable research information was gained; yet, in many respects, these designs followed technology routes rather disconnected from the emerging commercial wind turbine market. In contrast to this, in Denmark during the 1970s and 1980s, a gradual development of wind technology had occurred. This was a result of public pressures to develop renewables and to avoid nuclear energy combined with a lack of indigenous conventional energy sources. Wind turbine design development proceeded with incremental improvement of designs which

---

1 A historical review of wind technology (also written by the author) similar to the text up to this point may be found in the EWEA publication, *Wind Energy the Facts.*

were being maintained in commercial use and with gradual increase in scale into ratings of a few 100 kW. And, a much more successful technology resulted.

Just as the first-generation commercial Danish designs were emerging in the early 1980s, a combination of state and federal, energy and investment tax credits had stimulated a rapidly expanding market for wind in California. Over the period 1980–1995, about 1700 MW of wind capacity was installed, more than half after 1985 when the tax credits had reduced to about 15%.

Tax credits fuelled the indiscriminate overpopulation of various areas of California (San Gorgonio, Tehachapi and Altamont Pass) with wind turbines many of which were ill-designed and functioned poorly. It was the birthplace and graveyard of much more or less casual innovation. This created a poor image for the wind industry, which took time to remedy. However, the tax credits created a major export market for European, especially Danish, wind turbine manufacturers who had relatively cost-effective, tried and tested hardware available. The technically successful operation of the later, better designed wind turbines in California did much to establish the foundation on which the modern wind industry has been built.[2]

0.7 Basis of Wind Technology Evaluation

A summary of some of the key issues in the evaluation of new wind technology is presented here. These topics are addressed in more detail in Chapters 9 and 10.

0.7.1 Standard Design as Baseline

The most straightforward way to evaluate new technology is to set it alongside existing state-of-the-art technology and conduct a side-by-side comparison. This is particularly effective in the case of isolated components which are innovative and different in themselves from the standard solution but have little direct impact on the rest of the system. It is then reasonable to assume that all other components in the system have the same costs as in the standard design and conduct a cost of energy analysis in which only the new component is differentiated. Other innovations may be much more challenging. For example, a new rotor concept can have wide-ranging implications for system loads. In that case, one approach is to tailor certain key loads to be within the same level as the standard design and therefore to have no impact on the components designed by those loads. Another more challenging route is to develop analyses where the impact of loads on component cost is considered. The development of a baseline standard, state-of-the-art design will be seen as a key element in most of the evaluations of innovative technology.

0.7.2 Basis of Technological Advantage

If an innovative system is feasible in principle, the next obvious question is why is it better than anything that precedes it? Does it offer performance gains or cost reduction,

---

2 Capacity factor is the ratio of energy output produced over a period to that which would be produced if the system operated always at its full nameplate rating over the same period. Providing the wind turbine is reliable, capacity factor in the context of wind energy is primarily a measure of how good the site wind conditions may be.

does it enhance reliability? In the first instance, it is not a matter of assessing the level of merit or the realism of the claim so much as confirming that there is a core reason being offered why the system may have merit.

0.7.3   Security of Claimed Power Performance

Especially with radically new system designs, there may be a question mark over the likely level of performance. The evaluation of this is clearly critical for the system economics and a number of evaluations go no further than consideration of whether the proposed system has a sufficiently good power performance coefficient. This is particularly the case in systems that sacrifice performance for simplicity. Sometimes the illusion of a very simple and cheap system will persuade an uncritical inventor that an idea is very promising when, in fact, the system in its essential concepts sacrifices so much energy that the considerable capital cost savings that it may achieve are not enough to justify the concept and the cost of energy is high.

0.7.4   Impact of Proposed Innovation

Where can successful innovation make the greatest impact?

This is addressed by looking at the relative costs of components in a wind turbine system and any impacts they have on system productivity through efficiency or reliability. Innovation is disruptive and needs to offer sufficient benefit to be worth the trouble. The capital cost of a yaw system of a large horizontal axis wind turbine is typically around 3% or 4% of total wind turbine capital cost. About half to two-thirds of this cost is in the yaw bearing. This major component is generally not dispensable and so it is clear that no yaw system solution, however innovative, can make large capital cost savings in relation to wind turbine capital cost. On the other hand, if a new yaw system has improved reliability, its total value in terms of impact in cost of energy is enhanced.

0.8   Competitive Status of Wind Technology

After many years of battling to reduce costs, assailed by critics about the extent to which wind was subsidised, as if other energy supplies had not been, a breakthrough picture is emerging. The following article [6] by Paul Gardner, Global Segment Leader, Energy Storage at DNV GL, summarises the situation very well:

> In November, the UK Government published an updated version of its Electricity Generation Costs analysis. This is a rigorous assessment of Levelised Cost of Energy (LCOE) for a very wide range of generation options … a major benefit of this kind of study is that it compares competing technologies on the same basis, or at least on mutually consistent assumptions.
>
> …—the report has a section specifically highlighting… enormous reductions from the costs forecast in the previous issue (2013), for large-scale PV, onshore wind, and offshore wind.
>
> For projects commissioning in 2020, the cheapest options available at significant scale all have similar costs: H-class combined cycle gas turbines (CCGTs) at 78

EUR/MWh, onshore wind (projects larger than 5 MW) at 74 EUR/MWh, and large-scale ground-mounted solar at 79 EUR/MWh. By 2025, wind and PV are the clear winners at 72 and 74 EUR/MWh; CCGT costs increase to 96 EUR/MWh due to higher assumed gas and carbon costs.

A good result for renewables. In fact, by 2025 even large hydro, small building-mounted PV, and near-shore wind will be competitive with large CCGTs. However, on closer inspection of the figures there's a more important and perhaps surprising conclusion. A common criticism of costings of the variable renewables (wind, PV, and others) is that they don't include the costs for 'backup' generation to cover demand when needed. In the UK the worst case is an extended period of anticyclonic weather in winter, resulting in days or weeks of very low winds, low temperatures, and high electricity demand. This criticism is justified, though the assumption that every wind or PV project should be 'charged' the capital cost of fossil generation of the same capacity is overly simplistic. However, the UK figures show that even with this overly simplistic assumption, wind and PV still win.

How? Well, the cost forecasts include fixed and variable costs. For CCGTs in 2025, the fuel, carbon, and variable O&M costs alone total 86 EUR/MWh. This is significantly more expensive than the total costs for wind and PV. So, wind and PV projects could indeed afford to pay for the costs of CCGTs, to be treated as 'firm', and would still be the cheapest generation option available at scale. Or in other words, a CCGT operating in 2025 as 'baseload' will find it cheaper to buy the output of wind or PV projects, whenever available, in preference to buying and burning gas.

This marks the next stage in cost-competitiveness of renewables. First there is 'retail parity', where behind-the-meter wind or PV beats the retail price of electricity to residential, commercial or industrial consumers. Then there is 'whole-sale parity', where renewable costs compete on wholesale or spot markets. And now on the horizon we can see 'fuel parity', where renewables become cheaper than just the fuel (and carbon) costs of fossil generation.

In fact, the 'horizon' is not that far off: interpolating the UK figures for 2020 and 2025 shows that fuel parity is forecast for 2023. That is only 6 years from now. Companies currently developing potential new CCGT projects will no doubt be factoring this into their calculations.

References

1 Murray, H.J.R.A. (1913) *A History of Chess*, Oxford University Press, London; (1985) Benjamin Press, Northampton, MA. ISBN: 0-936317-01-9.
2 Eggleston, D.M. and Stoddard, F.S. (1987) *Wind Turbine Engineering Design*, Kluwer Academic Publishers. ISBN: 13: 9780442221959.
3 Watson, L. (1984) *Heavens Breath: A Natural History of the Wind*, Hodder General Publishing Division. ISBN: 0340430982 (0-340-43098-2).
4 Spera, D.A. (ed.) (2009) *Wind Turbine Technology: Fundamental Concepts in Wind Turbine Engineering*, 2nd edn, ASME Press, New York.

5 Hau, E. (2006) *Wind Turbines: Fundamentals, Technologies, Application, Economics*, Springer-Verlag. ISBN: 13 978-3-540-24240-6

6 Gardner, P. https://www.linkedin.com/pulse/renewables-track-beat-fuel-parity-combined-cycle-gas-turbines-paul?trk=v-feed&lipi=urn%3Ali%3Apage%3Ad_flagship3_feed%3BsXZmYFRfaHIrsHuo%2FxkBiQ%3D%3D (viewed February 2017).

Part I

Design Background

1

Rotor Aerodynamic Theory

1.1 Introduction

Theoretical background in energy extraction generalities and, more specifically, rotor aerodynamics of horizontal axis wind turbines (HAWTs) is developed in this chapter. Some prior knowledge of fluid dynamics in general and as applied to the analysis of wind turbine systems is assumed, in particular basic expressions for energy in a fluid flow, Bernoulli's equation, definitions of lift and drag, some appreciation of stall as an aerodynamic phenomenon and blade element momentum (BEM) theory in its conventional form as applied to HAWTs. Nevertheless, some of this basic knowledge is also reviewed, more or less from first principles. The aim is to express particular insights that will assist the further discussion of issues in optimisation of rotor design and also aid evaluation of various types of innovative systems, for example, those that exploit flow concentration.

Why focus much at all on theory in a book about innovative technology? Theory is often buried in more or less opaque computer code, which may generate loads of information that engineers can use in design. However, as is amplified in the following chapters, theory is in itself:

- Food for innovation and suggestive of methods of performance enhancement or alternative concepts;
- A basis for understanding what is possible and providing an overview appraisal of innovative concepts;
- A source of analytic relationships that can guide early design at a stage where many key parameters remain to be determined and there are too many options to subject each to detailed evaluation.

Prior to discussions of actuator disc theory and the BEM theory that has underpinned most practical engineering calculations for rotor aerodynamic design and determination of wind turbine loads, some discussion of aerodynamic lift is presented. This is intended particularly to highlight a few specific insights which can guide design and evaluation of wind energy systems. In general, a much more detailed understanding of basic aerodynamics is required in wind turbine design. This must cover a wide range of topics, 2D and 3D flow effects in relation to aerofoil performance, stall behaviour, aeroelastic behaviour, unsteady effects including stall hysteresis and induction lag, determination of suitable aerofoil data for wide ranges in angle of attack, and so on. References [1–10] are a sample from extensive published work covering some of these issues.

Innovation in Wind Turbine Design, Second Edition. Peter Jamieson.
© 2018 John Wiley & Sons Ltd. Published 2018 by John Wiley & Sons Ltd.

1.2 Aerodynamic Lift

The earliest wind turbines tended to use the more obvious drag forces [11] experienced by anyone exposed to wind on a windy day, and use of the potentially more powerful lift forces was almost accidental. Exploitation of the aerodynamic lift force is at the heart of efficient modern wind turbines, but surprisingly the explanation of lift has been quite contentious. Before entering that territory, consider first Bernoulli's equation which is derived in many standard sources on fluid mechanics. Ignoring gravitational, thermal and other energy sources and considering only pressure and kinetic energy, this equation becomes: $p + 1/2\rho U^2 = p_0$, where p is static pressure in a fluid element moving with a velocity of magnitude U, ρ is fluid density and p_0 is the total pressure which, in the absence of energy extraction, is constant along any streamline in the flow field.

Bernoulli's equation is essentially an energy equation expressed dimensionally in units of pressure and can be viewed as conservation of energy per unit volume of the fluid. In that connection, pressure can be regarded as the source potential energy (per unit volume) that drives fluid flow. This interpretation is discussed subsequently and is seen to be crucial to a clear understanding of how a wind turbine rotor works.

Returning to the issue of aerodynamic lift, one view of the explanation of the lift force has been that the fluid, should it have a longer path to traverse on one side of an aerofoil, will travel faster in order to meet the fluid flowing past the other side at the trailing edge of the aerofoil. With increase in velocity, the associated static pressure in that region will reduce in consequence of Bernoulli's equation. The pressure deficit on the side of the plate with the longer flow path is then considered the source of the lift force.

There are various problems with this as an explanation of the lift force. Firstly, a thin plate set at an angle in a uniform flow field will generate significant lift when, considering its shape, there is negligible difference between the upper surface and lower surface paths. Secondly, if an aerofoil with a shape with a noticeably longer flow path on one side is considered and the assumption that the flow on each side will traverse the length of the aerofoil in equal times (something that in itself can be challenged) is made, the difference in static pressure calculated on the basis of the implied velocities on each side of the aerofoil will be found quite insufficient to account for the observed lift force.

An apparently authenticated story relates to the efforts of the famous physicist Albert Einstein in aerofoil development. Einstein's effort, inspired by the path-length-related concept, was a miserable failure[1] and he later commented '*That is what can happen to a man who thinks a lot but reads little.*'

---

1 According to Carl Seelig (*Albert Einstein: A Documentary Biography* by Carl Seelig, 1960, pp. 251–252; Translated by Mervyn Savill, London: Staples Press, Bib ID 2263034), an accredited biographer of Einstein: 'It is not well known that … Einstein … undertook a new aerofoil design intended for serial production. Eberhard, the chief test pilot, treated the fruit of the famous theoretician's efforts with suspicion.' 'Ehrhardt's letter continues (EA 59–556, as quoted in Folly 1955): A few weeks later, the "cat's back aerofoil" had been fitted to the normal fuselage of a LVG biplane, and I was confronted with the task of testing it in flight. … I … expressed the fear that the machine would react to the lack of angle of incidence in the wing by dropping its tail and would thus presumably be obliged to take off in an extremely unstable attitude. Unfortunately the sceptic in one proved to be right, for I hung in the air like a "pregnant duck" after take-off and could only rejoice when, after flying painfully down the airfield, I felt solid ground under my wheels again just short of the airfield at Aldershof. The second pilot had no greater success, not until the cat's back aerofoil was modified to give it an angle of incidence could we venture to fly a turn, but even now the pregnant duck had merely become a lame duck'.

Considering the basic definition of lift as the force created on an object at right angles to the incident flow, it is evident that such a force, like all forces according to Newton's Second Law, will be associated with a rate of change of momentum in that direction. Thus, the magnitude of the lift force will, in principle, be unambiguously determined by integrating all the components of momentum in the flow field normal to the incident flow that result from the object causing deviation of the flow.

Whilst this explanation is pure and fundamental, it does not immediately shed light on why lift forces can be so large.

The explanation relating to Bernoulli's equation has some relevance here. Where flow is accelerated around a curved surface, the reduction in static pressure assists in maintaining attachment of the flow and contributes to large suction forces. As nature proverbially abhors a vacuum, strong suction on the boundary layer near a curved surface will induce a large deviation in the general fluid flow some distance from the surface, thereby giving a large overall change in fluid momentum and producing a strong lift force. Aerofoil design is very much about the extent to which such forces can be sustained as the curvature is increased and more severe changes of flow direction are attempted in order to increase lift.

An associated consequence of the Bernoulli equation is the so-called Coanda effect. Aerofoils with elliptical section were developed and used on the X-wing plane/helicopter design [12]. Such aerofoils will have only moderate lifting capability attributable to their shape alone. However, the discharge of a thin jet of air tangential to the surface near the trailing edge will attract the general flow to the jet and cause a much larger deviation in flow direction and consequently much enhanced lift.

The 'attraction' of the jet to the surface arises as the jet brings increased momentum into the boundary layer where the jet flow is next to the body surface. This overcomes the natural tendency of the (reduced momentum) boundary layer to separate under the adverse (rising) streamwise pressure gradient due to the aerofoil curvature. Due to the large curvatures involved, there is a noticeable pressure change across the jet, which can be calculated from the mass flow rate in the jet and the radius of curvature of the flow. The jet tries to entrain any fluid between itself and the wall (very efficiently because it is normally turbulent) and this entrainment keeps it attached to the wall. Then, because the streamlines are now curved, the wall pressure falls below the external ambient value. In fact, in the absence of external flow incident on the aerofoil, such a jet will almost completely encircle the aerofoil.

This phenomenon is often called the Coanda effect in recognition of Henri-Marie Coanda, who discovered it apparently through rather hazardous personal experience.[2] Controlling lift on an aerofoil section by blowing a jet tangential to the surface is often referred to as circulation control. It is a form of boundary layer control which has been considered for regulation of loads and control or performance enhancement of wind turbine blades [13].

Lift is intimately related to vorticity [14]. Associated with this is the Magnus effect, whereby a rotating cylinder (or sphere) can generate lift. This affects the flight of balls in many sports, has been employed in the form of the Flettner rotor [15] to power ships and

---

2 Henri Coanda was asked to devise a system to divert the hot jet discharges from an aircraft's engines away from the cockpit and fuselage. In blowing air to this end, the jet did exactly the opposite and attracted the hot gases to the fuselage surface with dangerous consequences.

has been exploited in at least two innovative wind turbine designs [16, 17]. Wikipedia [18] is quite informative on lift, vorticity and the Magnus effect and also provides a commentary on some popular incomplete views such as have been discussed. Finally, in the context of wind turbine systems, lift may also be involved in the performance of wind devices that have been casually categorised as 'drag' devices (see Section 13.2).

1.3 Power in the Wind

In ideal modelling of wind flow, it is usual to start with a wind field that is of uniform constant velocity everywhere, introduce an energy extraction system such as a rotor and examine the resultant flow field that is established in steady state. For subsequent clarity, the basic conservation laws of a particle as compared with steady-state flow are recalled in Table 1.1. Generally, in discussion of steady-state flows, it has been common to use the particle-related terminology rather loosely (e.g. talking about momentum theories where it is really momentum rate or force that is being considered or energy balances that are really power).

Referring to an axisymmetric enclosed surface bounding the flow volume that is continually passing through the rotor disc as the bounding streamtube, power flowing through the far upstream area at the source of that streamtube comprises kinetic power and, most importantly, also pressure power. The fundamental expression for power in the air is illustrated in Figure 1.1. In the notation adopted, m, U_0, V, p, A and ρ are respectively mass, air velocity, volume, pressure, area and air density.

The expression for source power in the wind is widely disseminated as $0.5\rho U_0^3 A_0$. However, this is incomplete being only the kinetic power and has led to many misunderstandings,[3] especially in the context of systems that aim to exploit flow augmentation and also in analysis of the rotating wake. Any volume, V, of gas at pressure, p, stores an amount of energy $E = pV$, which becomes a power $P = p\dot{V} = pAU$, if there is steady

Table 1.1 Conservation laws.

Particle	Steady flow process	
Mass	Mass flow rate	
Energy	Energy rate	⟶ Power
Linear momentum	Linear momentum rate	⟶ Force
Angular momentum	Angular momentum rate	⟶ Torque

---

3 The almost universal teaching that the power in the wind is proportional to the cube of velocity, without mention or inclusion of the pressure power, has led to terrible misunderstandings, especially with regard to systems that augment flow. The author is aware of a system (never actually manufactured) that enhanced flow by a factor of ~1.2 (eventually verified by computational fluid dynamic (CFD) modelling) but was marketed via business plans based on the assumption of $(1.2)^3$ power augmentation as opposed to ~1.2. This fundamental error was maintained in financial calculations with ownership of the technology changing hands over a period of ~30 years.

Figure 1.1 Power in the air.

$$\text{Total energy} \equiv \text{kinetic} + \text{potential} = \frac{1}{2}mU_0^2 + pV$$

$$\text{Pressure} \equiv \text{energy/unit volume} = \frac{1}{2}\rho U_0^2 + p$$

$$\text{Force} \equiv \text{pressure} \times \text{area} = \frac{1}{2}\rho U_0^2 A + pA$$

$$\text{Power} \equiv \text{force} \times \text{velocity} = \frac{1}{2}\rho A U_0^3 + pAU_0$$

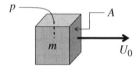

flow at velocity, U, through a surface area, A, of the volume, V. The correct expression for source power must therefore include both kinetic and pressure power and is:

$$\text{Source power in the wind, } P_0 = 0.5\rho U_0^3 A_0 + p_0 A_0 U_0 \tag{1.1}$$

The subscript, 0, denotes values far upstream prior to any energy extraction. The second equation of Figure 1.1 is Bernoulli's equation in the usual form for many wind energy analyses, where only energy associated with fluid pressure and velocity is considered. Clearly, there would be a fundamental inconsistency if the pressure-related term in the first equation of Figure 1.1 or in Equation 1.1 were missing. Relative to vacuum pressure, the pressure power term is huge,[4] but very small differences in it create the atmospheric wind resource and play a critical role in all wind power conversion systems. A quick perspective on the relevant pressure differences may be gained considering an extreme storm gust of $U = 70\,\text{m/s}$ which corresponds to a dynamic pressure of $0.5\rho U^2 \cong 3000\,\text{Pa} \cong 3\%$ of atmospheric pressure, while operation of any size of wind turbine at optimum aerodynamic performance in a wind of $10\,\text{m/s}$ corresponds to a rotor plane pressure difference of $<1\%$ of atmospheric pressure. All forces on a wind turbine rotor and its capability to extract energy arise in consequence of such small pressure differences.

Further it must be emphasised, as appears in the discussion in Section 1.5.4, that kinetic energy is never directly extracted in any physical process. It is always converted to another form such as pressure energy or heat. All aerodynamic machines exploit pressure energy changes to provide the forces that do work.

1.4 The Actuator Disc Concept

The actuator disc is a valuable concept that arose early on in the development of analyses of rotors and propellers. Without any specific knowledge of or assumptions about

---

4 Otto Von Guericke's famous experiment of 1654 demonstrated that two teams of eight horses could not pull apart a large pair of touching copper hemispheres from which most of the air had been evacuated.

the system that may extract energy flowing through an arbitrary area in a uniform flow field, consideration of energy and momentum conservation allow some basic information about the consequent flow and limits on maximum possible energy extraction to be established.

A simple analysis of a rotor (or other energy extraction device) in open flow (Figure 1.2) leads to Froude's theorem[5] and the well-known Betz limit. In addition, actuator disc theory is further presented in a recently developed, more generalised form that will also deal with a rotor in *constrained* flow (Figure 1.3). Constrained flow is defined as the situation in which the following happens:

- An object is introduced into a flow field which modifies at least locally an otherwise uniform flow field of constant velocity.
- No energy is introduced or extracted by that object (conservative system).

An energy extraction device may then be introduced into the constrained flow field, in principle anywhere but most usually in a region of *flow concentration* where there is a higher local velocity and hence higher mass flow through unit area normal to the flow than in the far upstream flow. Typical examples of constrained flow are where there is a hill, a duct or a diffuser.

In the context of evaluating innovation, the point of considering this more general situation is twofold. Understanding the limitations on power performance of wind farms in complex terrain (hills) is a mainstream concern. Although there are no mainstream large-scale commercial wind energy systems that exploit flow concentration systems, nevertheless such systems have long been considered, some developed to prototype stage and others are under development at present. So they continue to receive increasing attention among innovative wind turbine designs.

Figure 1.2 represents a rotor in open flow. The flow field in the absence of the rotor would be of constant velocity everywhere and parallel to the axis of the rotor. Figure 1.3 represents a rotor in a diffuser (toroid with aerofoil cross section, as indicated). This is an example of constrained flow. Even in the absence of the rotor and of any energy extraction, the flow in a region around the diffuser is altered[6] by its presence and is substantially non-uniform.

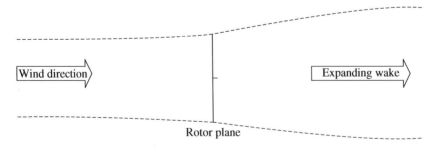

Figure 1.2 Open flow.

5 This is the result (for open flow) that the velocity at the rotor (energy extraction) plane is the average of the far upstream velocity and far wake velocity.

6 It may also be noted that the ground itself, even if completely level, constrains the flow. Although the ground effect extends in all directions to infinity, its constraint effect on the streamlines does exist locally near the wind turbine as if there is a mirror image of the turbine in the ground. It is not normally taken into account, but the effect is quite noticeable, for example, on the wake which because of its swirl lies at a small

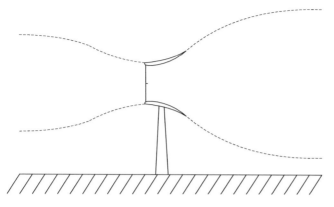

Figure 1.3 Constrained flow example (diffuser).

1.5 Open Flow Actuator Disc

1.5.1 Power Balance

The axial induction at the rotor plane is defined as the fractional reduction in far upstream wind speed local to the rotor. Thus (see Figure 1.4), the velocity through the rotor plane is;

$$U_1 = U_0(1 - a) \tag{1.2}$$

A key assumption in the actuator disc model of a wind turbine system (without wake rotation) that atmospheric pressure is restored in the far wake ($p_2 = p_0$). Conservation of mass in the steady-state process requires that there is the same constant mass flow rate $\rho A_0 U_0 = \rho A_1 U_1 = \rho A_2 U_2$ everywhere within the streamtube bounding the rotor plane. Thus, the pressure power, $p_0 A_0 U_0 = p_2 A_2 U_2$, is unchanged in the overall process, although it necessarily changes across the rotor plane (see Figure 1.6). Considering change in kinetic energy between far upstream and far wake, the power (rate of change

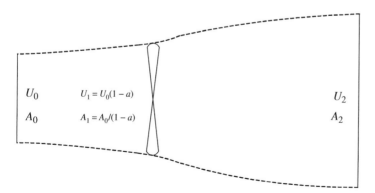

Figure 1.4 Open flow actuator disc model.

angle to the free stream. In contrast, the wind shear effect associated with the ground boundary layer is not due to the normal velocity constraint and extends everywhere independent of the presence of the turbine.

of kinetic energy) extracted P is

$$P = \frac{1}{2}\rho A_0 U_0^3 - \frac{1}{2}\rho A_2 U_2^3 \tag{1.3}$$

Since, $\rho A_0 U_0 = \rho A_2 U_2$:

$$P = \frac{1}{2}\rho A_0 U_0 (U_0^2 - U_2^2) \tag{1.4}$$

1.5.2 Axial Force Balance

The mass flow rate through the rotor plane is $\rho A_1 U_1$. The change in fluid velocity between upstream far wake is $(U_0 - U_2)$. Hence, rotor thrust as rate of change of momentum through the rotor plane is

$$T = \rho A_1 U_1 (U_0 - U_2) \tag{1.5}$$

and power, P is

$$P = TU_1 \tag{1.6}$$
$$= \rho A_0 U_0^2 (1 - a)(U_0 - U_2) \tag{1.7}$$

1.5.3 Froude's Theorem and the Betz Limit

From Equations 1.4 and 1.7

$$U_2 = U_0 (1 - 2a) \tag{1.8}$$

It can be seen that the far wake induction is thus twice the value at the rotor plane. This result was first derived by Froude [19]. Defining power coefficient C_p as the ratio of fraction of power extracted by the rotor to the amount of kinetic power that would pass through the rotor swept area with the rotor absent, then:

$$P = \frac{1}{2}\rho A_1 U_0^3 C_p \tag{1.9}$$

Using Equation 1.8 to substitute for U_2 in Equation 1.7 and also noting that $A_0 = A_1 (1 - a)$

$$C_p = 4a(1 - a)^2 \tag{1.10}$$

Differentiating Equation 1.10 to determine a maximum leads to $a = 1/3$ and to the Betz limit:

$$C_p = 16/27 \tag{1.11}$$

Investigations by Bergey [20] and van Kuik [21] indicated that Lanchester (1915), Betz (1920) and Joukowski (1920) suggested that all, probably independently and certainly by methods differing in detail, have determined the maximum efficiency of an energy extraction device in open flow. Later, Okulov and van Kuik [22] concluded that the attribution to Lanchester by Bergey was inappropriate. Thus, the Betz limit may apparently most properly be called the Betz–Joukowski limit, although for convenience the short reference as Betz limit is retained.

Although open flow actuator disc theory is over a century old, it is by no means done and dusted. The Betz limit and the ideal actuator disc have been the subject of extensive and continuing discussions. In real flow, external flow that does not pass through the rotor plane may assist in transport of the wake and therefore contribute additional energy to the system. However, this goes beyond the ideal actuator disc in inviscid flow, which cannot be expected to reflect behaviour at the tip of a real rotor with discrete blades or address gains and losses associated with viscous flow effects.

Although it is now confirmed both theoretically (in the following analysis, for example) and by numerical analyses based on vortex theory [23, 24], the validity of the Betz limit for ideal inviscid flow through an actuator disc had been questioned. Greet [25], considering a one-dimensional analysis, and Rauh and Seelert [26], considering 3D axisymmetric potential flow, arrived at the same conclusion that Froude's theorem and the Betz limit could not be rigorously be proved. They considered the problem to be a failure to account fully for streamtube forces. There were no analytical errors in their analyses, but they reached an impasse and formed false conclusions on failing to conduct a complete momentum balance considering external as well as internal forces on the streamtube. Energy extraction relates only to the magnitude of the streamtube areas far upstream, at the plane of the disc and far downstream, but the shape, pressure distribution and axial force on the streamtube boundary arises intrinsically from a balance of static pressure between internal and external flows. Thus, total streamtube forces will be indeterminate if an analysis does not consider a control volume as in Figure 1.5 that includes some external flow.

The streamtube is of course a virtual entity like a line or geometric figure. It has specific properties that no fluid and hence energy or momentum flows across its boundaries. However, for any chosen volume within a steady-state flow field, such as the volume, V, in Figure 1.5, it must be possible to demonstrate equilibrium of forces, mass flow, and so on. Otherwise, steady-state flow would not be maintained. The sum of axial streamtube forces is expressed by the integral over the whole surface enclosing the volume of the streamtube.

$$F_x = \oint p\widehat{dA}.\hat{x} = p_0 A_0 - p_2 A_2 + \int_{A_0}^{A_2} p dA \qquad (1.12)$$

In Equation 1.12, dA is an element of area normal to the axial direction with the vector direction of $\hat{x}$, a unit vector in the axial direction. The integral is then split into end forces

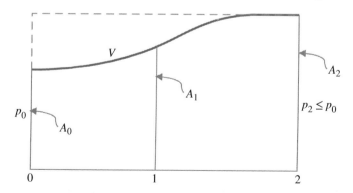

Figure 1.5 Axisymmetric control volumes for an actuator disc at plane 1.

at planes 0 (far upstream) and 2 (far downstream) plus the axial force on the curved surface. In the simple actuator disc model of Betz, the wake expands but is not rotating. The far downstream static pressure is then p_0, although the analysis presented considers the more general case where, if the wake is rotating, suction pressure is required to balance centrifugal force and hence the far downstream pressure, p_2, and indeed pressure anywhere inside the wake, will be less than p_0.

The momentum (axial force balance) equation for the ideal actuator disc, considered within the streamtube that bounds the disc, involves rate of change of fluid momentum, rotor plane thrust, T, and axial forces on the streamtube:

$$0 = \dot{m}(U_0 - U_2) - T + F_x \tag{1.13}$$

The sum of streamtube axial forces, F_x, is zero for the ideal actuator disc without wake rotation when $p_2 = p_0$. Sharpe [27] reasons that it can be deduced immediately observing that (in spite of local pressure variations on the curved surface) the whole streamtube that bounds the disc is ultimately immersed in fluid at atmospheric pressure. An explicit proof that $F_x = 0$ is now presented.

Consider the control volume, V, enclosed by the streamtube curved surfaces and the dotted lines. Mass flow enters axially through the annulus defined by the vertical dotted line and exits obliquely through the cylinder wall defined by the horizontal dotted line. Regarding the streamtube boundary as a 'wall', radial momentum is temporarily imparted to the flow (and later absorbed outside the control volume V when the streamtube curvature reverses). However, there is no energy extraction and hence no change in axial momentum of the external flow. Thus, the force associated with the fluid axial momentum rate at entry to the control volume, V, namely, $\rho(A_2 - A_0)U_0^2$, is exactly balanced by a similar term involving the integral of components of axial velocity over the exit area. The end force at plane 0 on the annulus represented by the vertical dotted line is then clearly $p_0 (A_2 - A_0)$. There cannot be any axial force on the parallel surface. Thus, the curved surface axial force, F_c is given as:

$$F_c = \int_0^2 p dA = p_0 (A_2 - A_0) \tag{1.14}$$

Considering Equation 1.12, this proves that $F_x = 0$ when $p_2 = p_0$ and so reduces Equation 1.13 to a balance of rate of change of fluid momentum with the thrust force as assumed by Betz.

1.5.4 The Power Extraction Process

The power balance for the ideal actuator disc is illustrated in Figure 1.6. A key assumption for the simplest case where the wake is not rotating is that general atmospheric pressure is recovered in the far wake so that $p_2 = p_0$. The actuator disc model requires continuity of the fluid axial velocity through the rotor plane. This is essential as the wind turbine removes energy from the air flow passing through it but does not remove any of the air itself! *Locally* at the rotor plane, with no change in axial velocity, there is no change in the associated kinetic power in the flow and *only pressure power is extracted.* Overall, considering source power far upstream and residual power far downstream, the assumption that the air pressure returns to ambient far downstream determines

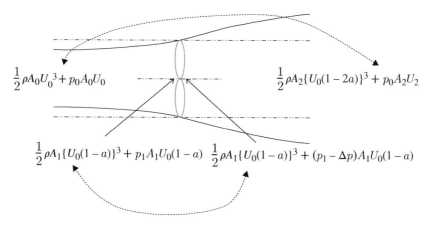

$$\frac{1}{2}\rho A_0 U_0{}^3 + p_0 A_0 U_0 \qquad\qquad\qquad \frac{1}{2}\rho A_2 \{U_0(1-2a)\}^3 + p_0 A_2 U_2$$

$$\frac{1}{2}\rho A_1\{U_0(1-a)\}^3 + p_1 A_1 U_0(1-a) \quad \frac{1}{2}\rho A_1\{U_0(1-a)\}^3 + (p_1 - \Delta p)A_1 U_0(1-a)$$

Figure 1.6 Power balance for ideal actuator disc (no wake rotation).

that there is no net change in pressure power ($p_2 = p_0$ and $A_2 U_2 = A_0 U_0$) and hence the overall process appears as an extraction of source kinetic power.

If locally the wind turbine rotor is not extracting kinetic energy and yet the system is producing power and therefore extracting energy from the fluid, what is the energy source? The answer is *potential energy*, which in this case is pressure energy. This is true of all fluid machines, fans, propellers, wind turbines and even gas turbines where thermal energy changes must be accounted but can only contribute to forces through creating or modifying pressure differences.

A wind turbine rotor produces power from the torque generated by the rotor blades. This torque arises from forces on blade elements, which in turn are the consequence of pressure differences on each side of the aerofoils. The wind turbine works by offering an appropriate resistance to the fluid flow slowing the fluid approaching the rotor. The reduction in fluid velocity occurs conservatively ahead of the rotor plane. Hence, considering Bernoulli's equation, a rise in static pressure occurs to provide conservation of energy per unit volume. The pressure difference across the rotor plane in conjunction with the through flow velocity is then the determinant of the energy extraction and, as was discussed previously, pressure is effectively potential energy per unit volume of fluid.

The basic equation for power at the rotor plane is then;

$$P = \Delta p\, A_1 U_1 \tag{1.15}$$

This defines power in the air flow through the actuator disc. Equation 1.15 is valid under all circumstances (open or augmented flow). Some of this power is not available to a rotary type of energy converter and remains in the air as the rotational kinetic power in the wake. In addition, some of the total power available to the rotor may not be converted usefully due to aerodynamic and drive train loss mechanisms.

1.5.5 Relativity in a Fluid Flow Field

According to basic physical laws established since the time of Galileo, who first described this principle in 1632 in his *'Dialogue Concerning the Two Chief World Systems'*, velocity is relative and there is no preferred inertial frame. Thus, provided there is no significant interaction with a ground boundary layer, driving a rotor at

Case A   Wind flowing through stationary wind turbine system

$U = U_0$ $U = U_0(1 - a)$ $U = U_0(1 - 2a)$

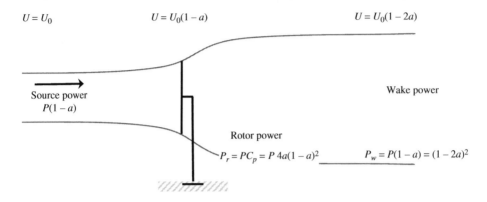

Source power
$P(1 - a)$

Wake power

Rotor power
$P_r = PC_p = P\,4a(1 - a)^2$ $P_w = P(1 - a) = (1 - 2a)^2$

Rotor + wake power $= P(1 - a)\{4a(1 - a) + (1 - 2a)^2\} = P(1 - a) =$ source power

Case B   Wind turbine system travelling(no ambient wind)

$U = 0$ $U = -aU_0$ $U = -2aU_0$

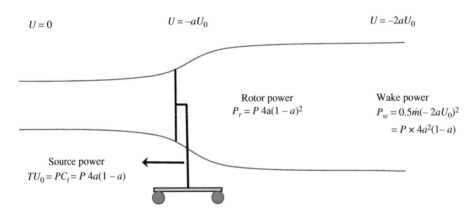

Rotor power
$P_r = P\,4a(1 - a)^2$

Wake power
$P_w = 0.5\dot{m}(-2aU_0)^2$
$= P \times 4a^2(1 - a)$

Source power
$TU_0 = PC_t = P\,4a(1 - a)$

Rotor + wake power $= P\,4a(1 - a)\{(1 - a) + a\} = P\,4a(1 - a) =$ source power

Figure 1.7 Power balance in reference frames of Cases A and B.

constant velocity U_0 through still air relative to a ground reference frame (Case A in Figure 1.7) must, *for the rotor*, be exactly equivalent to the rotor being mounted on a stationary support in a wind field of constant velocity U_0.

This has some practical significance. In a low wind region, it may take a long time to collect adequate data to characterise a power curve and driving a rotor through still air offers a valid, if not ideal, means of measuring rotor performance at least of small rotors. See Chapter 18 where this method is used in testing the Katru system. The analysis of this case shows that in different reference frames, velocities, momentum, energy and power measures change; but naturally energy and momentum are conserved and thus power also in a steady-state flow process. Also, with airborne systems (Chapter 8), it is often only in the reference frame of the moving airborne rotor that steady-state flow can be considered to exist and this may become a preferred frame for analysis.

Let $P = 0.5\rho\pi R^2 U_0^3$, where R is rotor radius. Let $\dot{m}$ be the mass flow rate of air in the streamtube bounding the rotor plane, as measured in the rest frame of the streamtube and wind turbine system.

Case A (Figure 1.7) is the normal situation where the wind turbine system is stationary relative to the ground and the upstream wind speed is U_0. In Case B (Figure 1.7), the upstream air is calm (no wind) and the wind turbine system is driven at speed U_0. If any effects related to ground proximity are ignored, then Galilean (also called Newtonian) relativity demands that the local velocity of air relative to the rotor plane is the same in each case, namely, $U_0(1 - a)$, that the physical process (power extracted by the rotor and shape of streamtube) is the same and that conservation laws are upheld.

There is superficially a paradox in that the work done by the rotor thrust, T, is $TU_0(1 - a)$ in Case A but TU_0 in Case B. This arises because the source power is different in each case. However, the rotor power is the same and energy conservation (power in steady-state flow) is satisfied as in both cases. The sum of rotor power and wake power is equal to source power.

1.6 Why a Rotor?

The actuator disc idea considers an arbitrary energy extraction system which need not be a rotor. Yet all present mainstream wind energy conversion systems rely on the rotor concept. Why? A wind energy system is not only, as is axiomatic, an energy conversion system turning fluid mechanical energy in the wind into electrical energy but is also an energy concentration system.

For example, a typical modern 1.5 MW wind turbine may have parameters as in Table 1.2. In a case when the wind turbine is producing its rated output (1500 kW in the generator output electrical cables) at 11.5 m/s rated wind speed, it has received wind energy over the swept area at a power density ~1 kW/m^2 and is transporting output after losses at a power density of around 1.6 GW/m^2. As the power passes through the system, it is concentrated first in the composite of the blades, then in the steel of the shaft, subsequently in the field of the generator and finally in the copper of the electrical cables.

It is vital to effect this massive concentration with as little cost as possible and the first major gain is made in the rotor itself. The rotor typically has a solidity of ~5% and hence blade frontal area is less than the swept area by a factor ~20. This is evident in the highlighted concentration factor (**19.1**, in Table 1.2).

This is the key factor in favour of the rotor concept. The rotor can confront all the extractable energy in the swept area with blades that may occupy only about 5% of the swept area. This is in direct contrast to a translating aerofoil or, say, an oscillatory wave energy device where, although the source energy density is usually much greater than for wind, a metre length of wave energy converter must confront each metre of wave front from which energy is to be extracted.

Thus, an efficient rotor is typically concentrating the extractable energy in the rotor disc by a factor of about 20 and thereby reducing the size and cost of the primary collectors (blades) compared with alternative systems such as an oscillating aerofoil that do not have this benefit. The answer to 'why a rotor?' is therefore not only the legitimate

Table 1.2 Power concentrations in a 1.5 MW wind turbine.

	D (m)	Area (m²)	Efficiency	Power (kW)	Power density (kW/m²)	Concentration factor	Cumulative concentration factor
Wind over swept area	70.5	3 900	1	3 640	0.93	1	1
Rotor blade input	70.5	204	0.44	3 640	17.83	**19.1**	19
Low speed shaft input	0.564	0.25	1	1 600	6 400	359.6	6 900
Gearbox input	0.564	0.25	0.98	1 600	6 280	1	6 900
Generator input	0.12	0.01	0.95	1 570	131 700	21.6	150 000
Electrical cables	—	0.001	1	1 490	1 490 000	10.7	1 600 000

common observation that mechanical energy in rotational form best suits conventional electricity generating systems but also that, in sweeping an area of the source energy flux that is much greater than the physical surface area of the rotor blades, the rotor effects a significant primary increase in energy density.

This is the main reason why the rotor concept is very hard to beat and why many of the alternatives such as oscillating or translating aerofoils that are perfectly feasible technically may struggle to be cost competitive.

1.7 Actuator Disc in Augmented Flow and Ducted Rotor Systems

1.7.1 Fundamentals

The ducted rotor system comprises a bare turbine rotor with an added surrounding structure intended to alter the inflow to the turbine in magnitude and or direction (diffusers and other types). It is commonplace in aeronautical and marine applications including ships and also tidal turbines.

The following statements are valid, but contrary views have been long standing.

a) The available (potentially extractable) power from any system with augmented flow (or not) is linearly proportional to the mass flow and to rotor plane pressure drop. In an augmented flow system, this is not as the cube of the velocity augmentation factor, although kinetic power through the rotor plane is augmented by this factor.

b) The maximum section diameter of a duct (exit area of a diffuser type) is a very obvious geometric parameter to consider, but *it has no unique relation to duct performance*. In particular, the view that ducted rotor performance is limited by the Betz limit as applied to the maximum duct diameter[7] is erroneous and misleading about the potential performance of ducts.

Regarding statement (a), ducted rotors in real flows may suffer adverse effects from friction losses and flow separation. However, they can also gain from viscous interaction

---

7 Sørensen [28] (pp. 22–23) provides an apparent derivation that Betz related to area ratio is a limit, but acknowledges that this relies on unverified assumptions.

with external flows. Thus, some energy that does not flow through the rotor may assist rotor performance by entraining the wake or in creating vortices that reduce pressure on the downstream side of the rotor plane. In some cases, essentially through increasing the pressure drop across the rotor (implying an ideal optimum $C_t > 8/9$), this can give a power gain that is somewhat better than linear with velocity augmentation. Such benefit has been observed in experimental testing by Phillips *et al.* [29] and Ohya and Kara-sudani [30], for example. In addition, much more elaborate systems can be developed to constructively involve the external flow involving boundary layer injection through slots, mixing and entrainment as developed by Werle and Presz [31] following practice in aircraft engine design.

Figure 1.8 shows three diffuser-type duct shapes among many more similarly analysed by McLaren–Gow [32] using a vortex ring model. Each duct has exactly the same area ratio (ratio of maximum to minimum duct diameter). The ducts were modelled as line ducts with no wall thickness and the C_p max for each design was determined by varying the rotor plane pressure drop. The inset table shows the ratio of $(C_p \text{ max}/1.161)$, the Betz limit factored by area ratio being $(16/27) \times (7/5)^2 = 1.161$. As an ideal inviscid analysis, this supports statement (b) showing that the area ratio is not a unique parameter. Evidently, the performance varies with duct shape and the performance is neither limited by nor related to the Betz limit as applied to the exit area. This refutes a common assertion that the ducted rotor can perform no better than a larger bare rotor with diameter equal to maximum duct diameter which, inappropriately, has led some parties to summarily dismiss ducted rotors as a concept. It will be shown (Chapter 18) that this is not simply the case for highly idealised inviscid flow conditions and that the limiting performance of some real systems probably exceeds the Betz limit as applied to the maximum duct diameter. While Figure 1.8 refutes the idea that Betz factored by area ratio is a valid limit, it is entirely consistent with the limit, 'C_p limit' in Figure 1.8, based on Equation 1.34. In the discussion around Equation 1.29, it is shown that C_p limit is in effect the Betz value applied to an area of streamtube section at a so-called reference plane. That area generally exceeds the maximum section area of the duct unless the duct is producing little augmentation and the wake does not expand beyond the duct exit. Seven duct shapes of the same exit area ratio (see later discussion of Figure 1.11) were analysed including the

Figure 1.8 Area ratio fallacy.

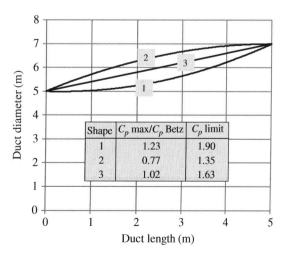

Shape	C_p max/C_p Betz	C_p limit
1	1.23	1.90
2	0.77	1.35
3	1.02	1.63

three of Figure 1.8. Those that were of a more concave shape than duct 2 of Figure 1.8 (as viewed from the axis of symmetry) had a maximum ideal power coefficient less than Betz factored by the area ratio and those more convex than duct 1 exceeded Betz so factored. This is further clarified in Section 1.7.2. For any duct design where the flow expands downstream to an area greater than the exit area, the reference plane is also downstream of the duct, of greater diameter than the exit diameter and consequently defines a limit greater than Betz factored by area ratio.

In the early 1980s, Oman *et al.* [33] conducted experimental work on the diffuser-augmented wind turbine (DAWT) concept, showing that power coefficients exceeding the Betz limit could be obtained. In 1999, Hansen *et al.* [34] published CFD results confirming that the Betz limit could be exceeded. Hansen noted that the increase in C_p was in proportion to the augmentation of mass flow achieved by the diffuser (as in the discussion of Figure 1.13), but that this did not explicitly define a limit for C_p.

1.7.2   Generalised Actuator Disc

Although the open flow actuator disc theory, which determines the Betz limit, has evidently been established for over 90 years and van Bussel [35] in a comprehensive review notes that diffuser research has been in progress over 50 years, the generalisation of actuator disc theory arises from analysis by Jamieson [36]. This work includes new relationships for limiting values of C_p and a preliminary validation has shown close quantitative agreement with Hansen's CFD results [34]. In many previous analyses of turbines in ducts and diffusers, speedup factors are introduced and definitions of C_p and C_t other than the standard ones have been employed. This is understandable in the historical context, but there is no longer need for it and there is some potential for confusion. The following analysis maintains standard definitions of axial induction, power and thrust coefficients.

Axial induction, a, at the rotor plane is defined exactly as before (Equation 1.2). Thus, if the flow is augmented at the rotor plane, a is negative. As in open flow, the power coefficient and thrust coefficient are defined with respect to the far upstream wind speed and referenced to the rotor swept area. They are, respectively:

$$C_p = \frac{P}{\frac{1}{2}\rho A U_0^3} \tag{1.16}$$

and

$$C_t = \frac{T}{\frac{1}{2}\rho A U_0^2} \tag{1.17}$$

From these basic definitions of the power coefficient, C_p, and the thrust coefficient, C_t, the power to thrust ratio can be expressed as in Equation 1.18.

$$\frac{P}{T} = U_0 \frac{C_p}{C_t} \tag{1.18}$$

However, considering also the basic definition of power as a product of force and velocity as applied at the rotor plane:

$$P = T\, U_0(1-a) \tag{1.19}$$

Hence,

$$\frac{P}{T} = U_0(1-a) \tag{1.20}$$

Hence, from Equations 1.18 and 1.20,

$$\frac{C_p}{C_t} = (1-a) \tag{1.21}$$

Equation 1.21 applies to an ideal rotor in open or augmented flow, where the local inflow is a fraction $(1-a)$ of the remote undisturbed external wind speed. A system is defined as the region in which axial induction is influenced between the freestream and the far wake. Energy extraction is considered to take place across a planar area normal to the flow and at a definite location within the system.

Let $f(a)$ be the axial induction in the far wake (Figure 1.9). At any plane of area, A within the system where there is a pressure difference, Δp associated with energy extraction, the thrust, T is given as;

$$T = \Delta p A = \frac{1}{2}\rho U_0^2 A C_t \tag{1.22}$$

Hence

$$C_t = \frac{2\Delta p}{\rho U_0^2} \tag{1.23}$$

Considering Bernoulli's equation, applied upwind of the extraction plane,

$$p_0 + \frac{1}{2}\rho U_0^2 = p_1 + \frac{1}{2}\rho U_0^2(1-a)^2 \tag{1.24}$$

and on the downstream side of the extraction plane.

$$p_1 - \Delta p + \frac{1}{2}\rho U_0^2(1-a)^2 = p_0 + \frac{1}{2}\rho U_0^2\{1-f(a)\}^2 \tag{1.25}$$

From Equations 1.23–1.25,

$$C_t = 1 - \{1-f(a)\}^2 \tag{1.26}$$
$$C_t = 2f(a) - f(a)^2 \tag{1.27}$$

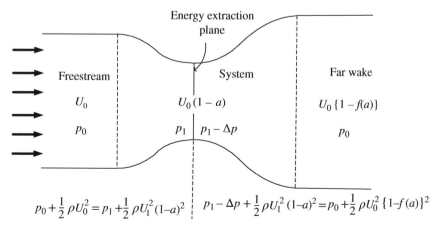

Figure 1.9 diagram labels:

Energy extraction plane

Freestream — U_0 — p_0

System — $U_0(1-a)$ — p_1 | $p_1 - \Delta p$

Far wake — $U_0\{1-f(a)\}$ — p_0

$$p_0 + \frac{1}{2}\rho U_0^2 = p_1 + \frac{1}{2}\rho U_1^2(1-a)^2 \quad \bigg| \quad p_1 - \Delta p + \frac{1}{2}\rho U_1^2(1-a)^2 = p_0 + \frac{1}{2}\rho U_0^2\{1-f(a)\}^2$$

Figure 1.9 General flow diagram.

Now consider that:

a) For energy extraction to take place, the velocity in the far wake must be less than ambient, that is, $f(a) > 0$
b) If the flow is augmented above ambient at the rotor plane, then purely from considerations of continuity, there must exist a reference plane of area A_{ref} downstream of the rotor plane where the induction is half that of the far wake, that is, $= f(a)/2$

Considering conservation of mass in the flow, then at the rotor plane (area A_r),

$$\rho A_r U_0 (1 - a) = \rho A_{ref} U_0 \left(1 - \frac{f(a)}{2}\right) \tag{1.28}$$

The factor of 2 in Equation 1.28 may initially seem arbitrary. However, the idea is to view the augmentation system as a disturbance ahead of the reference plane, which itself is in open flow at a location where Froude's theorem and the established formulae for C_t and C_p can apply if related to the axial induction there. Now for Froude's theorem to apply, the induction is required to be specifically $1/2$ of $f(a)$ as distinct from any other fraction. This also implies that *the Betz limit applied to the reference plane area* (as opposed to the duct maximum diameter) must be the valid ideal inviscid performance limit for the duct under consideration. In the absence of energy extraction, note that $f(a) = 0$.

Let the axial induction at the energy extraction plane be a_0. Then,

$$\rho A_r U_0 (1 - a_0) = \rho A_{ref} U_0 \tag{1.29}$$

In Figure 1.10, streamtube sections bounding the diffuser are illustrated for differing levels of rotor disc pressure drop. From Equation 1.29, the reference plane has invariant area $A_r(1 - a_0)$ and 'moves' from far downstream at zero disc loading towards the rotor plane as disc loading is increased.

Seven duct shapes, as in Figure 1.11, were analysed as axisymmetric line ducts by MacLaren–Gow *et al.* [32] using a vortex ring model (three of these appear in Figure 1.8).

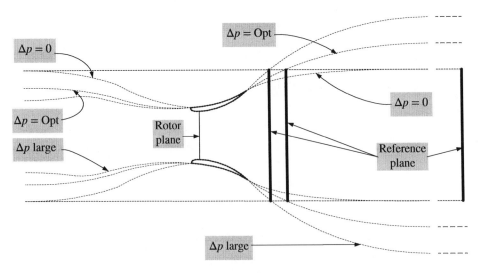

Figure 1.10 The reference plane in relation to disc loading.

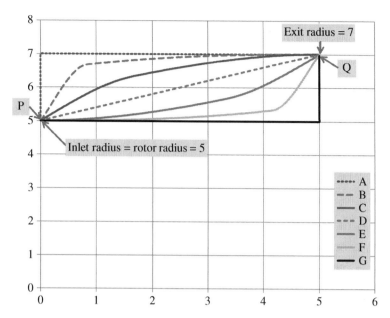

Figure 1.11 Seven shapes of line duct analysed as in inviscid flow.

Table 1.3 Limiting performance of ducts.

	A	B	C	D	E	F	G
Reference plane radius	6.116	6.204	6.383	7.187	8.033	8.778	9.433
C_p/Betz at reference plane	0.922	0.931	0.943	0.968	0.977	0.977	0.970
C_p/Betz at exit plane	0.707	0.734	0.787	1.025	1.292	1.542	1.768

The inlet radius (also the rotor plane radius) is always 5 units and the exit 7 units. As C_p increases from poorest duct A, to best G, the reference plane moves downstream from within the duct (with radius less than exit radius of 7 units) to values substantially exceeding the exit radius. In other words, flow expansion takes place beyond the duct exit.

The C_p values at the reference plane (Table 1.3) are all close to the Betz limit by amounts that can be related to C_t at maximum C_p. The C_p performance can, moreover, be seen to be quite unrelated to exit area ratio with values ranging from 29% less to 77% greater.

Returning to Figure 1.10, a detail reflects behaviour also observed in numerical modelling. Especially with higher loading on the disc, the inflow streamtubes, 'sensing' flow resistance ahead, start to expand (see 'Δp large' in Figure 1.10) as would happen for an open flow rotor before they contract due to the suction of the diffuser.

From Equations 1.28 and 1.29,

$$f(a) = 2\left\{\frac{a - a_0}{1 - a_0}\right\}$$

(1.30)

Now substituting for $f(a)$ in Equation 1.27, gives

$$C_t = \frac{4(a - a_0)(1 - a)}{(1 - a_0)^2} \tag{1.31}$$

And hence from Equation 1.21:

$$C_p = \frac{4(a - a_0)(1 - a)^2}{(1 - a_0)^2} \tag{1.32}$$

Differentiating Equation 1.32 with respect to a determines a maximum at $a = a_m$ of

$$a_m = \frac{1 + 2a_0}{3} \tag{1.33}$$

The associated maximum C_p is then:

$$C_{pm} = \frac{16}{27}(1 - a_0) \tag{1.34}$$

For the open flow rotor with $a_0 = 0$, Equations 1.31–1.34 correspond, as they must, to the established equations for open flow. The familiar results that the open flow rotor operates optimally when $a_m = 1/3$ and has an associated maximum power coefficient $C_{pm} = 16/27$ (the Betz limit) are evident.

A more striking result emerges out of the limit Equation 1.32. On substituting a_m from Equation 1.33 in Equation 1.32, it is found that

$$C_t = \frac{8}{9} \tag{1.35}$$

whereas a_m and C_{pm} have specific values for each system configuration, this result is now independent of a_0. Equation 1.35 is therefore a general truth for optimum energy extraction in an ideal system. This result was mentioned to the author in 1995 by K. Foreman, as an observed outcome (without theoretical explanation) of his extensive experimental work within Grumman Aerospace in the 1980s with the DAWT concept. It was proved more recently by van Bussel [35] and now directly as a consequence of the generalised limit Equations 1.31 and 1.33. Considering Equation 1.23, a corollary to Equation 1.35 is that the pressure drop across the rotor plane for optimum energy extraction is always $4/9U_0^2$.

Thus, in any flow field of uniform far upstream velocity, U_0, *regardless of what local flow augmentations are created (conservatively) within the system and wherever a rotor is located,* the rotor will, in optimum operation to maximise power extraction, experience the same loading in terms of thrust, T, thrust coefficient C_t and rotor plane pressure drop Δp. This does not in the least contradict statements in many sources (e.g. Lawn [37]) that a rotor in augmented flow must be 'lightly loaded'. Loading it at the same level of thrust as would be optimum in open flow, when the wind speed local to the rotor may be several times greater than ambient, amounts to very 'light' loading. The level of loading is independent of the level of flow augmentation achieved by the diffuser and will therefore appear all the lighter, the greater the flow augmentation.

Results are summarised in Table 1.4. Consider now Figure 1.12 where, instead of fixing the rotor swept area, the source flow area is fixed. With the same source flow area, the source mass flow rate and source power are the same in all three cases, namely, the general case with an arbitrary system, the particular case of a diffuser concentrator and the standard case in open flow.

Table 1.4 Summary results comparing open and constrained flow.

	Betz open flow	Generalised constrained flow
General operation		
Upstream wind speed	U_0	U_0
Wind speed at energy extraction plane	$U_0(1-a)$	$U_0(1-a)$
Far wake wind speed	$U_0(1-2a)$	$U_0\left(\dfrac{1-2a+a_0}{1-a_0}\right)$
Performance coefficient, C_p	$4a(1-a)^2$	$\dfrac{4(a-a_0)(1-a)^2}{(1-a_0)^2}$
Thrust coefficient, C_T	$4a(1-a)$	$\dfrac{4(a-a_0)(1-a)}{(1-a_0)^2}$
Pressure difference across rotor	$\dfrac{1}{2}\rho U_0^2 C_T$	$\dfrac{1}{2}\rho U_0^2 C_T$
Optimum performance		
Maximum C_p	$\dfrac{16}{27}$	$\dfrac{16}{27}(1-a_0)$
Associated axial induction factor	$\dfrac{1}{3}$	$\dfrac{1+2a_0}{3}$
Far wake axial induction factor	$\dfrac{2}{3}$	$\dfrac{2}{3}$
Associated thrust coefficient	$\dfrac{8}{9}$	$\dfrac{8}{9}$
Pressure difference across rotor	$\dfrac{4}{9}\rho U_0^2$	$\dfrac{4}{9}\rho U_0^2$

From Equation 1.34

$$C_{pm} = \frac{16}{27}(1-a_0)$$

Energy extracted by the rotor is, by definition:

$$E_1 = \frac{1}{2}\rho U_0^3 A_1 C_{pm}$$

From continuity of flow;

$$E_1 = \frac{1}{2}\rho U_0^3 \frac{A_0}{(1-a_m)} C_{pm}$$

Using Equation 1.35

$$E_1 = \frac{1}{2}\rho U_0^3 A_0 \frac{8}{9}$$

The power available to the rotor evidently is 8/9 of the kinetic power in the upstream source area. Note that the thrust coefficient is identically equal to the fraction of source kinetic power that is extracted. Note also that all of the preceding analyses relate to

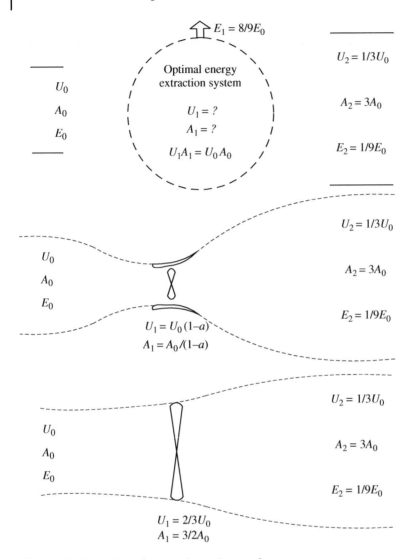

$E_1 = 8/9E_0$

$U_2 = 1/3U_0$

Optimal energy
extraction system

U_0

A_0

$A_2 = 3A_0$

E_0

$U_1 = ?$

$A_1 = ?$

$U_1A_1 = U_0A_0$

$E_2 = 1/9E_0$

$U_2 = 1/3U_0$

U_0

A_0

$A_2 = 3A_0$

E_0

$E_2 = 1/9E_0$

$U_1 = U_0(1-a)$

$A_1 = A_0/(1-a)$

$U_2 = 1/3U_0$

U_0

A_0

$A_2 = 3A_0$

E_0

$E_2 = 1/9E_0$

$U_1 = 2/3U_0$

$A_1 = 3/2A_0$

Figure 1.12 Comparison of cases with equal source flow areas.

ideal actuator discs as power extraction systems. This means that the wake does not rotate as it would with any real rotor system and that the assumption of wake pressure recovery to atmospheric pressure is valid. Wake rotation can, however, be extremely important for the aerodynamic performance of ducts in real flows. With a high tangential velocity component in the wake flow, the angle to horizontal of the resultant flow on its expanding spiral path is much less than the geometric expansion of the duct in the axial direction. This enables short ducts with high geometric expansion angles to perform effectively without premature flow separation.

In the open flow case, the 'system' is always 'ideal' in that the flow is unconstrained and free to flow through or around the rotor in a way that can vary with rotor loading. In all other cases, the system comprises some physical entity additional to the rotor which constrains the flow. In general, this system, whether hill, diffuser or other, because it is of fixed geometry, will not be ideal in every flow state and may not be ideal in any. Hence, even in terms of purely inviscid flow modelling, such systems, regardless of the efficiency of the rotor or energy extraction device, may not extract energy as efficiently as in open flow. Comparing an effective diffuser system with an open flow rotor that optimally extracts the same amount of energy, the rotor in the diffuser system can be much smaller in diameter and the critical design issue is whether this advantage can justify the cost of the diffuser system.

In the limit state (ideal device in ideal system):

- The design of the device is completely decoupled from design of the system, which is completely characterised by a_0.
- The thrust and thrust coefficient that corresponds to optimum rotor loading are independent of the system that includes the rotor and the thrust coefficient is always 8/9.

The conclusion from this is that for any system influencing the local flow through an energy extraction device, the induction factor, a_0, at the extraction plane with the device absent provides a characteristic signature of the system. This statement is probably valid for non-ideal energy extraction devices such as rotors with drag loss, tip loss and swirl loss provided the system influencing the rotor plane induction is ideal.

Consider now the operation of a rotor in constrained flow. The Internet is littered with websites where claims are made that some innovative system around a wind turbine increases the air velocity locally by a factor k and therefore the power as k^3.

In any area of flow augmentation prior to energy extraction, the flow concentrator does not introduce extra energy into the flow field. Therefore, increased local velocity and the associated increase in local kinetic energy are created conservatively. Hence, according to Bernoulli's theorem, increased kinetic energy is obtained at the expense of static pressure (atmospheric potential energy). It is perfectly true that the kinetic energy locally is increased by a factor k^3. This must be the case by definition. However, as has been strongly emphasised in Section 1.3, there is no extraction of kinetic energy at the rotor plane. It is the pressure difference at the rotor plane that drives energy extraction, and both the inlet and exit pressure of an energy extraction device in a region of concentrated flow are at sub-atmospheric pressure. This means that much less energy can be extracted than might be supposed.

Perhaps the simplest way to appreciate this is as follows. Consider a fixed area, A_0, represented by the dotted lines of Figure 1.13, where a rotor may be placed, but for the present in the absence of energy extraction. If the velocity is increased over the prevailing upstream value, U_0, by a factor, say, 3 in a flow augmentation device, the streamtube passing through it, by conservation of mass flow, will have an upstream source area that is three times greater than the extraction area, A_0. It is then clear that no more than three times the energy and certainly not 3^3 times can be extracted from this streamtube.

In Equation 1.15, $U_1 = U_0 (1 - a)$ as defined in Equation 1.2. If the rotor is in a concentrator, a will be negative of a magnitude related to the flow augmentation factor, k (which, it should be noted, will change with rotor loading). Noting the results of Table 1.1, it can be seen that, for maximum energy extraction, in open flow or constrained flow, Equation 1.15 is unchanged. Hence, the increase in power is only linearly as the increase in local velocity in the region of flow concentration.

Streamtube – open flow

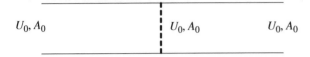

U_0, A_0 U_0, A_0 U_0, A_0

Streamtube – augmented flow

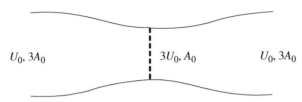

$U_0, 3A_0$ $3U_0, A_0$ $U_0, 3A_0$

Figure 1.13 Source area and energy gain with flow augmentation.

The generalised actuator disc theory implies that all rotors whether large or small, whether in open flow or in well-optimised diffusers or other concentrators will operate optimally in an optimal system with a similar pressure difference across the rotor plane. This pressure difference under such ideal circumstances is $4/9\rho U_0^2$ (see Table 1.1). However, in constrained flow fields, system inefficiencies (which importantly can arise from purely geometric aspects in addition to frictional losses) will, in general, further reduce the optimal pressure difference for maximum energy extraction.

1.7.3 The Force on a Diffuser

From Equation 1.28, the mass flow rate through the energy extraction plane (and elsewhere) is $\rho A_r U_0(1-a)$; and from Equation 1.30, the change in fluid velocity between far upstream and far downstream is $f(a) = 2(a - a_0)/(1 - a_0)$. Hence, the rate of change of momentum and total thrust force on the system is the product of these quantities.

$$T_{net} = \rho A_r U_0(1-a)\left\{ \frac{2U_0(a - a_0)}{(1 - a_0)} \right\} = \frac{1}{2}\rho A_r\, U_0^2 \left\{ \frac{4(1-a)(a - a_0)}{(1 - a_0)} \right\} \quad (1.36)$$

Now, thrust on the rotor is, by definition:

$$T = \frac{1}{2}\rho A_r U_0^2 C_t$$

And hence from Equation 1.31:

$$T = \frac{1}{2}\rho A_r U_0^2 \left\{ \frac{4(1-a)(a - a_0)}{(1 - a_0)^2} \right\} \quad (1.37)$$

Comparing Equations 1.36 and 1.37 it is clear that the force on the diffuser is,

$$T_d = \frac{1}{2}\rho A_r U_0^2 \left\{ \frac{-4a_0(1-a)(a - a_0)}{(1 - a_0)^2} \right\} = -a_0 T \quad (1.38)$$

This very simple result is important. The separation of total system thrust into the part that acts on the rotor or energy extraction device and the part that acts on the diffuser or flow concentrator is vital for an appropriate implementation of BEM theory to deal with modelling of system loads or optimisation of rotors in constrained flows. Note also that, as far as inviscid flow is concerned, although thrust on the diffuser may be several times that on the rotor, thrust on the diffuser only appears in association with rotor loading and is zero on the empty duct.

The simple form of Equation 1.38 provides insight, but applies only to an ideal diffuser. Sørensen [28] asserts, '*In most theoretical analyses, the authors have introduced different auxiliary variables, in order to derive general conclusions concerning maximum power output, etc. However this is not necessary…*'. His implication is that the variable a_0 is unnecessary. However, he instead introduces the auxiliary variable, which he calls T_{diff} the force of the diffuser. Equation 1.38 shows clearly that a_0 is related and not additional to $T_{diff} \equiv T_d$. The issue is simply that the level of augmentation and limiting performance of a diffuser differs, in general, for each possible diffuser geometry, and the limiting performance cannot be specified without introducing at least one integrated property of the diffuser. The introduction of a_0, in fact, provides a number of additional insights, among them relating velocity induction to duct force as in Equation 1.38.

Considering the diffuser as an axisymmetric aerofoil body, suction on the leading edge may, in general, produce a thrust component that is directed into the wind. The net force on the diffuser is then the sum of suction forces and pressure drag forces. When the diffuser is not ideal, the ratio of net force on diffuser to force on the rotor is increased and exceeds the prediction of Equation 1.38.

1.7.4 Generalised Actuator Disc Theory and Realistic Diffuser Design

The analysis presented here of ducts or diffusers in inviscid flow is only the starting point. Referring to this approach, which extends the open flow actuator disc theory to deal with augmented flows as 'generalised actuator disc (AD) theory', note that it

1) is a limiting theory (as is the Betz theory in open flow) that considers only inviscid flow,
2) shows clearly why flow concentration devices increase available energy linearly as increase of mass flow and not as the cube of the augmented velocity,
3) describes ideal systems, whereas real diffusers may be far from the limiting performance suggested; in general, their fixed geometry will only best suit one state of loading,
4) as an inviscid model, does not capture effects of flows external to the diffuser and rotor which can be used to augment performance through viscous interactions. Some diffusers are very much designed to exploit such effects as in the FloDesign wind turbine [38].

Generalised AD theory affords some new insights and provides energy limits for ideal systems, but there is still quite a gap between such ideal limiting theory and real-world design of flow augmentation systems.

The induction, a_0, of the ideal empty duct to be associated with a particular real duct can only crudely be estimated Jamieson [39] as $a_e \eta$ where a_e is the induction of the empty duct and η is the ratio of C_t associated with C_p max to 8/9. Moreover, determination of these parameters relies on detailed modelling of specific ducts. However, the generalised AD model refutes some fallacies about duct performance and does give a

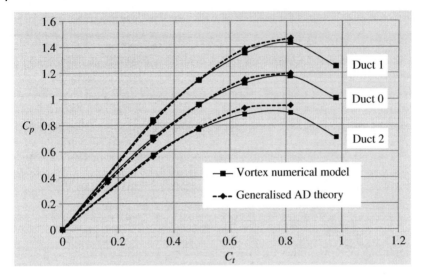

Figure 1.14 Performance characteristics of ducts.

reasonably self-consistent view of duct designs in inviscid flow (Figure 1.14) with some insights that are transferable to real duct design (Chapter 18). In Figure 1.14, results from the vortex ring model of McLaren–Gow [32] are compared with the generalised AD model based on Equation 23 of Jamieson [39].

1.8 Blade Element Momentum Theory

1.8.1 Introduction

BEM theory is the most widely used theory in practical design methods and computer codes for predicting loads and performance of wind turbines. In any balanced overview of wind turbine modelling reflecting current research directions, much attention would be devoted to vortex theories and CFD. These and numerical methods, in general, are not discussed. They may offer more accurate analysis of specific configurations, but they do not yield analytical relationships that can provide physical insight to guide parametric evaluations and concept design.

In BEM, the swept area of the rotor is considered as a set of annular areas (Figure 1.15) swept by each blade element. The blade is divided spanwise into a set of elements which are assumed to be independent of each other, so that balance of rate of change of fluid momentum with blade element forces can be separately established for each annular area. The basic theory is from Glauert [40], with the modern forms for numerical implementation in BEM codes having developed following the adaptation of Glauert's theory by Wilson *et al.* [41]. BEM theory is summarised here in order to preserve a self-contained account of some new equations that are developed from it.

1.8.2 Momentum Equations

Considering thrust as rate of change of linear momentum of the flow (overall axial velocity change, $2aU \times$ mass flow rate) passing through an annulus at radius r of width dr and

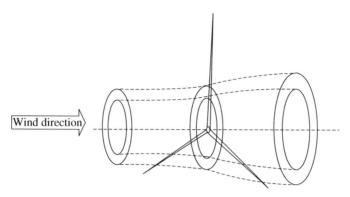

Figure 1.15 Actuator annulus.

denoting a tip effect factor (to be discussed) as F,

$$\text{Thrust} \quad dT = 4\pi \rho r U^2 a(1-a)F\,dr \tag{1.39}$$

and similarly considering torque and rate of change of angular momentum:

$$\text{Torque} \quad dQ = 4\pi \rho r^3 U a' \omega (1-a)F\,dr \tag{1.40}$$

In Equation 1.40, the tangential induction factor, $á$ is introduced. The development of tangential velocity in the air occurs at the rotor plane and is considered to be from zero as the air is non-rotating immediately upstream of the rotor rising immediately downstream to a value of $2á\omega$, which is the rotational angular velocity imparted to the wake by the torque reaction on the rotor. The torque reaction on the air grows from zero to maximum across the rotor plane and an average value of induction, $á$, rather than $2á$, is used in the flow triangle (Figure 1.16).

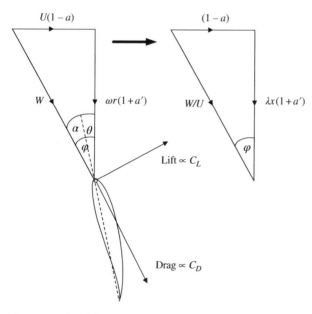

Figure 1.16 Local flow geometry at a blade element.

1.8.3 Blade Element Equations

Considering blade element forces on a blade element at radius r of width dr,

$$\text{Thrust} \quad dT = \frac{1}{2}\rho W^2 Bc(C_l \cos\varphi + C_D \sin\varphi)dr \tag{1.41}$$

$$\text{Torque} \quad dQ = \frac{1}{2}\rho W^2 Bc(C_L \sin\varphi - C_D \cos\varphi)rdr \tag{1.42}$$

Equations 1.39–1.42 allow dT and dQ to be eliminated, yielding two equations in the three unknowns, a, a' and φ. A third equation is given by considering the flow geometry local to each blade element at radius, r, that is at radius fraction, $x = r/R$.

From the flow geometry (Figure 1.16),

$$\tan\phi = \frac{U(1-a)}{\omega r(1+a')} = \frac{(1-a)}{\lambda x(1+a')} \tag{1.43}$$

Equating Equation 1.39 with Equation 1.41 and Equation 1.40 with Equation 1.42 to solve for the induced velocities a and a' (also making use of Equation 1.43) gives

$$\frac{a}{1-a} = \frac{\sigma(C_L + C_D \tan\varphi)}{4F \tan\varphi \sin\varphi} \tag{1.44}$$

$$\frac{a'}{1+a'} = \frac{\sigma(C_L \tan\varphi - C_D)}{4F \sin\varphi} \tag{1.45}$$

where σ, the local solidity, is defined as $\sigma = Bc/2\pi r$.

Usually, an iterative procedure is used to solve Equations 1.43–1.45 for each local blade element of width Δr. Hence, using Equations 1.41 and 1.42, the thrust and torque can be found on the whole rotor by integration.

The BEM analysis of Equations 1.39–1.45 has followed the widely used formulation of Wilson *et al.* [41], who suggested that drag should be neglected in determining the induction factors, a and a'. According to the PhD thesis of Walker [42]:

> ... it has been the assumption that the drag terms should be omitted in calcula-tions of a and a' ... on the basis that the retarded air due to drag is confined to thin helical sheets in the wake and (will) have negligible effect on these factors.

Although drag must be accounted for in determining the torque and power developed by a rotor, opinion is divided[8] about whether the drag terms should be included in evalu-ation of the induction factors. Neglecting drag leads to simpler forms for Equations 1.44 and 1.45 and can enable a closed-form solution (Section 1.8.7). It also simplifies the following analyses.

1.8.4 Non-dimensional Lift Distribution

From Equation 1.43,

$$\frac{a}{1-a} = \left(\frac{Bc}{2\pi r}\right)\frac{C_L \cos\varphi + C_D \sin\varphi}{4F\sin^2\varphi} \tag{1.46}$$

$$= \frac{B}{8\pi}\left(\frac{cC_L}{R}\right)\left(\frac{R}{r}\right)\frac{(\cos\varphi + C_D/C_L \sin\varphi)}{F\sin^2\varphi} \tag{1.47}$$

---

8 DNV GL use the formulation including drag as presented in Equations 1.44 and 1.45 in their commercial BEM code, Bladed.

The term (cC_L/R) represents a non-dimensional lift distribution where the chord distribution $c \equiv c(\lambda, x)$ is, in general, a function of radius fraction, $x = r/R$, and design tip speed ratio, λ.

Let $\Lambda(\lambda, x) = \frac{c(\lambda, x)C_L}{R}$, and let $k = \frac{C_L}{C_D}$:

Then

$$\frac{a}{1-a} = \frac{\Lambda B}{8\pi x F} \frac{[1 + (1/k)\tan\varphi]}{\sin\varphi \tan\varphi} \tag{1.48}$$

And

$$\sin\varphi = \frac{(1-a)}{\sqrt{(1-a)^2 + \lambda^2 x^2 (1+a')^2}} \tag{1.49}$$

Hence, after some manipulation:

$$\Lambda(\lambda, x) = \frac{8\pi a(1-a)}{B\lambda(1+a')\sqrt{(1-a)^2 + \lambda^2 x^2 (1+a')^2}} \frac{F}{\left[1 + \dfrac{(1-a)}{k\lambda x(1+a')}\right]} \tag{1.50}$$

The tangential induction factor, a', can be solved as in Equation 1.51 in terms of a using Equations 1.43–1.45 to eliminate φ.

$$a' = \frac{\{\lambda^2 k^2 x^2 + 2\lambda k x - 4ak[\lambda x - k(1-a)] + 1\}^{0.5} - (\lambda k x + 1)}{2\lambda k x} \tag{1.51}$$

Note that the elimination of φ in Equations 1.50 and 1.51 is only apparent as the lift-to-drag ratio, k, depends in general on the angle of attack, α, and $\alpha = \varphi(x) - \theta(x) - \psi$ where $\theta(x)$ is the blade twist distribution and ψ is the pitch angle of the blade. In the limit of zero drag when $k \to \infty$:

$$a' = \frac{(4a - 4a^2 + \lambda^2 x^2)^{0.5} - \lambda x}{2\lambda x} \tag{1.52}$$

Equation 1.52 also appears in Manwell et al. [43].
With further approximation:

$$a' = \frac{a(1-a)}{\lambda^2 x^2} \tag{1.53}$$

1.8.5 General Momentum Theory

When a rotating wake is considered in ideal inviscid flow, the angular momentum imparted to the air by torque reaction at the rotor plane is conserved and wake rotation is preserved through the whole of the wake. To balance the external pressure, the internal static pressure on the streamtube boundary of the far wake must be atmospheric. However, in order to maintain wake rotation by balancing centrifugal forces, the static pressure within the wake must reduce below ambient atmospheric. The power balance diagram is then similar to Figure 1.6; except that in the far wake, for any single stream tube annulus or for the average over the whole disc, the residual pressure power is now $p_2 A_2 U_2 = p_2 A_0 U_0$ where $p_2 < p_0$. Hence, unlike the ideal actuator disc with a non-rotating wake, there is a change in pressure power as well as kinetic power between far upstream and far downstream. The theory which considers this is usually referred to as 'general momentum theory' and has been the subject of extensive discussion [28, 40–42, 44]. Standard BEM models, such as those developed in Sections 1.8.1–1.8.4, consider wake rotation to the extent of relating the tangential velocity developed at the rotor plane and its associated induction factor to the rotor torque, but they ignore power terms associated with rotational kinetic energy and suction potential energy of the wake. An extensive review comparing BEM models including general momentum theory is available in Sørensen [28].

1.8.6 BEM in Augmented Flow

The generalised actuator disc results of Section 1.7 can be used to derive a generalised BEM that will assist in the optimisation of rotors in ducts or diffusers. In order to revise the BEM equations for generalised flow conditions, consider first the elemental thrust and axial momentum balance:

The mass flow rate through the rotor plane is $\rho(2\pi r dr)\ \{U(1-a)F\}$. The total change in flow velocity between far upstream and far wake (see Table 1.4) is:

$$dU = U\left\{1 - \left(\frac{1 - 2a + a_0}{1 - a_0}\right)\right\} = \frac{2U(a - a_0)}{(1 - a_0)} \tag{1.54}$$

However, considering the thrust force on the rotor alone (see Equation 1.41):

$$dT = 4\pi \rho r U^2 \frac{(a - a_0)(1 - a)}{(1 - a_0)^2} F\, dr \tag{1.55}$$

Equation 1.42 is unchanged and hence

$$dQ = 4\pi \rho r^3 U a' \omega (1 - a) F\, dr \tag{1.56}$$

Equations 1.41, 1.42 and 1.45 are also unchanged. The tangential induction factor a' may be approximated by neglecting drag and generalised as

$$a'(1 + a') = \frac{(1 - a)(a - a_0)}{(1 - a_0)\lambda^2 x^2} = \frac{C_t(1 - a_0)}{4\lambda^2 x^2} \tag{1.57}$$

The further development of the generalised BEM model is simplest if a_0 is assumed to be a suitably averaged constant value over the rotor disc and that is tacitly assumed in the following analyses. However, there is no requirement for this and a variation of $a_0 \equiv a_0\ (r, \theta)$ may be defined over the rotor disc. The equation system in the generalised flow case may be solved by iterative numerical methods in the same way as in standard BEM. The associated non-dimensional lift distribution is

$$\Lambda(\lambda, x) = \frac{2\pi C_t}{B\lambda(1 + a')\sqrt{(1 - a)^2 + \lambda^2 x^2 (1 + a')^2}} \frac{F}{\left[1 + \frac{(1 - a)}{k\lambda x (1 + a')}\right]} \tag{1.58}$$

For an optimum rotor, $a = a_m = (1 + 2a_0)/3$ and $C_t = 8/9$.

If the maximum lift-to-drag ratio of a chosen aerofoil section occurs at an angle of incidence $\alpha = \alpha_0$, then the optimum twist distribution is given as:

$$\psi(x) = \tan^{-1}\left\{\frac{2(1 - a_0)}{3\lambda x(1 + a')}\right\} - \alpha_0 \tag{1.59}$$

The optimum twist distribution will evidently vary with a_0 and hence may vary significantly according to the nature of the system affecting the rotor plane induction. Suppose there is substantial flow augmentation at the rotor plane. As $C_t = 8/9$ universally, the rotor is optimally loaded at exactly the same value of thrust coefficient and thrust as in open flow with no augmentation system present. This implies that the blade elements must be pitched much further into the flow direction so that a reduction in the lift component producing thrust exactly compensates for the potential increase in thrust due to the augmented local flow velocity. However, in this situation there is then a much larger lift contribution to rotor torque than in open flow. This corresponds to the increased power performance coefficient, which may exceed the Betz limit in proportion to the flow augmentation achieved.

The usual actuator disc theory, whether standard or generalised, considers only inviscid flow. In order to be more realistic and useful for design calculations, empirical modelling is introduced to represent the thrust coefficient in the turbulent wake state. Experimental validation for systems with concentrators is not yet available, and so the results derived represent no more than a consistent extension from the standard open flow model to the generalised actuator disc theory.

If a is the induction at the rotor plane in open flow, then the transformation, $a \to (a - a_0)/(1 - a_0)$ determines the value of axial induction at a plane (not the rotor plane) in constrained flow where the induction is half of that in the far wake. However, as is explained in Jamieson [36], the value of thrust coefficient, C_t, is independent of location in the system. Therefore, this transformation may be employed to determine an expression for thrust coefficient that is applicable at the rotor plane.

In open flow, various formulations are employed to modify the thrust coefficient equation of the inviscid flow actuator disc as the rotor approaches the turbulent wake state. DNV GL's commercial BEM software package, *Bladed*, defines thrust coefficient, C_t, as:

$$C_t = 4a(1 - a) \quad \text{for } 0 \leq a \leq 0.3539 \tag{1.60}$$

$$C_t = 0.6 + 0.61a + 0.79a^2 \quad \text{for } 0.3539 < a \leq 1 \tag{1.61}$$

In generalised flow states, applying again the transformation of Equation 1.90 results in the equations:

$$C_t = \frac{4(a - a_0)(1 - a)}{(1 - a_0)^2} \quad \text{for } 0 \leq a \leq a_0 + 0.3539(1 - a_0) \tag{1.62}$$

$$C_t = 0.6 + 0.61 \left\{ \frac{a - a_0}{1 - a_0} \right\} + 0.79 \left\{ \frac{a - a_0}{1 - a_0} \right\}^2 \text{for } a_0 + 0.3539(1 - a_0) < a \leq 1 \tag{1.63}$$

Equation 1.81 does not accurately accord with BEM theory as reflected in Equation 1.85, where the tip effect modifies the thrust coefficient. Thus, the method of application is to factor the C_t in the BEM solutions as a ratio of Equation 1.61 or 1.63 to the corresponding actuator disc Equations 1.60 and 1.62.

In open (unconstrained) flow, the thrust coefficient is essentially unique and optimum at 8/9 at least in the ideal inviscid flow case. However, as is elaborated in Jamieson [36], in constrained flow, the thrust coefficient is a system property. Irrespective of rotor efficiency, C_t in an ideal system is optimally 8/9 and maximises C_p at that value. If the system is not *ideal* in the optimal rotor loading state (which for a diffuser would mean that the diffuser is not fully optimised for the flow field that will develop in operation at a rotor thrust coefficient of 8/9), then the C_t that maximises C_p will be <8/9 and the associated maximum C_p will be less than it would be in an ideal system. On the other hand, any external mass flow (i.e. flow not passing through the rotor) that influences the overall energy exchange, for example, by assisting wake transport, may increase the optimum C_t at the rotor plane to above 8/9. Results of Phillips *et al.* [29] suggest that in a well-designed diffuser, an optimum C_t of around unity may be achieved.

Great care is required in applying generalised BEM to real systems, but the new theory offers a rationalised approach and parametric insight for the optimisation of rotor design in flow concentrators that has not been previously available. The general approach as in Jamieson [39] will be to replace a_0 with $\eta(a)\, a_0$, where $\eta(a)$ is a system efficiency function to be defined from empirical information, CFD analyses or otherwise. Also, the estimation of a_0 is not straightforward, as is discussed in Jamieson [39].

Nevertheless, the introduction of the variable a_0 characterising flow augmentation in conjunction with the indicated generalisation of the thrust coefficient indicates an extension of the BEM theory, which is simple to implement and can address the design of rotors in flow concentrators.

It is also important to be aware of systems where ducted rotors may be used, and the generalised BEM theory developed here will not be applicable as in the case of a tidal turbine where distances from the free water surface and sea bed are not large compared to the rotor dimensions. In such a case, Bernoulli's equation with only pressure and velocity terms (the usual basis of wind turbine actuator disc models) is insufficient and an adequate model [45] must account for buoyancy terms as the pressure drop behind the turbine will induce a local drop in sea surface level.

The analytical relationships developed in the foregoing discussion of BEM theory can be bypassed in the use of the usual numerical methods for BEM solutions. However, identification of explicit formulae is considered to be of great value in design development, facilitating preliminary parametric studies that provide insight into how some of the key variables in rotor design may influence performance. This will be revisited later in the context of specific case studies.

1.8.7 Closed-Form BEM Solutions

With the assumption advocated by Wilson *et al.* [41] that drag should be neglected in evaluating the induction factors, a closed-form solution to the BEM equations can be obtained. This has been previously established [43], but the following presentation provides a form that, with piecewise representation of aerofoil characteristics, could be adapted to represent aerofoils with nonlinear lift characteristics (stall). Neglecting the tip loss, F, the term with lift-to-drag ratio k on the assumption that $k\lambda$ is very large and, initially at least, also neglecting the tangential induction factor which is very small in effective operating states, Equation 1.50 reduces to

$$\Lambda(\lambda, x) = \frac{8\pi a(1 - a)}{B\lambda\sqrt{(1 - a)^2 + \lambda^2 x^2(1 + \acute{a})^2}}$$

which is equivalent to;

$$\frac{c(\lambda, x)\mathrm{Cl}}{R} = \frac{8\pi a \sin(\phi)}{B\lambda} \tag{1.64}$$

Let us now represent the lift coefficient, Cl(α), as

$$Cl(\alpha) = Cl_0 \sin(\alpha + \alpha_0) \tag{1.65}$$

This representation of lift coefficient provided by Equation 1.65 is effectively linear over the usual range of attached flow and the constant, α_0, can represent the effect of camber in providing positive lift at zero flow incidence. The lift slope Cl_0 may be represented as $2\pi k_0$, where k_0 is a correction factor to the ideal lift slope value of 2π. For example, $k_0 = 1.062$ for a NACA 63421 section that may typically be used on a large HAWT.

Relative to the rotor plane, a total equivalent pitch angle may be defined as;

$$\theta(x) = \theta_t(x) + \theta_s + \theta_p - \alpha_0 \tag{1.66}$$

In its most general form, θ_s is a set angle of the blade relative to the rotor plane at zero degrees pitch and θ_p is the applied pitch angle (relative to the set angle). From Equation 1.65, noting from Figure 1.16 that $\alpha(x) + \alpha_0 = \phi(x) - \theta(x)$ where ϕ is the inflow angle, the axial induction, a, may be expressed as

$$a(x) = \left(\frac{BCl_0 \lambda c(x)}{8\pi R} \right) \frac{\sin(\phi(x) - \theta(x))}{\sin(\phi(x))}$$

$$a = \frac{k_1 \lambda \sin(\phi - \theta)}{\sin(\phi)} \quad \text{where } k_1(x) = \frac{BCl_0 c(x)}{8\pi R} \tag{1.67}$$

In Equation 1.67, for convenience, the explicit dependence of variables has been removed. Now the flow triangles of Figure 1.16 provide a second equation for the axial induction factor, a.

$$a = 1 - \lambda x \tan(\phi) \tag{1.68}$$

In eliminating a from Equations 1.67 and 1.68, a quadratic equation in $\sin(2\phi)$ is obtained with the solution:

$$\sin(2\phi) = \left\{ \frac{\sqrt{(-c_0^2 + c_1^2 + 1)} - c_0 c_1}{c_1^2 + 1} \right\} \tag{1.69}$$

where $c_0 = \dfrac{(k_1 \sin(\theta) - x)}{(k_1 \sin(\theta) + x)}$ and $c_1 = \dfrac{(1 - k_1 \lambda \cos(\theta))}{(k_1 \sin(\theta) + x)\lambda}$

Hence, the inflow angle, angle of attack, axial induction factor and lift coefficient can be calculated at any radius fraction, x. To complete a BEM analysis, aerofoil drag may be represented as a function of angle of attack, as, for example, in Equation 1.70 (which is a curve fit representative of data for a NACA 63421 aerofoil).

$$Cd(\alpha) = 0.4147 \, \alpha^2 - 9.775 \times 10^{-4}\alpha + 5.388 \times 10^{-3} \tag{1.70}$$

The lift-to-drag ratio can then be computed for any given α and the tangential induction factor, $\acute{a}$, from Equation 1.52. Accepting some further approximation in results, the tip loss factor (such as in the Prandtl form of Equation 1.71 may also be added. In estimating the induction factors, drag is in effect set to zero in Equations 1.44 and 1.45. It must naturally be accounted for in later calculations of rotor torque, power and loads.

This 'closed-form' BEM analysis, at least for operating states not too far off design, gives only slightly different results from iterative solutions that include drag in estimating the induction factors such as DNV GL code Bladed. For use as computer code, it is probably no more efficient than the arguably more exact iterative solutions which can usually converge very rapidly. It is occasionally useful for parametric analysis in enabling analytic rather than numerical investigations.

1.9 Optimum Rotor Design

1.9.1 Optimisation to Maximise C_p

Optimum states are simpler to describe than general conditions. An analogy is that only three coordinates will define the summit of a hill whilst infinitely many may be required to characterise the whole surface. A natural assumption, having chosen a particular wind turbine system scale and rotor diameter, is to consider as optimum an aerodynamic design that will maximise power performance by maximising the rotor power coefficient, C_p.

In the optimum state, for typical rotors designed for electricity production with design tip speed ratios above 6, the tangential induction factor, a', should be small over the significant parts of span ($x > 0.2$). It may be neglected with little loss of accuracy in $\Lambda(\lambda, x)$ or calculated from Equations 1.51, 1.52, or 1.53.

The actuator disc result of Betz [46] establishes an optimum rotor thrust loading corresponding to a value of the thrust coefficient, $C_t = 8/9$. This implies an optimum lift force on each blade element, which in effect specifies the product, cC_L, in Equation 1.41. Referring to Equation 1.42, it is plausible that with cC_L fixed, performance is maximised if C_D is minimum and hence k is maximum. Thus, in the optimum operational state of a wind turbine rotor, each blade element operates at maximum lift-to-drag ratio and the only aerofoil data required to define this state therefore is the maximum lift-to-drag ratio, k, and the lift coefficient, C_L, associated with this maximum lift-to-drag ratio for each element over the span of a blade. Optimum performance at maximum lift-to-drag ratio is not exactly true (see later discussion around Equation 1.84), but is a satisfactory approximation for mainstream designs with design tip speed ratio above 6.

For a lift-to-drag ratio, $k = 100$, design tip speed ratio, $\lambda = 9$ and considering an optimum rotor with $a = 0.3333$ at mid span where $x = 0.5$, Equations 1.51 and 1.53 give values for a' of 0.01010 and 0.01097 a difference in the optimum rotor state of around 10% albeit in a rather small quantity compared to the axial induction, a.

The square bracketed term in the denominator of Equation 1.50 which contains the lift-to-drag ratio k is effectively unity over the significant region of span for typical modern large rotors with design tip speed ratio >6 and $k \geq 100$.

The lift produced by an aerofoil section can be associated with a bound circulation which is virtual over the span of the blade but becomes a real vortex at the end of the blade where there is no material to support a pressure difference. The strength of this vortex depends on blade number and blade solidity, and it is through models of this 'tip effect' that the effect of the blade number on rotor performance is expressed in BEM theory. Various tip effect models have been developed (see Section 1.10.3), the most rigorous by Goldstein [47]. The most commonly used model in BEM theory is from Prandtl (see Wilson [41] for example) and that model is adopted in the following analyses.

Adopting the Prandtl tip factor, $F = (2/\pi)\cos^{-1}(e^{-\pi s/d})$ where $d = 2\pi R(1-a)/(B\lambda)$ and $s = (1-x)R$:

$$F = \frac{2}{\pi}\cos^{-1}\left[\exp\left\{-\frac{(1-x)B\lambda}{2(1-a)}\right\}\right] \tag{1.71}$$

With the assumptions that $a = 1/3$ in optimum operation, that a is constant over the span and knowing the value of lift coefficient at maximum lift-to-drag ratio for the aerofoil section selected at each radial station, Equation 1.50 then defines the chord distribution of an optimum rotor as a function of radius fraction, x and design tip speed ratio, λ. If the approximations of neglecting a', neglecting the very minor effect of drag and neglecting tip effect are combined with the further approximation of neglecting $(1-a)^2$ in comparison to $\lambda^2 x^2$, Equation 1.50 then reduces to

$$\Lambda(\lambda, x) = \frac{8\pi a(1-a)}{B\lambda^2 x} = \frac{16\pi}{9B\lambda^2 x} \tag{1.72}$$

Equations similar to Equation 1.72 have appeared in various forms, in Gasch and Twele [48], Burton and Sharpe [49], and are easily understood intuitively. Lift per blade element is proportional to dynamic pressure and chord width. Dynamic pressure on a blade element is proportional to the square of the inflow velocity, which is predominantly the in-plane velocity when $\lambda > 6$. Thus, to maintain total rotor lift at the appropriate fixed optimum value, $\Lambda(\lambda, x)$ must approximately vary inversely as $B\lambda^2 x$.

Equation 1.50 (or its simplified forms such as Equation 1.72) allows the optimum chord distribution of a blade to be developed given a selection of aerofoil types that will then define at each radial station, x, the maximum lift-to-drag ratio, k, the associated design lift coefficient, C_L and the corresponding angle of incidence, α_0. An optimum blade twist distribution is then determined referring to Equation 1.43 and setting $a = 1/3$ as:

$$\theta(x) = \tan^{-1}\left\{\frac{2}{3\lambda x(1+a')}\right\} - \alpha_0(x) \cong \frac{2}{3\lambda x} - \alpha_0(x) \tag{1.73}$$

As far as optimal blade shape is concerned, the simplified Equation 1.72 predicts the chord distribution very well over the extent of span that most matters. It would be usual in real designs to round the tip in a way that may be guided by practical experience or CFD analyses and, for practical reasons associated with manufacture and/or transportation, to limit the chord to much less than ideal values inboard of say 20–25% span. With the chord being so limited, and the sections normally transitioning to a cylindrical blade root end, there is then no point to continue the twist distribution near the blade root to the very high angles that would be predicted by Equation 1.73. In that case, the approximate form of Equation 1.73 is a good estimate over the aerodynamically active part of the rotor.

As was discussed, the optimum lift force on each blade element specifies the product, cC_L, in Equation 1.42. This means that the chord width can be optimised structurally if aerofoils are available or can be designed with suitable values of design lift coefficient C_L. Thus, as is discussed further in Chapter 2, for the same blade design tip speed ratio, aerofoils with high or low design C_L can enable slender or wide optimum blades.

There is however constraint on having rapid changes in the spanwise variation of C_L. A basic principle is that it is generally undesirable to have rapid changes in section lift (specifically lift and not lift coefficient) along the span of the blade. The trailing vortices which generate induced drag (or the induction factor for a rotor) are proportional to the spanwise gradient of lift (actually circulation but effectively the same). A sudden change in blade chord is undesirable structurally and aerodynamically; this implies that C_L should not change abruptly since the relative inflow velocity varies only gradually with radial position. It is therefore usually assumed[9] that a constant C_L is desirable over most of span with a smooth reduction to zero at the tip.

Having characterised the optimum lift distribution as in Equation 1.50, expressions generally useful for parametric studies are now derived for the power coefficient, thrust coefficient and out-of-plane bending moment coefficient (to be defined). The torque coefficient, $C_q = \frac{Q}{0.5\rho U^2 \pi R^3}$, where Q is the rotor torque, is sometimes useful in evaluating the self-starting capability of wind turbines and is trivially related to the power coefficient as $C_q = \frac{C_p}{\lambda}$.

1.9.2 The Power Coefficient, C_p

Returning to Equation 1.42:

$$dQ = \frac{1}{2}\rho W^2 Bcr(C_L \sin\varphi - C_D \cos\varphi)dr \tag{1.74}$$

Hence elemental power is

$$dP = \frac{1}{2}\rho W^2 Bcr(C_L \sin\varphi - C_D \cos\varphi)\omega dr \tag{1.75}$$

$$= \frac{1}{2}\rho\left(\frac{W}{U}\right)^2 R^2 V^3 \Lambda(\lambda,x)B\left(\sin\phi - \frac{\cos\phi}{k}\right)\lambda x dx \tag{1.76}$$

and the rotor power coefficient is

$$C_p = \frac{B}{\pi}\int_0^1 \Lambda(\lambda,x)\left(\frac{W}{U}\right)^2\left(\sin\phi - \frac{\cos\phi}{k}\right)\lambda x dx \tag{1.77}$$

Substituting for W/U, $\sin\varphi$ and $\cos\varphi$ from Figure 1.16 and for $\Lambda(\lambda,x)$ from Equation 1.50:

$$C_p = \frac{B}{\pi}\int_0^1 \frac{8\pi a(1-a)F}{B\lambda(1+a')}\frac{\lambda\left\{(1-a)-\dfrac{\lambda x(1+a')}{k}\right\}x dx}{\left\{1+\dfrac{(1-a)}{k\lambda x(1+a')}\right\}}. \tag{1.78}$$

Considering a rotor without tip effect, assuming a constant over the rotor span, neglecting a' and neglecting also $\frac{(1-a)}{k\lambda x(1+a')}$:

$$C_p = 8a(1-a)\int_0^1 \left\{(1-a)-\frac{\lambda x}{k}\right\}x dx \tag{1.79}$$

---

9 The optimum spanwise distribution of circulation or lift on a fixed wing is a smooth elliptic variation (which was provided uniquely at all angles of attack by the Spitfire wing), but there may not be any simple theory to define the optimum shape for a rotor.

and hence:

$$C_p = 4a(1-a)^2 \left[1 - \frac{2\lambda}{3k(1-a)} \right] \tag{1.80}$$

Equation 1.80, as it must, tends to the actuator disc result of Betz as $k \to \infty$. With k finite, it is similar to a limiting case derived by De Vries [50] which arose from a quite different beginning in the context of BEM theory for vertical axis wind turbines. De Vries' equation is

$$C_p = 4a(1-a)^2 - \frac{BcC_D\lambda^3}{2R} \tag{1.81}$$

Noting that $\frac{BcC_D\lambda^3}{2R} = B \left(\frac{cC_L}{R} \right) \frac{\lambda^3}{2k}$ and also considering Equation 1.72, the Equations 1.80 and 1.81 have a similar form.

Returning to Equation 1.78, the general expression for C_p can be expressed as:

$$C_p(\lambda) = \int_0^1 \frac{8a(1-a)F[k(1-a) - \lambda x(1+a')]\lambda x^2}{[k\lambda x(1+a') + (1-a))]} dx \tag{1.82}$$

Equation 1.82 is a rigorous BEM relationship defining C_p first published, Jamieson [51], without derivation. As was mentioned in connection with Equation 1.51, the lift-to-drag ratio, k, is a function of angle of attack $\alpha(x) = \varphi(x) - \theta(x) - \psi$ and the flow angle $\varphi(x)$ is therefore implicitly present in Equation 1.82.

However, in the optimum rotor state, for typical design tip speed ratios above, say, about 6 and with typical aerofoil selections for large HAWTs, it is a very good approximation to assume that all the blade elements operate at their maximum lift-to-drag ratio, $k \equiv k(x)$ and, using Equation 1.51 or 1.52 to determine a', Equation 1.82 can then be directly integrated. Hence, maximum C_p may be expressed as a function of tip speed ratio and lift-to-drag ratio as in Figure 1.17. Although for convenience in the calculations presented in Figure 1.17, the lift-to-drag ratio is treated as constant over the blade span, there is no requirement for this to be the case. If k is defined as a function of x, Equation 1.82 can be used for a rotor with differing aerofoil characteristics over the span, as is usually the case on account of thickness-to-chord ratio decreasing from root to tip.

A chart such as Figure 1.17 has been published previously [52], but the results were determined by numerical solution of the BEM equations and not from an explicit formula. After such an exercise, Wilson et al. [41] fitted data with the formula (Equation 1.83) which was claimed valid for $4 \leq \lambda \leq 25$, and $k = C_l/C_d \geq 25$ but restricted to three blades maximum.

$$C_{p\,max} = \left(\frac{16}{27} \right) \lambda \left[\lambda + \frac{1.32 + \left(\frac{\lambda-8}{20} \right)^2}{B^{2/3}} \right]^{-1} - \frac{(0.57)\lambda^2}{\frac{C_l}{C_d} \left(\lambda + \frac{1}{2B} \right)} \tag{1.83}$$

Some general similarity between this empirical relation Equations 1.81 and 1.78 may be noted with the additional complexity in Equation 1.81 taking account of tip effects over the range of applicability.

It will be evident (Figure 1.17) that for any given blade number, B, and maximum lift-to-drag ratio, k, there is a unique optimum value of tip speed ratio, λ to maximise C_p. Figure 1.18 shows C_p max as a function of design tip speed ratio for a range of lift-to-drag

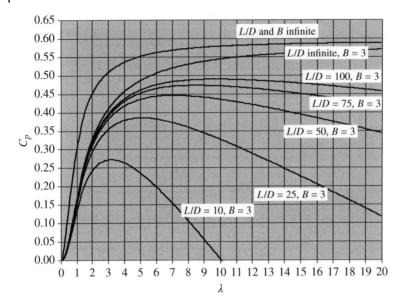

Figure 1.17 C_p max versus tip speed ratio for various lift-to-drag ratios.

ratios and blade numbers. As is confirmed in Figure 1.18 (and also in Figure 1.17), a typical state-of-the-art blade for a large wind turbine designed for a tip speed ratio of around 9 and with average equivalent maximum lift-to-drag ratio around 100 will achieve C_p max of ~0.5. While Figure 1.17 shows appropriate trends, without a full solution of the BEM equations, Equation 1.82 using the assumption of k constant at a maximum value for the chosen aerofoils will only predict C_p curves accurately in the region of C_p max.

Figure 1.18 clarifies an important point that to maximise the benefit from aerofoils that may achieve higher lift-to-drag ratios, it is important to design new (higher) optimum tip speed ratios. Figure 1.18 also provides a clear and immediate indication of how optimum one-, two-, three- or multi-bladed rotors will compare in power performance for any given choice of aerofoils, whilst Figure 1.17 clarifies the penalties that may apply in operating at non-optimum combinations of lift-to-drag ratio and design tip speed ratio.

Equation 1.82 shows that C_p is a complex function of the axial induction, a. The result that $a = 1/3$ results in maximum rotor C_p was based on simple actuator disc theory (Equation 1.11) and this clearly cannot be exactly true for Equation 1.82. It may also be noted that from Equation 1.74, it is only plausible and not rigorous that maximising $k = C_l/C_d$ will maximise the torque on each blade element as the elemental torque clearly also depends on flow angle, φ and how it may vary with k in the course of a full solution of the BEM equations. Considering Equation 1.76, a power coefficient can be defined local to each blade element as:

$$C_p(r, a, k) = \frac{dP(r, a, k)}{0.5\rho U^3(2\pi r\, dr)} \tag{1.84}$$

A strict optimisation according to the BEM theory presented will maximise $C_p(r, a, k)$ separately on each blade element. This results in a varying spanwise and k near to maximum but not absolutely maximum on each aerofoil section. For typical large

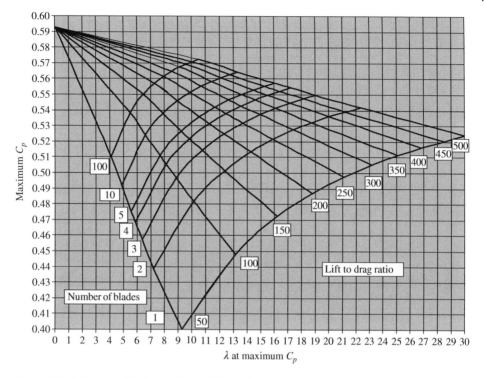

Figure 1.18 Influence of blade number and lift-to-drag ratio on maximum C_p.

electricity producing wind turbines with design $\lambda \geq 6$ and $k \geq 100$, in an ideal optimum blade design, a is a little different from 1/3 over most of the span and k is very close to the maximum for each aerofoil section. These effects are rather more significant, however, for a rotor based on aerofoils with low lift-to-drag ratio, for example, sailcloth blades or plate blades. The non-uniformity of an optimum distribution of axial induction is not of practical importance for optimum rotor design because of the limitations of BEM theory in its present form. Other issues appear in more accurate optimisation methods (see Section 1.10.4). However, based on the BEM equation system in a standard form, the non-uniformity of axial induction (i.e. not exactly constant at a value of 1/3 over the whole span) of an optimum rotor is a consistent outcome and is mentioned for that reason.

1.9.3 Thrust Coefficient

In a similar way to the derivation of maximum C_p from Equations 1.74–1.82, the associated thrust coefficient can be determined as

$$C_T = \int_0^1 8a(1-a)Fx\,dx \tag{1.85}$$

In the limit of no tip loss ($F = 1$), the familiar actuator disc formula, $C_T = 4a(1 - a)$, is recovered with $C_T = 8/9$ for an optimum rotor. Equation 1.85 has a much simpler form than Equation 1.82. The thrust coefficient is a *system* property dependent on rotor loading but independent of the efficiency of the rotor in power conversion. It is therefore unaffected by lift-to-drag ratio or the tangential induction factor and dependent only on the state of rotor loading characterised by the axial induction over the rotor plane which is naturally influenced by the tip effect.

1.9.4 Out-of-Plane Bending Moment Coefficient

The steady-state out-of-plane bending moment of an optimised blade in operation at its design tip speed ratio below rated wind speed and the introduction of pitch action may be derived from any standard BEM code and has a characteristic shape as in Figure 1.19. For typical large-scale electricity generating wind turbines, which are well optimised rotor designs with design tip speed ratios above 6, the shape is largely independent of design specifics and can usually be very well approximated by a cubic curve. Such representations have been convenient in studies developing blade designs embodying passive aeroelastic control (e.g. with flap-twist coupling as in Maheri [53]), where a simplified representation of blade loading is useful. BEM theory provides a simple derivation of the result as follows.

Defining a dimensionless bending moment coefficient at arbitrary radial distance, r, as $C_M(r) = \frac{M(r)}{(0.5\rho U^2 \pi R^3)}$ and following similar methods of analysis as for C_p in Equations 1.74–1.82 lead to:

$$C_M(r) = \frac{8a(1-a)}{B} \int_x^1 \frac{F(y)}{\left\{(1+a')\lambda y + \frac{(1-a)}{k}\right\}} \left\{\lambda y + \frac{(1-a)}{k}\right\} (y-x)y\,dy$$

$$(1.86)$$

Neglecting a' in comparison to unity gives

$$C_M(x) = \frac{8a(1-a)}{B} \int_x^1 F(y)(y-x)y\,dy \qquad (1.87)$$

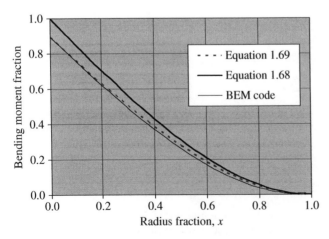

Figure 1.19 Out-of-plane bending moment shape functions.

Considering the case with no tip effect, where $F(y) = 1$:

$$C_M(x) = 4a(1-a)\left\{\frac{1}{3B}(2 - 3x + x^3)\right\} \tag{1.88}$$

And for an optimum rotor, taking $a = 1/3$:

$$C_M(x) = \frac{16}{27B}\left\{\frac{(x-1)^2(x+2)}{2}\right\} \tag{1.89}$$

For a typical conventional three-bladed wind turbine, Equation 1.89 has the appropriate cubic shape but, because the bending moment is most heavily weighted by the loading which is farthest outboard, the tip effect is very significant in relief of blade root bending moment.

Wilson (see Spera [52], Chapter 5, p. 261) and Milborrow [54] (a paper offering a variety of useful simplified parametric equations for blade loads) have previously derived an equation similar to Equation 1.89. Wilson noted that this relationship predicted values significantly greater than measured blade bending moment data [55] from the Mod-2 HAWT. He further observed that the Mod-2 blade design was far from an optimised configuration and described Equation 1.89 as 'an upper bound' which it is for an optimum rotor without tip loss. However, noting the more general form of Equation 1.88, the bending moment coefficient is unsurprisingly related to the thrust coefficient which is not maximised at $a = 1/3$ even in the ideal inviscid model without considering higher loadings that may result in the turbulent wake state. Thus, suboptimal rotors may have bending moment characteristics that exceed (or are within) the predictions of Equation 1.89.

A direct application of Equation 1.89 will overestimate the moment at shaft centreline of an optimum three-bladed rotor with a design tip speed ratio of $\lambda = 7$ by about 13%. Equation 1.87 is therefore not immediately appropriate for use in parametric studies without some adjustment to account for the tip effect.

Equation 1.90 is a useful approximation to Equation 1.83, which can represent the blade root moment quite accurately when (as is the case for mainstream electricity generating wind turbines) the product of blade number and design tip speed ratio, $B\lambda > 10$.

$$C_M(x) = \frac{16}{27B}G(B\lambda)f(x) \tag{1.90}$$

where

$$G(B\lambda) = 5.5744 \times 10^{-7}B^3\lambda^3 - 8.2871 \times 10^{-5}B^2\lambda^2$$
$$+ 4.4085 \times 10^{-3}B\lambda + 2.3245 \times 10^{-1} \tag{1.91}$$

and

$$f(x) = \frac{(x-1)^2(x+2)}{2} \tag{1.92}$$

In retaining the simple cubic function, $f(x)$, and providing an accurate match to the blade root bending moment, Equation 1.90 is somewhat conservative in blade out-of-plane bending moment estimates on the outboard blade. Accuracy in estimation of blade root bending moment is probably of greatest interest in parametric studies and, for a three-bladed wind turbine with $\lambda = 7$, Equation 1.90 gives $C_M(0)$ as 0.8809, whilst integration of Equation 1.86 using the Prandtl tip loss factor gives a corresponding value of 0.8827.

1.9.5 Optimisation to a Loading Constraint

As is discussed in Chapter 9, the most prevalent system optimisation criterion is to min-imise cost of energy (COE). Maximising rotor aerodynamic performance is certainly a natural design objective but may not be overall optimum in minimising COE. Snel [56] noted that at peak C_p (corresponding in the ideal case to an axial induction factor, $a = 1/3$) the thrust coefficient, C_t is rising rapidly so that it may be beneficial to back off maximum C_p a little, sacrificing a little power in order to reduce loads rather more substantially. This idea has been implemented in many past rotor designs but not to the extent of having aerodynamic designs targeting a very much lower induction, a concept more recently explored by Chaviaropoulos and Voutsinas [57]; and this idea was taken further including specialised aerofoil design for such a rotor in the Innwind.EU project [58]. The possibility of a COE benefit in a design for low axial induction and reduced C_p max was investigated. The aim is to have a larger and more productive rotor able to operate at similar load levels to one designed for C_p max at $a = 1/3$.

The essence of this can be derived very simply considering the ideal actuator disc equations for C_p and C_t. From Equation 1.88, the steady-state out-of-plane bending moment, M, at fixed wind speed U is seen to be proportional to the cube of radius, R, and to the thrust coefficient, C_t. Thus,

$$M \propto C_t R^3 \propto a(1-a)R^3$$

Suppose an optimum rotor design is developed to maximise power on the basis that a and R may vary while M is kept constant. Thus,

$$M \propto C_t R^3 \propto a(1-a)R^3 = k_m \text{ (constant)} \tag{1.93}$$

Equation 1.93 holds true for the out-of-plane bending moment at any blade radius when a is the associated local induction but, for the present analysis, consider M to be evaluated at the blade root. The power at the same wind speed, P, is given as

$$P \propto C_p R^2 \propto a(1-a)^2 R^2 \tag{1.94}$$

And thrust, T, is given as

$$T \propto C_t R^2 \propto a(1-a)R^2 \tag{1.95}$$

The variation of power and thrust with a and R may be considered. As a and R are now related through M being constant, $P(a)$, $R(a)$ and $T(a)$ may be determined. From Equations 1.93 and 1.94; $P = k\,a^{1/3}(1-a)^{4/3}$ (with k, another constant) and

$$\frac{dP}{da} = \frac{k}{3}a^{-2/3}(1-a)^{1/3}(1-5a) \tag{1.96}$$

Hence, for P to be maximum,

$$a = \frac{1}{5}$$

Applying the subscript s to values associated with a standard rotor design ($a = a_s = 1/3$), when P is maximum:

$$\frac{R}{R_s} = \left\{ \frac{a_s(1-a_s)}{a(1-a)} \right\}^{1/3} = 1.116$$

$$\frac{P}{P_s} = \frac{a(1-a)^2}{a_s(1-a_s)^2}\left(\frac{R}{R_s}\right)^2 = 1.076$$

$$\frac{T}{T_s} = \frac{a(1-a)}{a_s(1-a_s)}\left(\frac{R}{R_s}\right)^2 = 0.896$$

General trends of R, P, M and T relative to unit values of the standard rotor are presented in Figure 1.20.

The analysis indicates that a rotor designed for an axial induction factor of 0.2 that may be 11.6% larger in rotor diameter can operate with 7.6% increased power and 10% less thrust at the same level of blade rotor out-of-plane bending moment as the baseline design. The rotor diameter is larger by 11.6%, but this does not produce additional power proportional to the 25 % swept area increase because a lower optimum induction of $a = 0.2$ is intrinsic in the trade-off and the limiting rotor power coefficient is reduced from the Betz limit of 0.593 to a value $C_p = 4a(1-a)^2 = 0.512$.

A somewhat more detailed and realistic analysis will modify these numbers, but Figure 1.20 captures the essence of the low-induction rotor design concept. Designing for the same blade steady-state out-of-plane bending load level, say, at rated power is not necessarily the overall optimum solution for minimum system COE but it is a clear basis to investigate more deeply whether designs that target a significantly lower induction than 1/3 may reduce COE. There is reduced rotor thrust associated with maximum power (compared to a purely aerodynamic optimisation with $a = 1/3$) and, in consequence, wind farm wake losses may be reduced. A key issue for the low-induction design concept is whether the larger rotor can always defend against critical load increases over the full spectrum of operating conditions and hence avoid related cost increases.

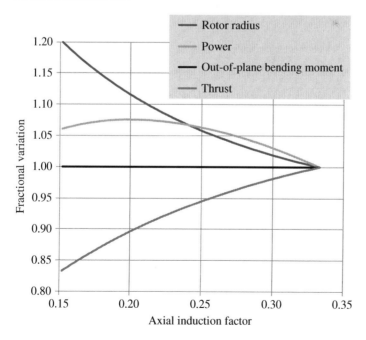

Figure 1.20 Design parameters related to axial induction.

1.9.6 Optimisation of Rotor Design and Hub Flow

Refinements to BEM have been considered [59, 60] in the light of analyses suggesting reduced inboard axial induction associated with reduced pressure near the wake core. These may include viscous effects supported by CFD analysis and are targeted at improving design methods to determine aerodynamically optimum platforms. The following very simplified analysis suggests, however, that such refinements may be pointless unless combined with a model of the flow around some specific hub shape. Flow augmentation benefit from effective design in the hub area may exceed any effects due to wake core suction, while losses in standard design arrangements may conflict with wake suction benefits.

Consider as a reference level an ideal rotor capable of achieving a local C_p on every blade element of 16/27 and hence a complete rotor C_p of 16/27. Now the central region of real rotors varies considerably. There is invariably at least a small nose cone over the hub region. Often, there is an exposed section of blade near the root which is cylindrical having negative aerodynamic performance and creating some drag but no lift. Often, the blade has no positive aerodynamic contribution until beyond 15% span and aerodynamic function is generally reduced inboard of maximum chord which is typically around 25% span.

Let us suppose then that there is a central region of the ideal rotor up to a radius r_a which contributes no aerodynamic power. Then the limiting C_p for the whole rotor is

$$C_p(r_a) = \frac{16}{27} \left(\frac{R^2 - r_a^2}{R^2} \right)$$

(1.97)

Consider instead having a large aerodynamically shaped nose cone of radius r_h covering all aerodynamically inactive hub and blade parts and which deflects the central flow constructively over the remainder of the aerodynamically active blade. The whole rotor C_p may now be represented as

$$C_p(r_h, k) = \frac{2}{R^2} \int_{r_h}^{R} \frac{16}{27} \left(1 + k\frac{r_h^3}{2r^3} \right) r\,dr$$

(1.98)

The explanation for Equation 1.98 defining $C_p(r_h, k)$ is as follows. If potential flow over a 2D object, say a cylinder, is considered, the flow concentration at right angles to the flow direction varies inversely as square of distance r. For a 3D object, say a sphere, the flow concentration varies as inverse cube of distance. The formula bracketed in Equation 1.98 is the standard potential flow solution for a sphere when the coefficient, k, is unity. Thus, a smooth displacement of the central flow outwards with the added mass flow varying as inversely as cube of the radial distance is represented with a coefficient, k, modelling the specific shape of cone as distinct from a spherical cone. According to Section 1.7 and generally accepted wisdom about flow concentration, local C_p will increase directly in proportion to the mass flow increase as is modelled in Equation 1.98.

Losing performance from an area within 20% span incurs ~4% reduction in rotor C_p, whilst deflecting the central flow over working parts of the blades is predicted to give a 1.7% gain for a hub cone of 20% span (assuming k arbitrarily as 0.5). Such a large hub cone seems quite extreme, but note that GE Wind has recently conducted experiments with very large hub cones (Figure 1.21) and with claims for a performance gain

Figure 1.21 GE Wind experimental hub flow system on a 1.7 MW wind turbine. Reproduced with permission of General Electric.

of ~3% [61]. They describe this system as ECO ROTR (Energy Capture Optimization by Revolutionary Onboard Turbine Reshape). The idea that the central flow may be more productively displaced radially outward also appears in studies of downwind turbines by ETH Zurich [62], supported by the Japanese downwind turbine manufacturer, Hitachi. Nacelle blockage preventing flow through the central region of the rotor is found to give small increases in performance.

1.10 Limitations of Actuator Disc and BEM Theory

1.10.1 Actuator Disc Limitations

In spite of the great practical value of his actuator disc concept, Froude was aware of unresolved issues especially in regard to what happens at the edge of the disc. In classical inviscid models, a singularity exists at the edge of the disc because a constant pressure difference across the disc is assumed to exist all the way from the centre to the edge. However, at the edge point, as viewed from outside the disc, the static pressure must in reality have a single unique value. Van Kuik and Lignarolo [23] shows that the iso-bars converge on the disc edge so that, in the inviscid mathematical model at least, the pressure is infinitely multiple-valued. As far as numerical modelling is concerned, this presents no fundamental problem in getting accurate results near the edge of the disc if the numerical resolution is high enough. It is a quite separate issue how best to model flow at the edge of a real rotor with discrete, finite blade number.

1.10.2 Inviscid Modelling and Real Flows

The actuator disc concept is not in itself limited to inviscid modelling and has been employed in many CFD studies [63], but BEM theory is basically inviscid except that some boundary layer effects are implicit in using aerofoil data accounting for drag and empirical add-ons may be incorporated to approximate modelling of unsteady effects like stall hysteresis. It is rather interesting that a full wake expansion is never observed in real flows due to entrainment and mixing of the external flow through viscous effects. Yet, the inviscid model as represented by standard BEM appears to predict power performance, for example, very well at least for aligned flow conditions in operating states avoiding stall. This BEM model is underpinned by the Betz actuator disc analysis; and after accounting for tip losses and finite aerofoil drag based on 2D wind tunnel data, the Betz limit remains as a very credible upper bound for performance of an open flow rotor.

1.10.3 Wake Rotation and Tip Effect

The so-called tip effect essentially differentiates rotors of similar solidity, aerofoil selection and design speed in terms of blade number. In the limit of an infinite number of infinitely slender blades travelling at infinite rotation speed, all the power is produced without a torque reaction or wake rotation.

Assuming a uniform wind field upstream of the rotor with no intrinsic rotating structures or initial angular momentum, the creation of angular momentum in the wake of the rotor is predicted for all real rotors with a finite number of blades and finite speed, which in turn implies a non-zero torque reaction.

This is not in question, but De Vries [50] and later Sharpe [27] have made the case for the view that wake angular momentum is associated with a reduction in wake core static pressure that arises conservatively from the blade circulation. In reviewing general momentum theory, it is apparent that the wake vortex is conservative as must be any vortex in steady state and in inviscid flow. However, this does not mean that it has no influence on available rotor power. The purely axial flow incident on the rotor provides available power to the rotor through the rotor plane pressure difference and also rotational kinetic power to the wake sharing the rotor plane pressure difference pressure difference and associated power in the ratio $(1 : \acute{a})$ between rotor and wake. Conventional BEM modelling as, for example, presented in Section 1.8 does not consider the effect of wake rotational kinetic power or wake suction power in the overall power balance. In CFD modelling, Madsen *et al.* [59, 60] has supported the De Vries interpretation. This is significant for the physical interpretations underlying BEM theory and for the accuracy of detailed design calculations on rotor aerodynamics. It little affects the top-level parametric analyses and formulae developed in Section 1.9 as applied to rotors with design tip speed ratios above about 6. It will matter especially in detailed aerodynamic design of rotors around the hub and tip areas especially and have substantial implications for rotors with very low design tip speed ratio.

There are a number of tip effect models (e.g. see Shen *et al.* [64]). The Prandtl model has been employed as the simplest available, purely for convenience, having in mind that differentiating these models or getting into accurate correspondence with real tip flows moves into territory where the simple BEM theory is generally inadequate. It should also be noted that there are also a number of different approaches in the application of the tip factors that differ in detail from Equations 1.39 and 1.40 including an elegant model from Anderson [65] which accounts for cyclic variation in the induction factors.

1.10.4 Optimum Rotor Theory

Optimum rotors produced by BEM theory differ a little from those developed with the ideal actuator disc assumption that the axial induction is 1/3 everywhere over the rotor span. These differences are considered unimportant because BEM theory is not accurate enough for them to be really meaningful. Work of Johansen *et al.* [66] on optimum rotor design following from the previous work [59, 60] also suggests that classical BEM solutions for optimum rotors will not be very accurately optimal. This is important both at a fundamental level and for practical detailed design of optimum rotors, but does not particularly undermine the value of equations such as Equation 1.72 for guiding parametric design investigations. As mentioned, using the ideal actuator disc model to optimise a rotor of fixed diameter in order to maximise power leads to the simple result that the axial induction should be 1/3. It is interesting to see that the corresponding optimisation at fixed blade bending moment without constraint on rotor diameter also leads to a simple, elegant but different result that the axial induction should be 1/5.

1.10.5 Skewed Flow

A major weakness of BEM theory is in modelling wind turbines in yawed flow. When the flow is oblique to the rotor plane, there are cyclic variations in angle of attack which can be important especially when flow angles approach stall. The strip theory assumption that the rotor can be analysed as annular elements that are independent of each other is less justifiable. Dynamic stall behaviour and stall hysteresis can have greater effect on rotor performance. Also, in yawed flow there are additional issues about the wake. Does it remain symmetric about the rotor axis or is it skewed in the wind direction? Experimental evidence and CFD analyses indicate the latter and skewed wake correction as, for example, based on Glauert [67] have been applied in using BEM to model yawed flow. BEM theory can adopt simplifying assumptions (such as taking account of the angle of the wind vector in the inflow calculations), can incorporate dynamic stall models and yield useful results. However, in yawed flow, there is much less certainty in basic calculations (even such as the determination of average rotor power) than in cases where the wind direction is normal to the rotor plane. In general, more sophisticated aerodynamic modelling using vortex wake models [9, 68], or CFD is desirable.

1.10.6 Summary of BEM Limitations

The limitations of BEM have been highlighted. CFD- and vortex-theory-based analyses may be more accurate in many circumstances. Nevertheless, although huge advances have been made in recent years and progress will continue, current CFD techniques do not yet solve the Navier–Stokes equations with the same objectivity as Mother Nature. Turbulence, transition and boundary layer modelling remain problematic. Some vortex wake models assume Froude's theorem and some CFD analyses are calibrated to reproduce actuator disc results. It is through a mixture of techniques and convergence of insights coupled with experimental feedback that progress is made.

References

1 Breton, S.-P. (2008) Study of the stall delay phenomenon and of wind turbine blade dynamics using numerical approaches and NREL's wind tunnel tests. Doctoral theses at NTNU, p. 171. ISBN: 978-82-471-1019-5.

2 Chaviaropoulos, P.K. and Hansen, M.O.L. (2000) Investigating three-dimensional and rotational effects on wind turbine blades by means of a quasi-3D Navier Stokes solver. *Journal of Fluids Engineering*, **122**, 330–336.

3 Raj, N.V. (2000) An improved semi-empirical model for 3-D post stall effects in horizontal axis wind turbines. Master of Science thesis in Aeronautical and Astronautical Engineering. University of Illinois at Urhana-Champaign, Urhana, IL.

4 Himmelskamp, H. (1947) Profile Investigations on a Rotating Airscrew. MAP Volkenrode Report and Translation No. 832.

5 Snel, H., Houwink, R., van Bussel, G.J.W. and Bruining, A. (1993) *Sectional Prediction of 3D Effects for Stalled Flow on Rotating Blades and Comparison with Measurements*. 1993 European Community Wind Energy Conference Proceedings, Lubeck-Travemunde, Germany, pp. 395–399.

6 Rasmussen, F., Hansen, M.H., Thomsen, K. *et al.* (2003) Present status of aeroelasticity of wind turbines. *Wind Energy*, **6**, 213–228. doi: 10.1002/we.98

7 Snel, H. (2003) Review of aerodynamics for wind turbines. *Wind Energy*, **6**, 203–211. doi: 10.1002/we.97

8 Ostowari, C. and Naik, D. (1985) Post stall studies of untwisted varying aspect ratio blades with NACA 4415 airfoil section part 1. *Wind Engineering*, **9** (3), 149–164.

9 Simoes, F.J. and Graham, J.M.R. (1990) *A Free Vortex Model of the Wake of a Horizontal Axis Wind Turbine*. Proceedings of the 7th BWEA Conference, Norwich, UK, pp. 161–165.

10 Schreck, S. and Robinson, M. (2002) Rotational augmentation of horizontal axis wind turbine blade aerodynamic response. *Wind Energy*, **5**, 133–150. doi: 10.1002/we.68

11 Hoerner, S.F. (1965) *Fluid Dynamic Drag*, Published by the Author.

12 Krauss, T.A. (inventor) (1986) X-wing aircraft circulation control. US Patent 4,573,871, March 1986.

13 http://www.rdmag.com/News/2010/01/Energy-More-power-circulation-control-to-alter-wind-turbine-design (accessed august 2010).

14 Lewis, R.I. (1991) *Vortex Element Methods for Fluid Dynamic Analysis of Engineering Systems*, Cambridge University Press, Cambridge, p. xix.

15 Windblat Enercon House Magazine, Issue 3 2008, http://www.enercon.de/www/en/windblatt.nsf/vwAnzeige/4DA5AEEBACEAEDFDC12574A500418221/$FILE/WB-0308-en.pdf (accessed 25 September 2017).

16 Di Maria, F., Mariani, F. and Scarpa, P. (1997) *Chiralic Bladed Wind Rotor Performance*. Proceedings of the 2nd European & African Conference on Wind Engineering, Palazzo Ducale, Genova, Italy, June 1997, pp. 663–670.

17 Holland, R. Jr. (1981) The Holland Roller Windmill. Investigation and Demonstration of Principles. Report No. DOE/R6/10969– T1, DE86 00478, US DOE, June 1981.

18 http://en.wikipedia.org/wiki/Lift_(force) (accessed August 2010).

19 Froude, R.E. (1889) On the part played in propulsion by differences in fluid pressure. *Transactions of the Institute of Naval Architects*, **30**, 390–405.

20 Bergey, K.H. (1979) The Lanchester-Betz limit. *Journal of Energy*, **3**, 382–384.

21 van Kuik, G.A.M. (2007) The Lanchester–Betz–Joukowsky limit. *Wind Energy*, **10** (3), 289–291.

22 Okulov, V.L. and van Kuik, G.A.M. (2011) The Betz-Joukowsky limit: a contribution to rotor aerodynamics by two famous scientific schools. *Wind Energy*, **15** (2), 335–344. doi: 10.1002/we.464

23 Van Kuik, G.A.M. and Lignarolo, L.E.M. (2015) Potential flow solutions for the energy extracting actuator disc flows. *Wind Energy*, **19** (8), 1391–1406.

24 McLaren-Gow, S., Jamieson, P. and Graham, J.M.R. (2013) *An Inviscid Approach to Ducted Turbine Analysis. June 2014.* European Wind Energy Conference & Exhibition 2013: Proceedings of a meeting held 4–7 February 2013, Vienna, Austria, New York, Vol. 1, pp. 931–938.

25 Greet, R.J. (1980) Maximum windmill efficiency. *Journal of Applied Physics*, **51** (9), 4680–4681.

26 Rauh, A., Seelert, A. and Betz, W. (1984) Optimum efficiency for windmills. *Applied Energy*, **17**, 15–23.

27 Sharpe, D.J. (2004) A general momentum theory applied to an energy-extracting actuator disc. *Wind Energy*, **7**, 177–188. doi: 10.1002/we.118

28 Sørensen, J.N. (2016) *General Momentum Theory for Horizontal Axis Wind Turbines*, Springer International Publishing, Switzerland, p. 23.

29 Phillips, D.G., Flay, R.G.J. and Nash, T.A. (1999) Aerodynamic analysis and monitoring of the Vortec 7 Diffuser-Augmented wind turbine. *IPENZ Transactions*, **26** (1), 13–19.

30 Ohya, J. and Karasudani, T. (2010) A shrouded wind turbine generating high output power with Wind-lens technology. *Energies*, **3**, 634–649. doi: 10.3390/en3040634

31 Werle, M.J. and Presz, W.M. (2008) Ducted wind/water turbines and propellers revisited. *Journal of Propulsion and Power*, **24** (5), 1146–1150.

32 McLaren-Gow, S., Jamieson, P. and Graham, J.M.R. (2013) *A Comparison Between Ducted Turbine Theory and Inviscid Simulation.* 2nd IET Renewable Power Generation Conference (RPG 2013), November 2013, pp. 1–44.

33 Oman, R.A., Foreman, K.M. and Gilbert, B.L. (1975) *A Progress Report on the Diffuser Augmented Wind Turbine.* 3rd Biennial Conference and Workshop on Wind Energy Conversion Systems, Washington, DC.

34 Hansen, M.O.L., Sorensen, N.N. and Flay, R.G.J. (1999) *Effect of Placing a Diffuser Around Wind Turbine.* European Wind Energy Conference, pp. 322–324.

35 van Bussel, G.J.W. (2007) The science of making more torque from wind: diffuser experiments and theory revisited. *Journal of Physics: Conference Series, Denmark*, **75**, 1–11.

36 Jamieson, P. (2008) Generalised limits for energy extraction in a linear constant velocity flow field. *Wind Energy*, **11** (5), 445–457.

37 Lawn, C.J. (2003) Optimisation of the power output from ducted turbines. *Proceedings of the Institution of Mechanical Engineers, Part A: Journal of Power and Energy,* **217**, 107–117.

38 http://www.flodesignwindturbine.org (accessed August 2010).

39 Jamieson, P. (2008) *Beating Betz–Energy Extraction Limits in a Uniform Flow Field.* EWEC 2008, Brussels, March 2008.

40 Glauert, H. (1963) Airplane propellers, in *Division L. of Aerodynamic Theory,* vol. **IV** (ed. W.F. Durand), Dover Publications, New York.

41 Wilson, R.E., Lissaman, P.B.S. and Walker, S.N. (1976) *Applied Aerodynamics of Wind Power Machines,* University of Oregon.

42 Walker, S.N. (1976) Performance and optimum design analysis/computation for propeller-type wind turbines. PhD thesis. Oregon State University.

43 Manwell, J.F., McGowan, J.G. and Rogers, A.L. (2008) *Wind Energy Explained,* John Wiley & Sons, Ltd, Chichester. ISBN: 13: 978-0-471-49972-5 (H/B)

44 van Kuik, G.A.M., Sørensen, J.N. and Okulov, V.I. (2015) Rotor theories by Professor Joukowski: momentum theories. *Progress in Aerospace Sciences,* **73**, 1–18.

45 Whelan, J., Thomson, M., Graham, J.M.R. and Peiro, J. (2007) *Modelling of Free Surface Proximity and Wave Induced Velocities Around a Horizontal Axis Tidal Stream Turbine.* Proceedings of the 7th European Wave and Tidal Energy Conference, Porto, Portugal.

46 Betz, A. (1927) NACA Technical Memorandum 474. Die Naturwissenschaften, Vol XV, No 46, November 1927.

47 Goldstein, S. (1929) On the vortex theory of screw propeller. *Proceedings of the Royal Society of London Series A,* **123**, 440–465.

48 Gasch, R. and Twele, J. (2002) *Power Plants: Fundamentals, Design, Construction and Operation,* Chapter 8,, James & James, ISBN: 10: 1902916379, ISBN: 13: 978–1902916378.

49 Burton, A., Sharpe, D., Jenkins, N. and Bossanyi, E.A. (2004) *Handbook of Wind Energy,* John Wiley & Sons, Ltd, Chichester.

50 De Vries, O. (1979) *Fluid Dynamic Aspects of Wind Energy Conversion,* AGARDograph, vol. **243**, NATO, Advisory Group for Aerospace Research and Development, ISBN: 92 835 1326–6.

51 Jamieson, P. (2009) *Lightweight High Speed Rotors for Offshore,* EWEC Offshore, Stockholm.

52 Spera, D.A. (ed.) (2009) *Wind Turbine Technology: Fundamental Concepts in Wind Turbine Engineering,* 2nd edn, ASME Press.

53 Maheri, A. (2006) Aero-structure simulation and aerodynamic design of wind turbines utilising adaptive blades. PhD thesis. Faculty of Computing, Engineering and Mathematical Sciences, UWE, Bristol.

54 Milborrow, D. (1982) Performance blade loads and size limits for horizontal axis wind turbines. Proceedings of the 4th BWEA Wind Energy Conference, Cranfield BHRA.

55 Boeing (1982) Mod-2 Wind Turbine System development Final Report, Vol. II. Detailed Report No. NASACR-168007, DOE/NASA/0002-82/2, NASA Lewis Research Centre, Cleveland, OH.

56 Snel, H. (2003) Review of aerodynamics for wind turbine. *Wind Energy,* **6** (3), 203–211.

57 Chaviaropoulos, P.K. and Voutsinas, S.G. (2013) *Moving Towards Larger Rotors – Is That a Good Idea?* EWEA Conference 2013, Vienna.

58 Deliverable 2.11. New Aerodynamics rotor concepts specifically for very large offshore Wind turbines http://www.innwind.eu/publications/deliverable-reports (accessed 25 September 2017).

59 Madsen, H.A., Mikkelsen, R., Øye, S. *et al.* (2007) A detailed investigation of the Blade Element Momentum (BEM) model based on analytical and numerical results and proposal for modifications of the BEM model. The Science of Making Torque from the Wind. *Journal of Physics: Conference Series*, **75**, 012016.

60 Madsen, A., Bak, H., Døssing, M. *et al.* (2010) Validification and modification of the blade element momentum theory based on comparisons with actuator disc simulations. *Wind Energy*, **13** (4), 373–389.

61 http://www.gereports.com/post/120795016210/how-does-a-wind-turbine-work-with-ges-new/as (viewed February 2017).

62 Kress, C., Chokani, N., Abhari, R.S. *et al.* (2016) Impact of flow inclination on downwind turbine loads and power, The Science of Making Torque from Wind (TORQUE 2016). *Journal Physics, Conference Series*, **753**, 022011. doi: 10.1088/1742-6596/753/022011

63 Réthoré, P.-E., Zahle, F., Sørensen, N.N. and Johansen, J. (2008) *Comparison of An Actuator Disc Model with Full Rotor CFD Model Under Uniform and Shear Inflow Condition.* 4th PhD Seminar on Wind Energy in Europe: Book of Abstracts, The European Wind Energy Association, pp. 125–126.

64 Shen, W.Z., Mikkelsen, R. and Sorensen, J.N. (2005) Tip loss corrections for wind turbine computations. *Wind Energy*, **8**, 457–475. doi: 10.1002/we.153

65 Anderson, M.B. (1981) An experimental and theoretical study of horizontal axis wind turbines. PhD thesis. Cambridge University, Cambridge, UK.

66 Johansen, J., Madsen, H.A., Gaunaa, M. *et al.* (2009) Design of a wind turbine rotor for maximum aerodynamic efficiency. *Wind Energy*, **12** (3), 261–273. doi: 10.1002/we.292

67 Glauert, H. (1926) A General Theory of the Autogyro. ARCR R&M No.1111.

68 Kloosterman, M.H.M. (2009) Development of the Near Wake behind a HAWT including the development of a free wake lifting line code. MSc thesis. TU Delft.

2

Rotor Aerodynamic Design

2.1 Optimum Rotors and Solidity

Equation 1.72 may be solved for the optimum blade chord width: Cl_d, the design lift coefficient, is the lift coefficient associated with maximum lift-to-drag ratio of the aerofoil section employed at radius, r. An associated optimum blade twist distribution is obtained from Equation 1.73.

$$c(\lambda, r) = \frac{16\pi R^2}{9B\ Cl_d \lambda^2 r} \tag{2.1}$$

In the context of innovative design, all of the parametric influences indicated in Equation 2.1 will be considered:

- Blade number, B (Chapters 11 and 22);
- Design lift coefficient, Cl_d (Section 2.4);
- Design tip speed ratio, λ – the high-speed rotor (Section 7.2).

The *solidity* of a wind turbine rotor is a term which loosely expresses the ratio of the surface area (on one side) of the blades to the rotor swept area. The area here is a 'planform' area and not the area projected on the rotor plane, which would, in general, be less due to the blade twist distribution.

A local solidity at radius, r, may be defined for an annulus of swept area of width Δr as

$$\sigma = \frac{Bc(\lambda, r)\Delta r}{2\pi r\Delta r} = \frac{Bc(\lambda, r)}{2\pi r} \tag{2.2}$$

and, hence, for the whole rotor as

$$\sigma_{Rotor} = \frac{B\int_0^R c(\lambda, r)dr}{\pi R^2} \tag{2.3}$$

The area around the rotor hub may be aerodynamically inactive and the question of whether to include or exclude it from the measure of rotor solidity then arises. There is no clear answer to this as it very much depends on what use is made of the solidity measure, however defined. Thus, in general, rotor solidity is an imprecise term. Nevertheless, it remains a valuable concept relating information about aerodynamic design choices, loads and structure.

The supposition that increasing design tip speed will have a benefit in increasing Reynolds number is occasionally encountered. In fact, for optimum rotor design, the

Innovation in Wind Turbine Design, Second Edition. Peter Jamieson.
© 2018 John Wiley & Sons Ltd. Published 2018 by John Wiley & Sons Ltd.

opposite is true. The Reynolds number on a wind turbine rotor varies from blade root to tip, with both the inflow resultant velocity and blade chord being functions of local radius. The Reynolds number, Re, at a specific location, say 70% span, may (arbitrarily) be taken as representative of the rotor as a whole. In Equation 2.4, a basic definition of Re is combined with Equation 2.1. This shows that, since each chord width of an optimum rotor varies inversely as the square of design tip speed ratio, Re will decrease with increasing design λ. Equation 2.4 also shows that, for an optimum rotor operating at constant tip speed ratio in wind speeds below rated, Re is approximately constant over the blade span and proportional to ambient wind speed. The constancy of Re over the blade span is directly related to the aims for an ideal optimum rotor to have constant lift and constant lift coefficient over the span.

$$Re = \frac{\rho Wc(\lambda, r)}{\mu} \cong \frac{\rho \omega rc(\lambda, r)}{\mu} = \frac{16\pi \rho RU}{9B\,\mathrm{Cl}_d \mu \lambda} \qquad (2.4)$$

An optimum rotor design may be determined for any chosen tip speed ratio. However, Equation 1.82 shows that, for given aerofoils with specific maximum lift-to-drag characteristics and any chosen blade number, there is a unique design tip speed ratio for optimal overall performance. This is now considered in the context of variable speed rotor design.

2.2   Rotor Solidity and Ideal Variable Speed Operation

The ideal of variable speed operation is to preserve constant optimum flow geometry over the rotor by matching rotor speed to wind speed. Thus, considering Figure 1.16, in an optimum state, the axial induction, a, will have the optimum value of 1/3 and the relative flow angle φ at each radial station will correspond to that value of a. As wind speed, V, changes, rotor speed, ω, will be varied by the control system (usually through electronic control of the generator reaction torque) to remain in a fixed proportionality to the wind speed and the design tip speed ratio, λ, will thus be held constant. State-of-the-art aerofoils commonly used on large wind turbine blades have maximum lift-to-drag ratios around 100–125. Consulting Figure 1.18, this implies that an overall maximum will be obtained with a design tip speed ratio of about 9.

If rated wind speed (the lowest wind speed at which the rotor can produce rated power) is typically around 11 m/s, then to operate at a tip speed ratio, $\lambda = \omega R/U = 9$, would imply a tip speed of $\omega R = \lambda U = 99$ m/s. This is a much higher tip speed than is usual and would cause problematic acoustic emission levels[1] for rotors operating on land-based sites. However, reducing the design tip speed ratio would have two adverse consequences:

1) A slight reduction in the peak C_p attainable as 9 is the overall optimum tip speed ratio.
2) A requirement for higher rotor solidity (wider blade chords) potentially adding to blade mass and cost (see Section 3.6) and to transportation difficulties.

---

1 Acoustic noise emission from blades is a very complex phenomenon. Sound power is considered to vary as the fifth power of tip speed. This should at least make it obvious that a design that is close to the acoustic limits for a particular application cannot easily tolerate any increase in design tip speed.

The slight reduction in peak C_p that would result from choosing a lower than optimal design tip speed ratio follows from Figure 1.17. As is indicated in Equation 2.1, a lower design tip speed ratio implies higher solidity unless the aerofoil section selection is changed to employ sections with higher design lift coefficient (see Section 2.4).

In practical rotor design, the problems suggested in the previous discussion are readily resolved by restricting the range of variable speed. In general, the wider the speed range, the greater the energy capture but the higher is the cost in power electronics. Much of the energy benefit is achieved with a speed range of about 1.5 : 1 and this is typical of the very popular design solution for variable speed operation using a doubly fed induction generator (DFIG).

The typical steady-state operating schedule of a 1.5 MW wind turbine is illustrated in Figure 2.1.

The design solution of Figure 2.1 can be considered to harmonise a number of aspects:

- Zone of operation at optimum tip speed ratio for maximum aerodynamic performance.
- Use of state-of-the-art aerofoils but avoiding too high solidity and excessive blade surface area.
- Limiting the variable speed range and hence limiting cost of the power converter and avoiding acoustic problems that would be associated with a very high maximum tip speed.

Although the optimum aerodynamic design is predominantly determined by the design tip speed ratio corresponding to variable speed region in Figure 2.1, the operating schedule is now relatively complex with two additional regions of fixed speed below the rated wind speed. A more rigorous optimisation of the planform design (chord and twist distribution) to maximise annual energy return at a specified design site is then best achieved by coupling an optimising routine to blade element momentum (BEM) software. The predictions of Equation 1.50 or 1.72 and the related optimum

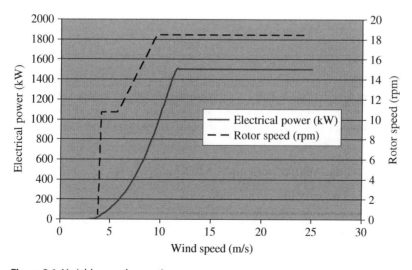

Figure 2.1 Variable speed operation.

twist distribution (Equation 1.73) can still be very useful[2] in providing a starting point for such optimisations. Of course, integrated blade design optimisation will address more than energy capture. The process, in general, will involve constraints (e.g. on maximum tip speed or on key loads) and other cost/value objective functions like cost of energy rather than simply energy if a sufficiently elaborate model is constructed to evaluate that.

2.3 Solidity and Loads

The design load envelop of a wind turbine system is commonly specified in international design standards [1–5]. The wind turbine system is then evaluated in a wide range of load cases that may arise in a variety of wind conditions and rotor states (comprising usually normal operation, idling in standby, stopped for maintenance and fault conditions). The methodology of evaluating new concepts often demands a comparison of design load impacts, and issues of loads and load cases are discussed further in Section 10.7.

From the basic actuator disc theory (Section 1.5), a rotor when operating optimally achieves a thrust coefficient of around 8/9 regardless of details of the aerodynamic design. Hence, rotor solidity can have no strong direct effect on the steady-state operational loads of an optimum rotor. Also, a rotor with low solidity (and, therefore, relatively higher design lift aerofoil sections) will have lower extreme storm loading (rotor parked or idling) from drag on the blades but much the same potential to pick up lifting loads as a high solidity rotor.

The effect of rotor solidity on dynamic (fatigue) loads and on a variety of extreme load cases is far more subtle. Reduced solidity will often imply reduced mass, reduced rotational inertia and usually greater structural flexibility. All these factors can in appropriate combination lead to wide-ranging load reductions to be associated with lower solidity blades. This is illustrated in the discussion in Section 7.2.

2.4 Aerofoil Design Development

Aerofoil design development, even exclusively in the context of wind technology, is a huge subject and this review considers it summarily in the context of how aerofoil selection interacts with blade structure. An optimum rotor design for a given design tip speed ratio requires only the product of chord and design lift coefficient (i.e. the lift coefficient corresponding to the angle of attack that maximises the ratio of lift-to-drag coefficient) at each blade radius to be defined. This is the basis of the non-dimensional function, $\Lambda(\lambda, x) = c(\lambda, x)Cl/R$ defined in Equation 1.72, being unique. Many different aerofoil selections and rotor planforms that maintain the appropriate value of the product $c(\lambda, x)$ Cl at each radial station can therefore satisfy an optimum rotor design for a given tip speed ratio. This design freedom has contributed to interest both in 'low-lift

---

2 A useful technique in many such optimisations (that can save calculation time substantially) is initially to freeze the twist distribution at the theoretical values determined by Equation 1.73, optimise the chord distribution and then free the twist distribution to complete a joint optimisation at a later stage in the process.

aerofoils' and 'high-lift aerofoils' which, respectively, allow relatively wider or narrower chord widths for any given design tip speed ratio.

Low-lift aerofoil development started at the Solar Energy Research Institute (SERI), the precursor of the National Renewable Energy Laboratory (NREL) in the mid-1980s. The main objective in the development of 'low-lift' aerofoils was not particularly to have low lift at the design operating point and therefore wider chords (say, for greater structural comfort). Thus, such 'low-lift' aerofoils are not directly thematic in terms of targeting a low design Cl. Nevertheless, they represent a most interesting development in wind turbine rotor aerodynamics. The intention with these so-called low-lift aerofoils was to restrict maximum lift and, most crucially, to *reduce the difference between the optimum lift design point and maximum lift*, thereby restricting the capability of the aerofoil to pick up high loads in off-design operation. It is notable that this is diametrically opposite to the objective in aircraft design[3] where a large difference between the design point (highest lift-to-drag ratio) and stall (potentially disastrous for an aircraft) represents a safety margin and an additional lifting capability for low-speed operation at take-off and landing.

Tangler and Somers [6, 7] developed aerofoils to this 'low-lift' design criterion. Comparing the SERI S805 aerofoil with the NACA 63418 (Table 2.1), at similar design lift coefficient, Cl_{des}, and associated angle of incidence, α_{des}, the S805 has an appreciably lower maximum lift, Cl_{max} and a reduced range $(\alpha_{max} - \alpha_{des})$ of incidence between the design point and stall.

These issues continued to be explored, for example, in the paper by Giguère *et al.* [8] in 1999. By regulating loads through reduced maximum lift, it was feasible to retrofit rotor designs with rotors of larger diameter and greater energy productivity without exceeding the original load design envelop. The design remit was subsequently enlarged and a new aerofoil family (NREL aerofoils) was developed [9–11], for a range of applications encompassing inner and outer rotor sections and different modes of operation. These have subsequently been used successfully on a number of wind turbine designs.

The ideas behind development of high-lift aerofoils are also diverse. A strong theme in the industry has been to develop higher lift sections for the inboard rotor. In practical blade design, because very little power production is at stake, inboard of about 20% span, aerodynamic design is largely abandoned and a structural transition is effected between the aerofoil section at 20% span and the (usually) cylindrical blade root end. However, optimum rotor theory predicts large chord widths and high angles of blade twist on the inboard blade and the maximum chord tends to be very large, much wider

Table 2.1 Design lift characteristics.

	Cl_{max}	Cl_{des}	α_{des}	α_{max}	$(\alpha_{max} - \alpha_{des})$
S805	1.20	0.70	4°	12°	8°
NACA 63418	1.62	0.75	4°	16°	12°

---

3 Wind turbine aerodynamics began with a perception that surely the aircraft industry that had honed aerofoil design through two World Wars and many years of civil aircraft operation must 'know it all'. Initial aerofoils choices were inappropriate, for example, the performance of the NACA 230xx sections widely used in helicopter technology was far too sensitive to roughness accretions from desert dirt and insects.

than is usefully structurally and also problematic for handling and transportation. Since a higher (design) lift section will have a relatively smaller associated optimum chord, this is one possible solution and many higher lift sections have been developed with that objective by FFA [12], Risø [13] and Delft [14].

An alternative solution to realise a thick section on the inboard blade is to truncate aerofoils.

As early as 1984, Howden Wind Turbines Ltd. of Glasgow, Scotland developed the design of a wind turbine blade for a 750 kW wind turbine based on progressively increased truncation of the NASA LS 21% section. Thus, although the sections remain 21% thick relative to the original complete aerofoil chord, their thickness relative to the truncated chord is much increased. The truncated aerofoil sections facilitated structural design and manufacture of the blade from mid span inboard to maximum chord without compromising the expected aerodynamic performance. There is a strong revival of interest in truncated aerofoils, now often described as 'flatback' aerofoils [15], for structural benefit and slight energy gains in use near the root of blades.

In 1993, pursuing a completely different objective investigating a device (SLEDGE – sliding leading edge) to control power and limit rotor overspeed [16], wind tunnel tests were conducted at Imperial College London on various 'D' sections comprising the first 20–30% chord of the NASA LS 21% section (see inset sketch of Figure 2.2). Test data such as in Figure 2.2 showed surprisingly high lift coefficients for the D sections approaching a value of 3. This extreme truncation of up to 80% of chord, as measured from the trailing edge, is of course far beyond a level that would be used on a power producing inboard blade section. However, it informs of the high lift characteristics of D sections, explaining why high lift can be expected of truncated sections or equivalently that truncation often has little effect on lifting capability. This also enlightens why rotors like the Savonius type (Chapter 13) are not pure drag devices.

Work on specific aerofoil design targets such as low lift for stall-regulated rotors and high lift for inboard aerofoil sections soon expanded into tailored aerofoil designs for

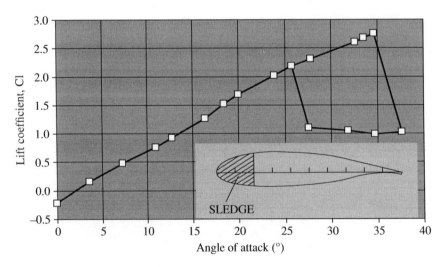

Figure 2.2 Lift characteristics of a 'D' section.

Table 2.2 Overview of HAWT aerofoils.

General aviation aerofoils	Wind-turbine-dedicated aerofoils
NACA 63-4xx	S8xx (NREL, USA)
NACA 63-6xx	FFA W-xxx (FOI, Sweden)
NACA 64-4xx	Risø-A1-xxx (also B, P series, Risø, Denmark)
—	DU xx-W-xxx (Delft, Netherlands)

HAWT, horizontal-axis wind turbine.

Figure 2.3 Overview of DU aerofoils.

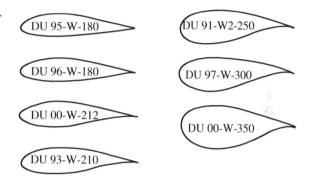

the whole rotor. Sections developed at Delft University, for example, are now very widely used in the industry.

Table 2.2 and Figure 2.3 are from van Rooij [14]. The DU sections of Figure 2.3 have been used by GE Wind, Repower, DeWind, Suzlon, LM Glasfiber, NOI Rotortechnik, Fuhrlander, Pfleiderer, Euros, NEG Micon, Umoe blades, Ecotecia, and others.

Much aerofoil development has been done at Risø. The Risø-B1 aerofoil family was developed for megawatt-size wind turbines with variable speed and pitch control. Seven aerofoils were designed with thickness-to-chord (t/c) ratios between 15% and 53% to cover the entire span of a wind turbine blade. The aerofoils were designed to have high maximum lift and high design lift to allow a slender flexible blade while maintaining high aerodynamic efficiency. This aim can be seen to be exactly in line with the issues discussed in Section 2.2.

High lift could clearly be useful in reducing the maximum chord width, often a problem for transport and for efficient structural transition to the blade root. However, high lift can also reduce the solidity of the outer blade and increase taper of the blade. Whilst a high-lift section may lead to small chord widths that are structurally inefficient in conventional material systems (glass polyester or glass epoxy), it may well suit higher strength systems using carbon efficiently and avoiding buckling reinforcement demands by having reduced blade surface area.

In the mid-1990s, partly assisted by UK government funding, GL Garrad Hassan conducted research exploring the potential of high-lift aerofoils [17] for use on wind turbines. The Garrad Hassan, High Lift 2, GHP_HL2 aerofoil is one of two high-lift aerofoils developed in this project. The NASA LS18, a high-performance aerofoil used on wind turbines with good insensitivity to roughness, was taken as a starting point for design

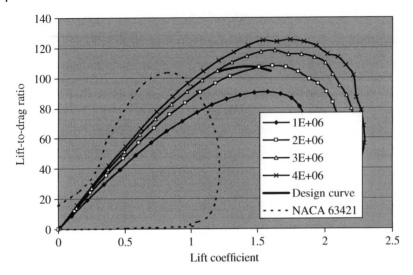

Figure 2.4 Lift-to-drag ratio comparisons.

and higher lift sections were developed by increasing trailing edge camber. The XFOIL code [18], developed by Mark Drela in the mid-1980s, was used for aerofoil section design and the predicted 2D performance was confirmed in wind tunnel tests at Imperial College London. Although this work was not taken further to develop a family of aerofoils for use on a wind turbine blades, some illuminating results emerged.

Figure 2.4 shows the variation of lift-to-drag ratio of the GHP_HL2 aerofoil over a Reynolds number range from 1 000 000 to 4 000 000. A lift to drag polar of the NACA 63421 aerofoil at $Re = 3\,000\,000$ is also included for comparison.

As intended by design, the lift coefficient at maximum lift-to-drag ratio of the GHP_HL2 is high, ~1.5–1.7 depending on Reynolds number. Suppose a design using the GHP_HL2 aerofoil is based on a Reynolds number, Re, of 2 000 000 and design lift coefficient of 1.5, this will determine an optimum chord c_{opt} to satisfy the optimum lift distribution $\Lambda(\lambda, x)$, as defined in Equation 1.50. Increasing or decreasing chord beyond the local optimum value will increase or decrease the local Re with little change to maximum lift to drag. This is illustrated in the series described as 'design curve' in Figure 2.4. The end result is that a considerable variation in chord width (to allow an optimum choice for structural design) is possible without any significant degradation of rotor performance. The variation in design lift coefficient in Figure 2.4 from around 1.2 to 1.6 (~±14%) allows about 30% total variation in chord width selection (see Equation 2.1) whilst keeping changes in maximum lift to drag within about 4%, which will have little impact on maximum C_p (see Figure 1.18).

Having a flat-topped lift to drag polar in the region of maximum lift to drag is therefore beneficial in aerofoil design. The compensatory relationship between chord and Re, which results in the design point characteristic being flatter than the polar characteristic at any fixed Re, is a further significant benefit. This is not, in principle, specific to high-lift aerofoil design, although it was an interesting property of the specific high-lift GHP_HL2 design. In addition to allowing design flexibility in terms of chord width selection, this will reduce the sensitivity of power performance when tracking maximum C_p to off-design operation.

Vertical-axis wind turbine (VAWT) aerofoil design has also been reawakened with the development of special thick aerofoils (Section 13.6.2).

2.5 Sensitivity of Aerodynamic Performance to Planform Shape

The design of a rotor having regard to aerodynamic performance, structure and loads, acoustic performance, manufacturing practicalities, and so on is extremely complex and increasingly employs sophisticated optimisation design tools [19]. Before approaching the general aspects of structural design, some issues around aerodynamic design and energy capture are explored. A structurally optimum beam for multi-directional wind loading would be more like a tapered tube than a conventional blade, whereas the most effective aerofoils may have rather vulnerable trailing edge sections and optimally be much thinner in t/c ratio than is desirable structurally. Since an optimal aerodynamic design is almost certainly not optimal structurally, the sensitivity of optimum chord width to power performance and energy capture may be considered for some selected designs. The idea is to explore what freedom there is to vary chord width from a strict aerodynamic optimum for potential structural benefit. Consider the following question:

> Without a net loss over the whole rotor of more than 0.5% of annual energy at a specified design site, how much variation in chord width from a strictly optimum design can be allowed, considering designs with different aerofoil families?

Results (Figure 2.5) obtained using software based on standard BEM theory are perhaps surprising in that quite large variations in solidity above ±10% can be admitted.

Planform designs are presented (Figure 2.5) for a 100-m diameter rotor designed for variable speed operation. The designs are based on NACA 634xx aerofoil sections with a

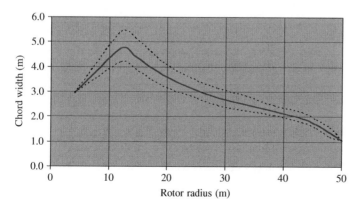

Figure 2.5 Planform envelopes based on the NACA 634xx aerofoils.

Table 2.3 Tolerance of aerofoil types to percentage solidity variation.

NREL S816/7/8 (%)	NACA 634xx (%)	High lift GHP_HL2 (%)
21	24	25

typical range of t/c values over the span of the blade. Apart from the usual modification in the transition to the blade rotor area and some reduction in maximum chord width below optimum, the solid middle line of Figure 2.5 represents the ideal optimum plan-form as may be determined from the standard BEM theory embodied in Equation 1.73. Irregularities are due to variations in aerofoil section properties across the span and would be smoothed out in a final blade design. The other lines represent boundaries within which the whole planform of the blade may be varied without an overall loss of energy exceeding 0.5%.

From a structural standpoint, this allows blades with a range of taper ratios or signif-icant widening of the planforms near mid-span if that area is found to be stress critical.

The results of Table 2.3 were obtained from analyses of the energy sensitivity to plan-form variations. Any energy calculation is based on assumed site conditions, in this case a site with annual mean wind speed of 8.5 m/s and Rayleigh wind distribution. The inter-pretation of the results in Table 2.3 is that solidity variations of a range from 21% to 25% (depending on chosen aerofoil type) corresponding to roughly $\pm10\%$ to $\pm12\%$ relative to a strictly aerodynamically optimum design are possible without loss of more than 0.5% of annual energy.

There is no decisive reason to choose 0.5% as a trade-off limit except that energy is the prime value of a wind turbine system and only quite small energy sacrifices are likely to benefit overall cost of energy. Moreover, the results presented in Figure 2.5 are specific to particular aerofoil choices and operational parameters. Nevertheless, it is generally the case that the energy capture capability of rotor designs can have sub-stantial tolerance to planform variations. This was partly explained in that the 'design curve' (Figure 2.4) has less sensitivity to off-optimum (Cl/Cd maximum) performance than any Cl/Cd polar at a specific Reynolds number. This tolerance is very helpful for blade design. It allows trade-off between loads, energy and blade (or other component) costs and can be exploited in an overall optimisation of cost of energy to meet design objectives other than purely maximisation of energy capture.

2.6 Aerofoil Design Specification

Some of the issues around aerofoil design and rotor optimisation have been discussed. A qualitative definition of requirements for an aerofoil developed for use on a variable speed wind turbine pitching in the direction of attached flow (i.e. making no direct use of stall) is presented in Table 2.4. This aerofoil design specification reflects a number of fea-tures, some essential for energy capture and some for structural efficiency, some impor-tant for dynamic performance and limitation of fatigue loading. Not all of these criteria can easily be targeted in design or simultaneously satisfied. The aerodynamic aspects have greater weight further outboard and the structural ones on the inboard rotor.

Table 2.4 Aerofoil design specification.

	Design demand	Reason
1	Design C_L (C_L at C_L/C_D max) chosen to suit structure	For optimal blade section structure
2	High C_L/C_D max ~120	For maximum energy capture
3	Roughness insensitivity of C_L/C_D near design C_L	For maximum energy capture
4	Effective t/c >90% of maximum t/c	For structural efficiency
5	Flat C_p curve around C_p max	For maximum energy capture
6	Avoid large difference in α at design C_L and α at C_L max	Limit fatigue loading
7	Benign stall	Reduction of fatigue loading
8	Stall hysteresis (loops help alleviate negative lift slope)	Reduction of fatigue loading

Reviewing Table 2.4, the first design demand results from an approach where a structurally determined local solidity will imply a required design value of C_L. This applies in the most integrated type of design approach where aerofoil design (or at least selection) is interactive with structural design. Regarding items 2 and 3, demand for high lift-to-drag ratio is fundamental to almost all aerodynamic machines and the values reflect what is currently achievable with aerofoils that retain a significant amount of turbulent flow and avoid undue roughness sensitivity. Roughness sensitivity is rather less consequential for non-stall-regulated designs as most of that sensitivity is manifested as separation progresses towards maximum lift unless the aerofoils are designed to have substantial laminar flow.

Demand 4 reflects the fact that the specific shape of the aerofoil, and not simply its maximum t/c, has a direct bearing on its structural efficiency. The effective t/c ratio, $(t/c)_{eff}$, is defined and discussed in Section 3.4.

The remaining criteria reflect aspects of dynamic performance of the wind turbine. Demand 5 is important for good C_p tracking (or, rather, providing tolerance to imperfect C_p tracking) of a variable speed wind turbine. Demands 7 and 8 are not primary design considerations, but it is certainly desirable to avoid any particularly adverse stall characteristic. It may be possible to conduct more or less generic design studies on the sensitivity of machine fatigue to stall characteristics. It will, however, be extremely difficult to tailor these characteristics with any accuracy as a part of the aerofoil design process.

2.7 Aerofoil Design for Large Rotors

There are issues in aerofoil design unique to very large rotors. As manufacturers develop wind turbines in the multi-megawatt range, Reynolds numbers, as may be verified substituting values typical of a 10 MW wind turbine in Equation 2.4, rise to orders beyond 10^6. The DTU 10-MW RWT (reference wind turbine) [20] has Re in a range $6–12 \times 10^6$. Higher Re will often tend to increase the maximum lift and maximum lift-to-drag ratio of aerofoils. Such increased lift-to-drag performance can increase C_p or be traded for extra aerofoil thickness to improve the structure.

However, at high Re the transition behaviour and roughness sensitivity may become more problematic in terms of maintaining stable performance. This is a general problem in aerofoil design for very high Re, which is now highlighted considering specifically the low-induction rotor concept explored in Section 1.9.5. The design of a low-induction rotor for, say, 10 MW rated power requires both attention to the mentioned high Re issues and must also address achieving a low design lift coefficient, Cl_d, so that, considering Equation 2.1 for example, the blade chord width is not too small for an efficient blade structure.

Chaviaropoulos *et al.* [21] consider the design of new aerofoils for large low-induction rotors [22]. New profiles are assessed using high-fidelity computational fluid dynamics (CFD) tools with the main uncertainty arising in transition modelling. A design objective is defined as maximisation of lift-to-drag ratio (Equation 2.5) over a range of moderately low Cl_d, considering only a selection of designs with a weighting for performance in turbulent flow in a range biased from 70% to 90%. In Equation 2.5, W_l and W_t are, respectively, weighting factors for laminar and turbulent flow operation. The maximisation over ranges in Cl_d, and over changes in flow state aims to introduce conservatism reducing performance sensitivities that may result in undesirable loading and production variations. Optimising blade structure and aerodynamic function requires a wide range in section t/c, and designs were developed for t/c ratios ranging from 15% to 40%. The design focus was in comparing a set of aerofoils designed with the weighting 30–70% ($W_l = 0.3$ and $W_t = 0.7$), with a second set with weightings 20–80% or in some cases 10–90%. More consistent results were found for the second set.

$$\text{Maximise} \left[\int_{cl_{d1}}^{cl_{d2}} \left\{ W_l \left(\frac{C_l}{C_d} \right)_l + W_t \left(\frac{C_l}{C_d} \right)_t \right\} dC_l \right] \quad \text{with } (W_l + W_t = 1) \quad (2.5)$$

Aerofoil performance predictions were made using MaPFlow (CFD solver), Foil2w (viscous–inviscid interaction solver) and XFOIL, with quite similar results around the design Cl of 0.8. Figure 2.6 shows only the XFOIL results in predicting performance of

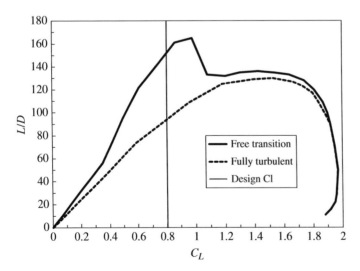

Figure 2.6 Performance (L/D) of the 18% low-lift 10–90 airfoil for transitional and fully turbulent flow conditions. Fixed transition locations were taken from XFOIL using the e^N model with $N = 4$.

the 24% t/c low-lift aerofoil with 10% laminar and 90% turbulent weighting. A substantial difference between free transition and fully turbulent models is evident. However, a design objective of avoiding an unacceptably low L/D in the fully turbulent state is met.

Overall, comparing the low-induction rotor design developed in Innwind [22] with the reference 10 MW wind turbine [20], an increase in production (capacity factor) of 7.5% was predicted, 4.5% being attributed to the low-induction rotor enabling a larger diameter and 3% from the dedicated low-lift aerofoils that, maintaining the same t/c everywhere, replaced the high-lift, FFA-W3-xx of the reference design.

References

1 International Standard IEC 61400-1. (2005–2008) *Wind Turbines – Part 1: Design Requirements*, 3rd edn, International Electrotechnical Commission, Geneva.

2 International Standard IEC 61400-3. (2009) *Wind Turbines – Part 3: Design Requirements for Offshore Wind Turbines*, International Electrotechnical Commission, Geneva.

3 Germanischer Lloyd (2010) Rules and Guidelines, IV – Industrial Services, Part 1 – Guideline for the Certification of Wind Turbines. Edition 2010.

4 DIBt (2004) Schriften des Deutschen Instituts für Bautechnik, Reihe B Heft 8, Richtlinie für Windenergieanlagen – Einwirkungen und Standsicherheitsnachweise für Turm und Gründung, Fassung März 2004 einschl. Korrektur Dezember 2006 (Writings of the German Institute for Construction Technology, Series B Issue 8, Guideline for wind turbines–effects and stability analysis for tower and foundation, version March 2004 including correction December 2006).

5 DNV Offshore Standard DNV-OS-J101. (2007) *Design of Offshore Wind Turbine Structures*, Det Norske Veritas, Hovek, Norway.

6 Tangler, J.L. and Somers, D.M. (1985) Advanced airfoils for HAWTs. Wind Power 85, San Francisco, August 1985.

7 Tangler, J.L. and Tu, P.K.C. (1988) *Peak Power and Blade Loads on Stall Regulated Rotors as Influenced by Different Airfoil Families*. IEA Meeting, R&D WECS Joint Action on Aerodynamics of Wind Turbines, Lyngby, November 1988. Also SERI report TP-3334.

8 Giguère, P., Selig, M.S. and Tangler, J.L. (1999) Blade Design Tradeoffs Using Low-lift Airfoils for Stall Regulated Horizontal Axis Wind Turbines. NREL/CP-500-26173, National Renewable Energy Laboratory.

9 Tangler, J.L. and Somers, D.M. (1995) NREL Airfoil Families for HAWTs. NREL/TP-442-7109.

10 Tangler, J., Smith, B., Jager, D. and Olsen, T. (1990) *Atmospheric Performance of the Special-Purpose SERI Thin-Airfoil Family: Final Results*. SERI/TP-257-3939, European Wind Energy Conference, Madrid, Spain, September 1990, .

11 Tangler, J.L. and Somers, D.M. (1987) *Status of the Special Purpose Airfoil Families*. SERI/TP-217-3264, Solar Energy Research Institute, Golden, CO, December 1987.

12 Björck, A. (1990) Coordinates and Calculations for the FFAW1-xxx, FFA-W2-xxx and FFA-W3-xxx Series of Airfoils for Horizontal Axis Wind Turbines. Report FFA TN 1990-15, Stockholm, Sweden.

13 Fuglsang, P., Bak, C., Gaunaa, M. and Antoniou, I. (2004) Design and verification of the Risø-B1 airfoil family for wind turbines. *Journal of Solar Energy Engineering*, **126** (4), 1002. doi: 10.1115/1.1766024

14 Timmer, W.A. and van Rooij, R.P.J.O.M. (2003) Summary of the Delft University wind turbine dedicated Airfoils. *Journal of Solar Energy*, **125**, 488–496.

15 Chen, X. and Agarwal, R. (2010) *Optimization of Flatback Airfoils for Wind Turbine Blade*. ASME 2010 4th International Conference on Energy Sustainability (ES2010), Phoenix, Arizona, 17–22 May 2010.

16 Jamieson, P. (1993) SLEDGE – A Novel Braking Device for Horizontal Axis Wind Turbines. Report WN 5119 for ETSU.

17 Jamieson, P. and Rawlinson-Smith, R. (1994) *High Lift Aerofoils for Horizontal Axis Wind Turbines*. Proceedings of 16th BWEA Conference, Stirling, June 1994, pp. 85–90.

18 http://web.mit.edu/drela/Public/web/xfoil (accessed August 2010).

19 Fuglsang, P. and Madsen, H.A. (1999) Optimization method for wind turbine rotors. *Journal of Wind Engineering and Industrial Aerodynamics*, **80** (1–2), 191–206.

20 Bak, C., Zahle, F., Bitsche, R. *et al.* (2013) *The DTU 10-MW Reference Wind Turbine*, Danish wind power research.

21 Chaviaropoulos, P.K., Beurskens, H.J.M. and Voutsinas, S.G. (2013) *Moving Towards Large (r) Rotors-Is That a Good Idea*. EWEA Conference 2013.

22 Chaviaropoulos, P.K., Sieros, G., Prospathopoulos, J., Diakakis, K. and Voutsinas, S. (2015) *Design and CFD-Based Performance Verification of a Family of Low-Lift Airfoils*. EWEA Conference 2015.

3

Rotor Structural Interactions

3.1 Blade Design in General

The emphasis, as elsewhere, is on a top-level parametric view to get basic insights, in this case, about the materials and structures of wind turbine blades. Although blade design is only discussed at an outline level, a new focus on integrated rotor design and dealing with the interactions between structural and aerodynamic design is presented. As a corrective to the simplistic overview at the level of parametric study and to reveal the underlying complexity of detailed blade design and manufacture, a snapshot of aspects of industrial blade technology (Section 3.6) is also presented.

The growth in scale of wind turbine technology is spectacularly obvious[1] considering rotor blades. In Figure 3.1, Vestas blades are being transported by sea near Didima in Greece. It has taken many years for each manufacturer to master the technology, and blade manufacturers have generally become substantially committed[2] to particular material systems and particular processes. This is quite understandable given the demand to develop ever larger blades, the complexity of the manufacturing process, the need to test thoroughly and the need to maintain quality assurance with every development.

In that context, the wider, more general but also most relevant question of what is the best material system for a rotor system to achieve minimum cost of energy (COE) is not usually addressed. An attempt is therefore made here to examine that issue with simple top-level parametric analyses. The aim is to illumine the important interactions between the aerodynamic design, which serves the prime functional output of the wind turbine – energy production – and structural design which impact on blade cost and system cost via impacts on rotor loading. As with many of the parametric analyses presented in this book, there is no claim to any final answer through such investigations. Their purpose is to shed light and guide technology development choices ahead of more intensively analytical detailed design stages.

---

1 In contrast to the latest giant blades, the author can recall (during a visit in 1984 to a site on the island of Mull in the west of Scotland) transporting an Aerostar 3.5-m blade (salvaged from a failed rotor) on the roof rack of a car.

2 More radical developments are more or less enforced by technology advances from time to time, as, for example, when Aerpac switched from hand lay-up using polyester resin to epoxy resin infusion.

Innovation in Wind Turbine Design, Second Edition. Peter Jamieson.
© 2018 John Wiley & Sons Ltd. Published 2018 by John Wiley & Sons Ltd.

Figure 3.1 Rotor blades manufactured by Vestas Blades in transportation. Reproduced with permission of Vestas Technology R&D.

3.2 Basics of Blade Structure

Blade design is a marriage of aerodynamics and structures, the former requiring slender aerofoil sections of minimum drag, the latter seeking circular sections that are robust under omni-directional loading. Thus, a compromise is needed and is set quite naturally by structure dominating the inboard blade (where less power is produced but the greatest bending loads are sustained) and aerodynamics ruling on the outer 50% of span. The situation is summarised in Figure 3.2. The blade root end poses one of the classic structural challenges – how to engineer an effective transition between metallic and composite structures. The primary strength in the blade composite is in longitudinal fibres, and a popular solution is to have tapered studs connecting to as many fibres as possible with internal threads accommodating bolts securing the blade end to the hub flange. The *most* popular root connection has been cross-bolt or IKEA type, where a nut is located some distance from the root face in a cylindrical hole drilled through the laminate thickness forming a deep anchor for a bolt through the face. It is much used because it is largely free of intellectual property restrictions. Compared with the bonded-in tapered studs, the spacing of the bolts has to be relatively wide so that the strength of the root for a given diameter is lower, overall making for a less economical blade root and hub.

Just beyond the blade root area, which is almost universally circular in modern designs, transition occurs to the start of the aerodynamic blade around 25% span. The load paths are complex in this region and the blade surface area per unit length is greatest, introducing concerns about buckling of skins. Moving further outboard, aerodynamic considerations become increasingly important and a battle is in progress to maintain aerofoil thickness without much aerodynamic performance compromise.

A typical distribution of power with blade span (see Figure 3.3, based on analysis of a 3 MW wind turbine in operation below rated wind speed), shows clearly the bias in importance of the outer blade. The power produced increases approximately linearly with radius up to a point where the tip effect attenuates it. Note also the drop off in power

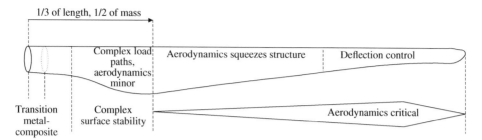

Figure 3.2 Blade structure zones.

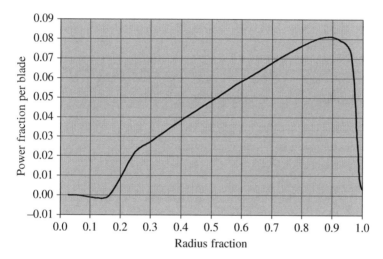

Figure 3.3 Blade power distribution.

inboard of maximum chord around 25% span and even very small negative contributions from drag near the blade root end. Outline COE estimates (9.6) inform that the rotor blades' capital value may be about one-tenth of lifetime cost on land-based sites and less offshore. Thus, each percent of energy capture is 10 or more times more valuable than a percent of blade set cost and any trade-off in aerodynamic efficiency for structural comfort (as has been discussed in Chapter 2) must be very carefully considered.

While very little power production is associated with the innermost parts of the blade, recent work [1] suggests that there is more useful benefit than may be expected from simple blade element momentum (BEM) analysis (as represented in Figure 3.3) in improving rotor aerodynamic design in the vicinity of the hub. This may be because flow is diverted from the hub centre by the spinner. Computational fluid dynamics (CFD) studies [2] make the case for there being augmented flow and negative induction near the centre of rotation.

Around 50% span of the blade structure may often be critically designed by fatigue loading. At this point, the bending moments are relatively large in relation to the section modulus that can be installed without unsatisfactory compromise on aerodynamic performance. In the outermost blade region, loads are minimal and the structure can readily accommodate the most efficient thin aerofoils. A very small proportion of blade mass and cost is associated with this region of the blade.

Stiffness can be critical in blade design. In the desire to make the larger blades as light and economical as possible, it is hard to avoid increased flexibility. This can be an asset leading to load reductions (see Section 7.2), but flexibility is certainly limited in conventional upwind rotor designs by the key requirement to avoid tower strike by the blade tips in the most critical load case.

3.3 Simplified Cap Spar Analyses

In many blade designs, the principal load-bearing structure of a wind turbine blade is a spar which functions essentially as a beam (box beam or *I* beam depending on the arrangement of shear webs). Extracting the cap spar from a typical blade design, it may be represented as a simple beam (Figure 3.4). This assumes that a flapwise bending load (extreme or fatigue) drives the blade design, but that is generally realistic.

The notation of Figure 3.4 is as follows:

b	Spar width
d	Spar thickness
L	Length of spar
M	External bending moment
E	Young's modulus
I	Second moment of area
t	Aerofoil section thickness = spar depth
t_i	Internal spar depth

Consider now two possibilities:

1) *The design is driven by a deflection limit*: Thus, even if the blade could in isolation sustain a greater deflection without damage, it must obviously be prevented from deflecting far enough in the most critical load case for the blade tip to strike the wind turbine tower.
2) *The design is strength limited*: In this case, the blades may be flexible (perhaps in a downwind rotor configuration) or so stiff that they would break before being deflected enough to strike the tower.

The question is now addressed how in each case the material and dimensional properties are related if a beam of minimum mass is the target. Such a beam will be of minimum

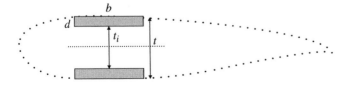

Figure 3.4 Cap spar parameters.

cost in any given material system, but the analysis will also allow the most basic level of comparison of different material systems to be established.

3.3.1 Design for Minimum Mass with Prescribed Deflection

Prescribing the deflection, δ, and introducing a constant, k, which depends on details of the beam section:

$$\text{Deflection,} \qquad \delta = \frac{kML^2}{EI}$$

$$\text{Allowable stress,} \qquad \sigma_a = \frac{Mt}{2I} = E\varepsilon_a \quad \text{where } \varepsilon_a \text{ is allowable strain}$$

$$\text{Mass of spar,} \qquad m = 2\rho bdL$$

$$\text{Second moment of area,} \quad I = \frac{bdt^2}{2}$$

From which it may be deduced that:

$$m = \frac{M\rho\delta}{kLE\varepsilon_a^2}$$

Comparing materials with fixed δ, a given length of beam L and defined external loading M, the relevant parameter is then:

$$\mu = \frac{\rho}{E\varepsilon_a^2}$$

$$\text{For carbon } \mu_c = \frac{1530}{138 \times 0.269^2} = 153$$

$$\text{For } E \text{ glass } \mu_g = \frac{1673}{26.8 \times 0.358^2} = 487$$

These values are almost exactly in the ratio $3:1$. Thus, a carbon-fibre-reinforced plastic (CFRP) blade that is a 'carbon copy' (same surface geometry) and designed for stiffness may be typically about one-third the mass of a glass blade.

3.3.2 Design for Fatigue Strength: No Deflection Limits

As in the case with prescribed deflection, it is easy to show that:

$$m = \frac{2M\rho L}{\sigma_a t}$$

The mass, m, decreases as t increases with no unique limit, but

$$Z = \frac{2I}{t} = bdt \quad \text{must remain constant}$$

Thus, as t increases, the product (bd) must decrease with various options for b and d. This is a familiar scenario where the beam in theory becomes ever lighter with larger separation of the caps and with thinner caps, but in fact limits are set by stability (buckling) or shear lag [3].

A limit to increasing t is set specifically by the following:

1) Buckling or impact if d is too thin, and
2) Inadequate edgewise stiffness if b is too small.

If rules are developed to define limits on d and b, for various materials, then the minimum mass for different material systems can be compared.

For example, if 2 mm minimum thickness of glass is compared with 1 mm minimum thickness of carbon, then taking $t \alpha 1/b$ indicates that

$$\lambda = \left(\frac{\rho d}{\sigma_a}\right) = \left(\frac{\rho d}{E \varepsilon_a}\right) \quad \text{is the relevant parameter}$$

$$\text{For carbon } \lambda_c = \frac{1530 \times 1}{138.2 \times 0.269} = 41.2$$

$$\text{For glass } \lambda_g = \frac{1673 \times 2}{26.8 \times 0.358} = 348.8$$

Ratio of mass is 8.5 : 1 (glass:carbon). This is rather extreme but only unidirectional properties have been considered. However, it is credible (supported by a parametric study involving outline blade designs discussed in Section 3.5) that a mass ratio of the order of 5 : 1 (glass to CFRP) may be achieved in a comparison, where each planform design is optimised to suit the material. It is critical to maintain fibre alignment to realise the full strength of the fibre and many manufacturers are shy of using much carbon in blades. The situation is exacerbated by uncertainties in supply and increase in costs. Nevertheless, the potential to have much lighter blades using much less material in a blade with very high carbon content remains.[3]

The foregoing analyses are extremely simplistic, but to some extent endorsed by practical experience of blade building. The French company, Atout Vent, made small, essentially all-carbon, blades of the same aerodynamic design as glass blades of the period and, at more or less neutral cost advantage, realised exactly the 3 : 1 mass reduction predicted for 'carbon copies' versus glass. It is therefore credible that even larger mass reductions may result comparing glass and CFRP blades in a scenario where rotor diameter and energy capture are the same but planform designs differ and are optimised to suit the chosen material system. This is supported in a parametric blade design study discussed in Section 3.5.

3.4 The Effective *t/c* Ratio of Aerofoil Sections

The thickness of an aerofoil is the maximum diameter that is normal to the chord line, usually expressed as a percentage of chord width. It can be assumed that, in order to preserve sufficiently good aerodynamic performance, the aerofoil specification will restrict the maximum allowable effective thickness to chord (t/c) ratio. A typical distribution of t/c values over the blade span is illustrated in Figure 3.5. This shows the dominant influence of aerodynamic requirements over the outer span of the blade with structural requirements taking over on the innermost one-third of the blade.

Whilst the aerofoil section t/c value is a rough indicator of section depth and structural efficiency of the aerofoil section in transverse bending, it does not distinguish between

3 Vestas uses more carbon fibre than any other manufacturer including other industries such as aerospace, and thus many years ago had reached the dominant position where carbon of the type that wind systems need (primarily cheap!) is produced to suit. The same is true of 'R' glass, about 12–15% stiffer than E glass, which LM has pioneered in wind applications. Some LM blades are nearly all R glass.

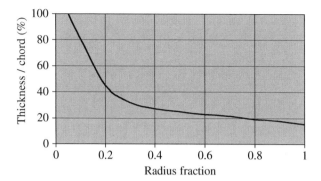

Figure 3.5 Typical spanwise distribution of thickness-to-chord ratio.

aerofoils with a wide region of near maximum (t/c) and aerofoils with a much more localised maximum.

In parametric blade design studies, it can be useful to define an *effective t/c*. This is defined by the I value (second moment of area value) of a cap spar (of defined chordwise extent and of unit thickness) based on the actual blade shape as compared to one which is uniform over the same width at the value of maximum t/c. The cap spar needs to be located in relation to desired section shear centre and mass centre so that an aerofoil with maximum thickness too far back from the leading edge may have reduced structural benefit. Torsional stiffness is essential for blade function in accepting the power-generating edgewise loading. Torsional stiffness decreases as blades become relatively lighter and of lower solidity. The location of the twist centre is then not only important for stability but can also be used to induce twist and offload some of the peak loads. This also affects edgewise stiffness and involves other structural and manufacturing considerations.

For typical cap spar widths, a demand for a value of $(t/c)_{eff}$ above 90% may be an appropriate input to the aerofoil design/selection process. This is achieved by the most structurally favourable aerofoils in use on wind turbines, typically the National Advisory Committee for Aeronautics (NACA) and National Aeronautics and Space Administration (NASA) series.

The smooth profile of the NACA 63-221 (Figure 3.6) may be contrasted with the Solar Energy Research Institute (SERI) aerofoil, S818 (Figure 3.7) which, on account of a more exotic shape with higher curvature around maximum thickness, has somewhat reduced structural efficiency $(t/c)_{eff} \sim 85\%$.

3.5 Blade Design Studies: Example of a Parametric Analysis

Blade design, and indeed the design of most wind turbine components, is a complex process. Firstly, a general specification must be developed considering the site class and environment, and following that a load specification typically based on detailed simulations of wind turbine operation with load cases specified and load calculations performed according to approved design standards. This provides the inputs to a structural design phase that, in general, will involve iterations with aerodynamic design, loads, closed-loop control and supervisory operational strategies and performance.

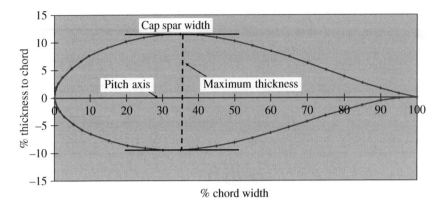

Figure 3.6 Effective *t/c* of a typical aerofoil section (NACA 63-221). Reproduced with permission of Airfoil Tools.

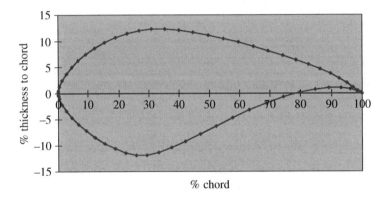

Figure 3.7 S818 aerofoil. Reproduced with permission of Airfoil Tools.

The following scheme describes a top-level investigation directed at gaining insights into material systems and impacts on blade design issues, especially mass and cost. The information is presented in relative terms and relates to blade design for systems around 1 MW rating.

The principal objective was to identify the materials systems, aerofoil selections and operational strategies most likely to allow a minimum cost blade. A plan of work was developed with the following stages:

1) Optimisation of aerodynamic designs for different aerofoil types: National Renewable Energy Laboratory (NREL), NACA, high lift;
2) Generation of a set of candidate planforms covering the ranges of aerofoil types and allowable planform variations;
3) Outline load specification;
4) Outline structural design;
5) Blade property evaluations – mass, stiffness, cost.

In stage 1, aerodynamically optimised designs were developed using each class of aerofoil types. The focus here was to restrict the study and compare aerofoils with more or less generic differences. The aerofoil types selected were SERI[4] aerofoils (high aerodynamic performance but not the most accommodating structurally), NACA aerofoils (good aerodynamic and generally good structural shapes) and high-lift aerofoils which could realise much more slender optimum blades for any given design tip speed ratio.

Stage 2 followed the concept illustrated in Figure 2.5. For each aerofoil type, a number of different planforms (about 60 in the whole study) were generated representing the most tapered and least tapered blade shapes, and so on within planform boundaries that sacrificed no more than 0.5% of annual energy at the chosen design site.

For stage 3, firstly, a comprehensive set of loads were generated for a baseline (state-of-the-art) wind turbine, all subsequent designs being compared at the same diameter and rated power. For each new rotor design, only a few load cases that were judged to be design drivers for the blade structure were considered. Extreme loads could be scaled according to the changes of planform solidity, and fatigue loads were scaled according to blade flexibility[5] in an iterative process (as an outline structural design was required to estimate blade stiffness).

In stage 4, an outline design of the blade for each planform was developed according to the chosen material system. The structural design concept comprised a cap spar in E glass or carbon, glass skins, stabilising cores of balsa and glass shear webs. A stress analysis was conducted to determine a required lay-up of materials (unidirectional and/or bidirectional fibres as appropriate) at each blade station that would meet the design loads with appropriate safety margins.

From the designs of stage 4, the key outputs (stage 5), mass and total blade cost could then be estimated. Manufacturing costs were added to cost of materials based on general experience of the assumed blade manufacturing processes.

The results in terms of relative blade mass are presented in Figure 3.8. The relative deflection is the ratio of deflection of each new design to a baseline design in a specific load case. There is often conservatism in the level of torsional stiffness provided in blade design, perhaps related to concerns about flutter stability, although a minimal level of torsional stiffness is of course essential in order to transmit the lift that provides rotor torque. The study explored the benefit in shedding some bi-axial material, reducing torsional stiffness but still preserving levels that were likely to be safe. The decrease in mass (relative again to a baseline design) at unit relative deflection reflects the reduction of bi-axial material without impact on flapwise stiffness.

The use of carbon spars with low-solidity planforms based around aerofoils with higher design lift coefficients can evidently allow very large overall weight reductions. Considering that these are not wholly carbon designs, the weight reduction ratio of around 4 : 1 as compared to the glass designs is quite consistent with the elementary material systems comparisons of Section 3.3.

The most extensive data set based on NREL aerofoils with E glass form a coherent curve, the form of which is to be explained shortly. Blades made with carbon spars (high-lift carbon) and only strength and not deflection limited allow very low mass flexible blades, as is predicted from the cap spar analyses of Section 3.3. The high-strength

---

4 SERI, the Solar Energy Research Institute, is now NREL, the National Research Energy Laboratory of the US Department of Energy.
5 This exploited a prior study on the influence of blade flexibility on dynamic loads.

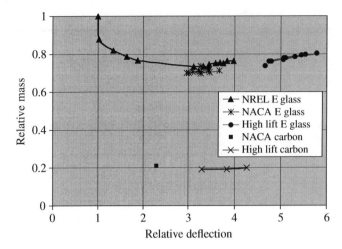

Figure 3.8 Relative mass of blades.

material suits the slender blades that result from using aerofoils with high design lift coefficient; whereas in glass, such blades are ultra flexible and not of minimum mass. The blades designed using NACA profiles can be expected to form a curve similar to the NREL set but of slightly lower mass on account of increased structural efficiency of the aerofoils (see Figures 3.6 and 3.7).

Transforming the data of Figure 3.8 from a mass to estimated cost, the data for carbon and glass designs converge considerably. The results of Figure 3.9 clearly depend on relative costs of glass and carbon, but suggest that blades with substantial amounts of carbon could be very cost-effective if designed around that material and not as 'carbon copies' of conventional blades. This makes most sense for highly flexible downwind rotor blades as are proposed for high-speed rotors (Section 7.2) where an increase of design

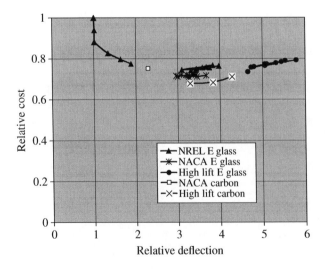

Figure 3.9 Blade design for minimum cost.

tip speed ratio rather than the use of high-lift aerofoils will naturally lead to more slender blade planforms.

This study dated to a period (circa 1999) when the cost of carbon fibres suitable for use in wind turbine blades was reducing. Over some years up to 2010, the carbon market was in a transitional state and supplies, availability and cost to wind turbine users were partly dictated by the vagaries of demand from the aerospace sector. Now, wind turbine industry demand for carbon from Vestas especially has reached a dominant volume, with that material supply better tailored to the wind turbine application (see Footnote 3). However, with carbon being much more expensive than glass, it tends to be used only where it is obviously beneficial. This is in spar caps but not in shear webs (usually) nor skins, and also not near the root where there is adequate section. Thus, the overall weight benefits are much smaller than may be realised in a predominantly carbon blade. The parametric study suggests that a near all-carbon blade could be competitive and even advantageous in an integrated design, optimally suiting the material system. Such a design would be downwind to allow load alleviation, mass and cost reduction from a low-solidity flexible blade without the problems of tower strike that would be much more acute in a flexible upwind design. The intrinsic stiffness of the material makes it easier to provide sufficient torsional stiffness. The cost penalty of carbon is mitigated by the large reduction in material mass compared with a stiff upwind design in glass. At the parametric level, there is no evaluation of acoustic noise or dynamics of blade or wind turbine system and insufficient consideration of stability and buckling. So the conclusion is not to rush into all-carbon blade design, but rather consider it for a deeper level of design and evaluation.

Discounting the specifics of cost, which are obviously crucial at any given point in time for the current blade design practice, the underlying principle, independent of whether carbon plays any role in a proposed design, remains in seeking an integrated design where material systems, aerodynamic design and the turbine operational concepts are well matched and none of these elements are assumed as a rigid starting point.

The data sets in Figure 3.8 conform approximately to a characteristic type of curve. Initially, some reduction in solidity (blade chord widths) sheds mass in the skins and leads to greater flexibility in the structure. Eventually, very slender flexible blades will struggle to maintain safe stress levels, and material must be added to the internal structure causing a rise in mass.

The shape of this curve can be explained by returning to the cap spar analyses of Section 3.3. If the planform solidity and hence local chord width, c, is changed, then the cap spar width, b, will change proportionately as will the aerofoil section thickness, t. In order to preserve blade surface stress at the allowable limit σ_a under the same applied design moment, M, the spar thickness d must also change.

$$\text{Cap spar thickness, } d = \frac{M}{\sigma_a bt}$$

Assuming $b \propto c = k_1 c$, say, then, the cap spar mass:

$$m_{spar} = 2\rho bdL = \frac{2M\rho L}{k_1 \sigma_a c} \quad \text{that is varies inversely as chord}$$

The blade skin mass is however roughly proportional to blade surface area and skin thickness (ignoring lightweight materials like balsa or foam that may be used to stabilise the skins) each of which is proportional to chord.

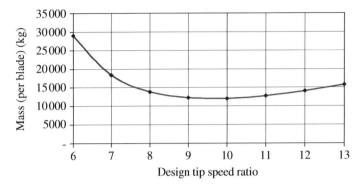

Figure 3.10 Blade mass related to design tip speed ratio.

Thus, skin mass

$$m_{skin} \propto c^2 L$$

Ignoring more minor terms,[6] this gives a rough model for total mass in the form:

$$m_{total} = \frac{2M\rho L}{k_1 \sigma_a c} + k_2 c^2 L \tag{3.1}$$

where k_2 is a constant to be determined from design analysis.

Recalling that chord width of an optimum blade design is inverse as the square of design tip speed ratio (Equation 2.1), Equation 3.1 may easily be recast in terms of design tip speed ratio. Figure 3.10 was derived from a simplified model of blade structural design involving equations similar to Equation 3.1 applied to the blade design of a 100-m diameter 3 MW wind turbine with unidirectional E glass as the principal structural material.

Each material system and aerofoil class will have its own characteristic curve of a type similar to that in Figure 3.10. Skin buckling concerns add mass to the highest solidity blades. Equally at some point as solidity reduces, the design becomes unfeasible in the given material system and design stress limits cannot be achieved. The most critical blade sections are then becoming almost solid and, with much of the material near the neutral axis, the design is very inefficient in terms of resistance to bending loads. For CFRP spars as compared to E glass, the curve shifts to much lower values of mass but also towards more flexible blades, a result that is logically accounted for in the cap spar analyses but may initially be unexpected considering that glass has a much higher strain limit than carbon.

In commercial blade design, in spite of the promise of CFRP for highly efficient designs, manufacturers will, in general, be quite cautious because of issues of availability of supply, cost and increased demands on manufacturing quality to ensure accurate alignment of fibres. Compression loads, especially in fatigue, are critical as with the non-flatness of the fibres (which will be present at some scale) the resin/fibre interface is stressed and could progressively fail. This is considerably more critical for carbon

---

6 The mass of gel coat or paint will be area related and the mass of glue or resin will tend to be area related if thicker laminates are used on larger blades or volumetric if more laminates of the same thickness are used.

fibres and therefore requires closer manufacturing control. This is also the reason carbon is generally run at lower strains than glass.

As blade flexibility is increased, the question then arises whether it is practical to remain with an upwind rotor. Downwind rotors are comparatively rare and the case is presented in Section 7.2 that it is only sensible to adopt this configuration with low-solidity flexible rotors when the rotor is high speed and the whole system can be lightweight. The Japanese wind turbine manufacturer, Hitachi, has nevertheless developed downwind designs with no particular emphasis on rotor flexibility but rather because, with a suitable combination of shaft axis tilt, they can better suit up-flows as occur on sloping terrain.

Increasing design tip speed ratio as in Figure 3.10 implies smaller chords and increasing flexibility. Design tip speed ratio was not varied in Figure 3.8, but increasing flexibility resulted from decrease in chord width through variation in aerofoil selection or materials selection. Thus, in terms of a characteristic shape, Figure 3.10 accords with the results presented in Figure 3.8.

Figure 3.10 relates to Figure 2.1 in a significant way. From Figure 3.10, it is evident that there is a substantial penalty in blade mass (and cost) for reducing design tip speed ratio, say from 9 to 7. If the maximum allowable tip speed is around 75 m/s, as may apply on land-based sites where acoustic noise emission is a concern, a penalty in energy capture must apply in employing a design tip speed ratio of 9 as a range of constant speed operation below rated wind speed is required (Figure 2.1) to avoid exceeding the tip speed limit.

Curves in the form of Equation 3.1 have been validated for a variety of blade designs in various materials systems including glass, carbon spar and wood-epoxy blades.

The preceding discussion has aimed to give a broad introductory picture of blade design issues considering aerofoil selection, materials and structures. A snapshot of aspects of industrial blade production is now presented.

3.6 Industrial Blade Technology

3.6.1 Design

Wind turbine design commences with a design specification embodying the target market requirements which will define site class and environmental constraints. Within the concept design of the whole system, an operational strategy is developed defining design tip speed limit and the range of speed variation including overspeed limits, and so on. The safety system, which defines hardware-related fail-safe actions for system protection in fault cases, will also be developed and followed by the supervisory control system, which embodies software control of all the normal states of the turbine (stopping, starting, yawing, power production), and various levels of fault detection (assuming that the control system itself is healthy). Within the supervisory control system lie closed-loop control systems such as control of blade pitching, for example. These cannot be detailed until the blade aerodynamic and structural properties and general system dynamics are determined.

Preliminary aerodynamic designs will be generated possibly coupled to structural design tools. Aerodynamic design, estimates of structural properties (blade mass and

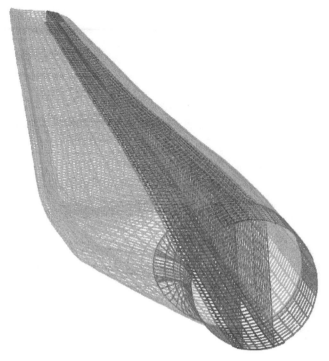

Figure 3.11 Rotor blade finite element mesh. Reproduced with permission of WINDnovation Engineering Solutions GmbH.

stiffness distributions in particular), dynamic loads and control system design are all strongly interactive and many iterations may be required to refine the design to the point where the design driving loads, the external blade surface and structural estimates are well defined, allowing detailed blade design to proceed. Aerodynamic design is of increasing sophistication and LM Wind, for example, the world's leading independent blade manufacturer, have made a huge investment to have an in-house wind tunnel facility [4]. The tunnel has a closed return loop with a contraction ratio of 10:1 and a cooling system that can cope with the maximum fan power of 1 MW. At a maximum flow speed of 105 m/s, a maximum Reynolds number of over 6 million is achieved with an aerofoil model chord width of 900 mm.

Joint areas are generally critical in blade design; and buckling, especially in areas around maximum chord, is usually a major concern. CFD is increasingly employed in verifying aerodynamic aspects of blade designs and finite element analysis (FEA) for anisotropic composite materials is now commonly used. A typical blade finite element mesh comprising blade surfaces and shear webs is illustrated in Figure 3.11.

3.6.2 Manufacturing

Modern blade production started around 1980 with hand lay-up of random oriented glass fibres wetted with polyester resin, a crude labour-intensive process as was common in boat building. As the wind turbine market grew increasingly sophisticated, manufacturing developed, and manufacture using resin infusion has become common and use of

Table 3.1 Blade technologies.

Comparison in 2001 of new technologies for a medium term: 40- to 50-m blade

	Wood–epoxy load-bearing shell and light webs 2001 current	Infused wood, carbon load-bearing shell and light webs New	Infused glass-reinforced epoxy load-bearing shell and light webs Established	Infused glass carbon epoxy load-bearing shell and light webs New	Prepreg glass load-bearing shell and light webs New	Prepreg glass-carbon load-bearing shell and light webs New	Prepreg glass light shell and load-bearing spar Established
Materials % weight	Wood–glass–epoxy 66%–8%–22%	Wood–glass–carbon–epoxy 50%–14%–3%–18%	Glass dry–wet epoxy 50%–40%	Dry glass–dry carbon–wet epoxy 26%–24%–40%	Pre-impregnated glass 90%	Pre-impregnated glass pp–carbon pp 45%–35%	Pre-impregnated glass epoxy 90%
Process type	Low-temperature hand lay-up and wet out	Low-temperature vacuum infusion	Low-temperature vacuum infusion painted finish	Low-temperature vacuum infusion	High-temperature vacuum press	High-temperature vacuum press	High-temperature vacuum press, spar made on rotating mandrel
Construction	Strong skin, integral root, single web	Strong skin with thin carbon lay, integral root, single web	Strong skin, separate root, double web	Light skin, separate root, double web with carbon caps	Strong skin, separate root, single web	Light skin, separate double with carbon caps	Light skin, strong spar, separate root
Material cost (%)	100	100	112	125	143	163	143
Production							
Cycle time (h)	32	10	10	10	12	12	12
Blade properties							
Weight (%)	100	67	67	51	67	51	66
Stiffness (%)	100	80–100	60–100	70–100	50–70	60–100	50–70
Environmental impact	Good	Best	Average	Average	Average	Average	Average
Development issues							
Development time to prototype set (mo)		6	12	12	18	18	24
Development time to production (mo)		12	24	24	36	36	48

prepregs has grown. Automated blade manufacture has been under consideration. For example, anticipating continuing growth in blade demand, MTorres (Torres de Elorz, Spain), in collaboration with Gamesa, has exploited a strong aerospace background in bringing some of the most advanced automated technology into blade production. They have developed a gantry-based blade production concept, the core system in their vision of fully automated blade plants. However, automation for the larger parts is extremely capital intensive and requires much faster deposition rates than in aerospace applications. This is perhaps the reason why after many years of research no major blade manufacturing unit is using it for the larger parts.

3.6.3 Design Development

As part of a decision process in 2001 regarding the direction for future blade developments, NEG Micon (now merged with Vestas) conducted a top-level review of blade manufacturing alternatives to develop 40- to 50-m-long blades over a medium-term timescale. The main choices for materials, processes, resultant blade properties and development time to new production are summarised in Table 3.1, giving a clear view of the most critical aspects of new blade development.

Some notes to Table 3.1 from NEG Micon applicable in 2001 were as follows:

- Costs were difficult to pin down exactly depending on procurement volumes, place of manufacture and future time of applicability. For example, the cost of carbon had gone up and down due to the projected demand from Airbus A380.
- Cycle times vary depending on the quality of equipment and the level of tooling investment or cure time acceleration assumed.
- Structural designs in this study assumed the same blade shape for each material and technology, whereas more COE differentiation may be possible optimising around different shapes.

This last point is the main theme of this chapter.

Although this is now a dated study, it shows the methodology. Infused glass, in particular, has improved through manufacturing control, allowing higher design strains to be exploited. R glass is used (primarily by LM) with typically 12% more stiffness. Pultrusions are used more extensively and more lately with tougher polyurethane resins at no greater cost. Progress in rotor blade development over the 30 years since around 1980, when wind set course to become a commercial source of electrical power on a national scale, can be measured by the quantity, quality and sophistication of modern blade manufacture, by increases in wind farm productivity (see Section 8.1) and also by relative mass reduction in wind turbine blades (see Figure 4.5).

References

1 Kooij, J.F. (2003) *One-Dimensional Variations: Blades. Report on Task 5 of Seminar.* Dutch Offshore Wind Energy Converter Project (DOWEC), September 2003.
2 Johansen, J., Madsen, H.A., Gaunaa, M. *et al.* (2009) Design of a wind turbine rotor for maximum aerodynamic efficiency. *Wind Energy,* **12** (3), 261–273. doi: 10.1002/we.292

3 Cox, B.N., Andrews, M., Massabo, R. *et al.* (2004) *Shear Lag and Beam Theories for Structures.* Structural Integrity and Fracture International Conference (SIF'04), Brisbane, September 26–29, 2004, pp. 63–69.

4 http://www.lmwindpower.com/Blades/Technology/Test/Aerodynamic%20test.aspx (accessed August 2010).

4

Upscaling of Wind Turbine Systems

4.1 Introduction: Size and Size Limits

For many recent years the big question has been – how big will wind turbines be? To the early pioneer looking back on the modest beginnings of commercial wind energy in the 1980s, the most recent offshore wind turbines are of awesome scale and are now the largest examples of rotating machinery in the history of mankind.

Cost models [1, 2] that may predict the economics of upscaled wind turbine systems have been developed and projects are under way to develop or evaluate multi-megawatt designs even as far as 20 MW rating.[1] The essence of the presentation of scaling issues in this book is that high-level, more or less complex empirical models that may encompass design variations and technology advances must be anchored to physical models of components and fundamental rules of scaling with similarity. Scaling in the present climate of wind technology almost invariably means *extrapolation* to larger sizes than any yet produced. It is in that context especially that empirical models, which may effectively describe data within the historical compass (*interpolation*), cannot be trusted without a clear underlying rationale that accounts for how the designs vary from upscaling with similarity.

Historically, considering designs that appeared through the twentieth century, the growth of turbine size has been quite irregular. This, however, was largely when wind technology (and renewable energy in general) was a fringe research interest.

A remarkable early wind turbine design, which has a lot to inform about the successful implementation of innovation, was the Smith–Putnam wind turbine (1942). This design brought together some of the finest scientists and engineers of the time (aerodynamic design by von Karman, dynamic analysis by den Hartog). The wind turbine, rated at 1.25 MW (a rating level not surpassed by wind turbines in commercial production for about another 50 years) with a 53-m-diameter rotor, was the first ever to deliver power into a public electricity network. It ran successfully and longer than some megawatt machines of the 1980s, from October 1941 to 1943, when a main bearing failure occurred. A subsequent blade failure in 1945 and steel shortages caused by the war resulted in the turbine being dismantled. With steel as the only well-established structural engineering material, the problems of engineering cantilevered blades to cope

---

1 The UPWIND sixth Framework project of the EU addressed questions of the viability of 10 and 20 MW wind turbine systems based on the standard three-bladed concept. This project also created cost models where intrinsic scaling effects are logically separated from the mass and cost reductions that arise from technology progress.

Innovation in Wind Turbine Design, Second Edition. Peter Jamieson.
© 2018 John Wiley & Sons Ltd. Published 2018 by John Wiley & Sons Ltd.

Figure 4.1 Blade hinge mechanism of the Smith–Putnam wind turbine. Photograph from the archives of Carl Wilcox. Reproduced with permission of Paul Gipe.

with self-weight-induced bending moments were acute and the design solution was to hinge[2] the blades. The massive structure associated with this is shown in Figure 4.1.

In 2000, Henrik Stiesdal,[3] then technical director of the long-established Danish supplier, Bonus AS (now owned by Siemens), had observed [3] a consistent exponential growth in turbine size considering the size of wind turbine at the centre of the company's production in successive years. Although larger turbines are presently (2011) under consideration, around 2004 the exponential trend stopped[4] with Siemens/Bonus and then also [4] for the industry as a whole (see Figure 4.2 and note the logarithmic scale of the 'y' axis). Stiesdal also compared the growth of wind turbine technology with aircraft technology and a similar basic pattern was observed (Figure 4.3).

Both technologies have in common an underlying square-cube law (explained further in Section 4.2) reflected in the ratios of rotor swept area (wind turbines) and lifting wing area (aircraft) to system volume and mass. Both technologies exhibited an extended initial period of exponential growth until a limiting size was approached. Moreover, these limits are probably more fundamentally economic than technological. Figure 4.3 shows that, in the aircraft industry after a period of exponential growth from the DC3 of 1935 to the Boeing 747 of 1970, there have been 30 years of no substantial growth in size. The Airbus A380 of 2008 (shown as a separate point isolated from the general data in Figure 4.3) bucks the trend a little.

---

2 Another early wind turbine with steel blades of 15 m diameter was designed by John Brown Engineering and erected on Costa Hill (a very severe site with high turbulence and mean wind speed ~11 m/s) in Orkney (island off the north of Scotland). Again, the hinged blade design route was selected because of the difficulties in making a large cantilevered blade in steel.

3 Stiesdal was 'in with the bricks' in the development of the Danish wind industry from farmyard experiments in the 1970s to what is now a significant component of the modern power generation industry.

4 The data of Figure 4.2 is a recent update of information published by Stiesdal and Kruse [3] and is published here courtesy of H. Stiesdal of Siemens Wind Power GmbH.

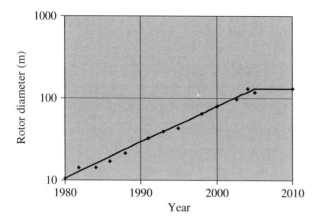

Figure 4.2 Growth in size of wind turbines.

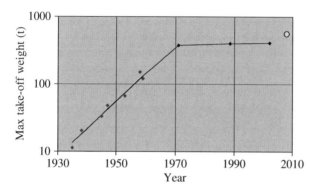

Figure 4.3 Growth pattern in the aircraft industry.

The flat top (Figure 4.3) has endured for 30 years and indicates with reasonable certainty that few much larger aircraft will appear. The momentum to develop larger scale wind turbine systems is now mainly in offshore applications. In 2009, Clipper announced funding for their Britannia project [5] to develop a 10 MW offshore wind turbine and American Superconductor are licensing a 10 MW design, Sea Titan [6]. Clipper is no longer trading and while the Sea Titan 'hangs in' in terms of a web presence, major progress is still awaited. Meanwhile, alongside the Vestas V164 (8 MW, 164 m diameter) are the Siemens SWT 80-154 (8 MW, 154 m diameter) and the Adwen AD-180 (8 MW, 180 m diameter) with the longest blades in the world, each of 88.4 m length. The gap between the REpower 5 MW of 2004 and these 8 MW turbines is filled with new turbines of Dongfang/Hyundai, Min Yang, GE, Sinovel, Senvion and Enercon. The largest floating offshore wind turbine (see Figure 6.10), MHI 7 MW, is installed offshore at Fukushima. Bullish predictions about commercial 10 MW technology by 2020 have been made.

However, the only secure conclusion about wind turbine growth is that, a few years into the current millennium, the growth trend ceased to be exponential.

The traditional reasons for upscaling wind turbines that generate into an electrical network have been as follows:

1) Electrical utilities are used to power in multi-megawatt scale units.
2) In a wind farm, a number of infrastructure items (beyond the turbine units themselves and also including maintenance) reduce in cost per megawatt of installed capacity, the larger the capacity of the wind turbine units. Larger unit capacity implies fewer numbers of the turbine units to realise a given total capacity.
3) Larger turbines can often use land and wind more effectively.
4) In public funding of wind energy, size has tended to be regarded as a metric of technology progress.

Large wind turbines have positive and negative environmental aspects. Often, the visual impact of a single large turbine may be more acceptable than many small ones. However, larger turbines have a greater range of visibility. The issues of item 2 are even more consequential offshore, where the infrastructure costs in foundations, interconnection of turbines and electrical connection to shore make the case for the very largest turbines.

In order to investigate in an objective way the implications of upscaling wind turbine technology, scaling fundamentals are now reviewed.

4.2 The 'Square-Cube' Law

The primary value of a wind turbine system is in the energy it produces, which is proportional to the swept area of the rotor and therefore the square of the rotor diameter. Costs, on the other hand, relate to the mass of material in the system, which is proportional to volume and therefore to diameter cubed. This relationship has been referred to as the square-cube law. It implies that, for the wind turbine system alone, excluding the infrastructure of a wind farm, the capital cost per unit capacity will increase linearly with wind turbine scale. Very little cost data for wind turbine systems is readily available, although over 10 years from 1997, the magazine *Windkraftanlagen Markt* published list prices. Careful analysis of that data [7], after considering impacts of higher tip speeds (of the more recent offshore wind turbines) and varying tower heights, supported the expectation of an increase in cost per rated kilowatt for wind turbines (the wind turbine hardware alone excluding installation and all balance of plant (BoP)) in the megawatt range.

The mass and hence cost of material in a system is in turn determined by the design loads that must be endured. Considering how loads scale and also various influences which are scale independent, the square-cube rule is subject to many modifications when detailed modelling of a system is considered, but it is essentially true. As is discussed later, it is often apparently, but for the most part only apparently, contradicted by commercial data.

4.3 Scaling Fundamentals

For the term 'scaling' to have much meaning implies a similarity between a set of objects so that their main difference is size alone. It is evident that a homogeneous solid object with a characteristic dimension D will have a surface area scaling as D^2, while volume and mass will scale as D^3.

How do aerodynamic properties scale when that object is in a fluid flow field? The most basic rule to preserve similarity in upscaling a wind turbine system is to keep the design tip speed ratio constant. If the rotor is upscaled with geometric similarity, preserving tip speed preserves the flow geometry over the blades. Strictly, this is only approximately true as the Reynolds number (Re) changes with scale and therefore the flow is not absolutely similar. For large wind turbines, Re is typically in the millions and the effect of scaling on flow similarity is very minor. With the usual definition of Re:

$$Re = \frac{\rho V c}{\mu} \tag{4.1}$$

where ρ is fluid density, V velocity, c a characteristic dimension[5] of the object interacting with the flow and μ the kinematic viscosity of the fluid. The variation in Re (directly as c) is in consequence of the fixed scale of fluid viscosity.

In the upscaling of any real design, strict similarity is always violated. This is because of the existence of absolute scales with characteristic dimensions that affect wind turbine behaviour but will not ever change with the scale of the turbine. Such absolute scales are as follows:

1) Atomic scale,
2) Scale of the earth's seas and terrain,
3) The related scale of atmospheric structures (including wind turbulence),
4) Human scale.

Surely the atomic scale is irrelevant for the macroscopic systems of a wind turbine? Not exactly. Viscosity is determined by the mean free path of molecules and the atomic scale sets the dimensions of molecules and their related parameters. Viscosity is the determinant of two boundary layers that affect wind turbine performance. In the earth's boundary layer, there is usually a gradient of wind speed with height. Thus, a large wind turbine at a given site will not see the same wind as a smaller one. Similarly, because of viscosity and the associated influence of Re on the aerofoil boundary layer, there is not exact flow similarity between large and small wind turbines even under identical wind conditions. Again, tracing the story back to viscosity and the mean free path of molecules, this arises because molecules refuse to upscale along with the wind turbine!

$$\text{Mean free path: } \bar{s} = \frac{1}{\sqrt{2\pi}\, n d^2}$$

where d is molecular effective diameter and n the number of molecules per unit volume.

$$\text{Viscosity: } \mu = \frac{\rho \bar{c} \bar{s}}{2}$$

where $\bar{c}$ is mean molecular speed and ρ density.

Atmospheric structures, climate and longer term wind variations are related to the source solar energy, the differential thermal effects on land and sea areas and the Coriolis forces due to the earth's rotation. Locally, turbulence is as much determined

---

5 An indicative value of Re relevant to a wind turbine rotor may be based on the chord width of a blade around 70% span.

by land topography or sea state, the thickness of the earth's boundary layer being related to surface roughness. Although turbulent wind variations are random, there are identifiable statistics, spectral characteristics and characteristic length scales. These properties are critical for the modelling of fatigue and extreme loads of wind turbines. The theory of wind turbulence is discussed in [8] and the models used for wind turbine load calculations are defined in leading international standards [9]. The interconnection between turbulence and viscosity is profound and famously described in a poem by Lewis Richardson (1922):

> Big whorls have little whorls;
> Which feed on their velocity;
> And little whorls have lesser whorls;
> And so on to viscosity.

The bottom line in terms of scaling implications is that wind turbulence has an absolute scale, and so the spatial and temporal variation of wind conditions across a rotor disc will not only be random temporally but in overall statistical properties also vary with turbine size.

The 'human scale' is another absolute scale that applies in the sense that different manufacturing and handling methods are required as systems grow in scale, whilst the size of personnel access hatches and other dimensions do not change. Therefore, there is some violation of similarity on account of the (approximately) fixed scale of human beings.

Scaling trends are often approximated by a single power-law expression with a fractional exponent, when in fact there are several terms with different dependencies. For example, with a painted cube of side D of solid steel, it is clear that the mass of the steel scales as D^3 while the mass of paint scales as D^2, provided the coating is always applied at the same thickness. A power law for total mass with a single fractional exponent between 2 and 3 may fit a particular range of 'painted cube' data quite well, but it will extrapolate erroneously. The larger the cube, the more the mass will tend to increase as D^3 – the familiar algebraic result that the highest power of x will eventually dominate as $x \to \infty$.

4.4 Similarity Rules for Wind Turbine Systems

4.4.1 Tip Speed

In upscaling a wind turbine system, a consistent basis of comparison requires that a representative tip speed is constant. This preserves the flow geometry[6] in terms of the relationship between rotor speed and wind speed at any given operating point. Maintaining a given tip speed at any given wind speed implies that in upscaling, rotor angular velocity, ω, must vary inversely with diameter, D, and decrease with increasing turbine size. Referring to the diagram on the right-hand side of Figure 1.7, it is clear that with axial induction set at an optimum value around one-third, then the optimal flow geometry at each radius fraction, x, will be preserved by fixing the value of the tip speed ratio, λ. Thus, scaling with similarity will be scaling at constant design tip speed ratio and at constant design tip speed.

---

6 Strict similarity is not possible because the Reynolds number increases with turbine size, but the impact of this is very minor for turbines rated above about 50 kW and can be neglected in differentiating between megawatt and multi-megawatt scale systems.

4.4.2 Aerodynamic Moment Scaling

The aerodynamic moments of a wind turbine blade about any axes and at any radial station can be shown to scale as D^3.

These moments arise as a sum of products of local air pressure, elemental areas and moment arms. The local air pressure at any given fraction of rotor radius is independent of scale being $(1/2)\rho W^2$, where ρ is air density and W is the resultant local inflow velocity. In any consistent geometric scaling of a blade, the product of any corresponding areas and moment arms are necessarily proportional to D^3. Hence, aerodynamic moments scale as D^3.

4.4.3 Bending Section Modulus Scaling

The bending section modulus of a beam is a product of width and thickness squared and hence is cubic with characteristic dimensions and scales as cube of diameter for any given cross section of a wind turbine blade. With cubic scaling of both applied aerodynamic moments and section modulus, in a pure geometric scaling, bending stress will be independent of scale.

4.4.4 Tension Section Scaling

Each section of a wind turbine blade is usually under tension due to centrifugal force associated with blade rotation. This force is proportional to $M\omega^2 R$, where M is the mass, ω is the angular velocity of the rotor and R is the radius. As M obviously scales cubically and ω scales inversely with diameter (see item 1), the centrifugal force term scales overall as diameter squared. It acts on a section area which clearly also scales as diameter squared. Hence, under scaling similarity, blade tension stresses are scale independent.

4.4.5 Aeroelastic Stability

Given that the centrifugal force scales is D^2, if a blade element is displaced through an angle α out of plane, then the moment arm of the restoring centrifugal force will be proportional to $D \sin \alpha$ and the resultant moment to $D^3 \sin \alpha$. An out-of-plane aerodynamic loading causing this displacement will scale as $D^3 \cos \alpha$. Equating these terms shows that $\tan \alpha$ and hence α is invariant with scale. This is equivalent to the Lock number (see Equation 17.4 and Section 17.6) being invariant with scale and implies that aeroelastic stability will not be scale dependent.

4.4.6 Self-Weight Load Scaling

It is obvious, however, that a blade bending load due to self-weight will be proportional to the product of a mass scaling as D^3 and a moment arm scaling as D. Hence, blade bending moments due to self-weight will scale as D^4. Thus, if the blade is so large that self-weight moments are design drivers, then blade mass (for a given material system) must increase by more than D^3 to preserve constant bending stress. In these circumstances, similarity may possibly be preserved in the surface geometry of a wind turbine but not in the internal structure.

4.4.7   Blade (Tip) Deflection Scaling

Considering only aerodynamic loading, given the previous discussion about bending moment and section modulus, the blade tip deflection will scale linearly as D. With tower dimensions and blade tip clearance scaling similarly, upscaling poses, in respect of aerodynamic loading, no problem for tower clearance. However, since the self-weight blade bending moment will couple into dynamic deflections of the blade and will scale as D^4, tower clearance will become more problematic with upscaling.

4.4.8   More Subtle Scaling Effects and Implications

4.4.8.1   Size Effect

There is a so-called size effect relating to the fracture mechanics of materials. For homogeneous materials, the larger the sample size, the greater the probability of a critical flaw existing in a given sample. This implies that at some level of upscaling, it is not adequate to design for constant stress. Instead, a reducing stress allowable must be adopted. This is equivalent to mass scaling more than cubically.

4.4.8.2   Aerofoil Boundary Layer

There are two fundamental violations of scaling similarity in wind turbine technology. These relate to fluid boundary layers. One concerns the boundary layer around the aerofoil sections of the blade. The Reynolds number characterises the effect and it is found that the flow around an aerofoil and its associated lift and drag characteristics depend not only on the shape of the aerofoil but also on its size. As has been mentioned, this is a minor effect, almost negligible for medium- and large-size wind turbines.

4.4.8.3   Earth's Boundary Layer, Wind Shear and Turbulence

Much more significant than the aerofoil boundary layer is the effect of the earth's boundary layer where friction between the atmospheric air and the ground leads to a strong gradient in wind speed (wind shear) with wind increasing significantly with height above ground. This picture represents a common situation, but it is a gross simplification as the wind shear gradient depends, in general, on atmospheric stability and may be negative (wind speed decreasing with height) in the presence of inversion layers.

The earth's boundary layer is associated with ground roughness and air viscosity. The roughness depends on surface topographical features differing, for example, with short cropped vegetation, trees, buildings, undulating hills, and so on. This in turn affects the turbulence in the wind and confers an absolute scale on the wind statistics. Loads due to wind turbulence are often design drivers for wind turbines. In general, this loading worsens with increasing turbine size. Characteristic turbulent length scales are comparable to the scale of large turbine rotors, which will then see a much greater differential loading across the rotor disc than across a small rotor.

Wind shear implies that upscaling will realise an increase in power from the wind turbine greater than as swept area or diameter squared because, as hub height increases, the associated wind speed is also increasing. However, it is then often carelessly asserted that this substantially impacts on the 'square-cube law' (to be discussed) forgetting that, in principle, the increase in wind speed with height will also imply an increase in loads.

The power, P, from a wind turbine rotor of radius, R, in a steady wind field of velocity, V, is proportional to $V^3 R^2$ and, assuming geometric similarity, the hub height, h, is proportional to R. In the presence of wind shear characterised by a wind shear exponent, α, V is proportional to h^α. Hence, power P is proportional to $R^{2+3\alpha}$. The bending moment on a blade, M, is proportional to $V^2 R^3$ and a similar analysis establishes that M is proportional to $R^{3+2\alpha}$. Hence, in the absence of wind shear ($\alpha = 0$), the ratio of power to load is as $1/R$ (the same as the area to volume or mass) and, in the presence of wind shear, is as $1/R^{1-\alpha}$.

Thus, wind shear may in principle confer a small benefit in the upscaling of conventional wind turbines associated with a power gain that slightly outweighs load increases. However, in reality, this gain may be offset by the differential loading in turbulent wind conditions increasing with wind turbine size. Design driving loads depend much more on the statistics of turbulence than on quasi-deterministic features of the wind field. Sometimes design driving loads may arise from such deterministic characteristics as wind shear (or extreme wind direction changes), but these are artificial load cases and such extreme wind shear values demanded for loading calculations are not available in real operation as a potential energy gain for an upscaled wind turbine.

4.4.9 Gearbox Scaling

If the gearbox output speed is held constant, then the gear ratio must increase and this, in conjunction with a cubic scaling of input torque, leads to a more than cubic scaling of gearbox mass and cost (to be offset by a less than cubic scaling of the generator). If the generator input speed is held constant with upscaling, as is usual in conventional wind turbine design, then the generator torque is varying as power leading to generator mass and cost is proportional to the square of diameter. A simpler picture is evident with the direct drive generator, where the generator is the whole of the power train and scales approximately cubically (i.e. as input torque).

4.4.10 Support Structure Scaling

The support structure (tower) of a conventional wind turbine is similar to a blade in respect of being a wind-loaded cantilevered beam and the tower can be expected to scale cubically for the same reasons. It may be noted, however, that with tower head mass scaling essentially cubically, this will contribute to a buckling moment that increases as D^4.

4.4.11 Power/Energy Scaling

The primary factor affecting power and energy capture is installed rotor area. However, the presence of wind shear or other systematic structure in the wind introduces some complication.

In the present work, a standard wind shear profile with a wind shear exponent of $1/7$ (0.143) has been adopted. This is applicable to a typical land-based site. Offshore surface roughness, turbulence intensity and wind shear (all interconnected) are typically less than for land-based sites and a wind shear exponent for offshore design of around 0.1 may be more representative.

4.4.12 Electrical Systems Scaling

In some cases, quite detailed cost models exist for scaling of electrical parameters. Nevertheless, scaling of electrical generators and system components is another major topic in its own right which has not been much explored from a fundamental standpoint. The assumption that is used in the cost of energy approach of this book is that electrical machines scale fundamentally as torque rating. The torque rating is equivalent to current rating provided a fixed output voltage is considered. It also implies (again considering constant voltage) that for systems upscaled with a fixed generator input speed (another violation of similarity) electrical generator costs will scale as power rating.

4.4.13 Control Systems Scaling

The preservation of tip speed in upscaling implies a corresponding decrease in rotational speed or rotor frequency when a design is upscaled with similarity. The reduction of frequencies and lengthening of time constants is a general feature of the upscaled turbine. Yaw and pitch rates decrease in proportion to diameter increase. Because of the complexity of control systems and the tuning of them to suit the specific dynamics of the drive trains and rotors, it is rather difficult to determine simple scaling rules even under quite idealised assumptions.

The optimal mode gain is a control parameter which is set to ensure the appropriate torque tracking (following an appropriate torque-speed characteristic to maximise rotor power) in the operation of a variable speed wind turbine below rated rotor speed. It is defined as

$$K_{opt} = \frac{\pi \rho R^5 C_p}{2 \lambda^3 G^3}$$

where ρ is air density, R rotor radius, C_p power coefficient at λ, λ desired operating tip speed ratio and G gearbox ratio.

In a design with strict similarity, the gearbox ratio G would be held constant. Thus, in a direct drive system ($G = 1$), K_{opt} (and rotor rotational inertia) increases as D^5; whereas in a typical conventional system with gearbox output speed in a fixed range, G will increase as D and K_{opt} will upscale only as D^2.

4.4.14 Scaling Summary

It can be concluded from Sections 4.4.1–4.4.3 that scaling with geometric similarity largely works from a stress point of view provided the wind turbine is not of such size or manufactured from such materials that self-weight loads become design drivers. Physical principles dictate cubic scaling of mass of similar objects. Hence, scaling of mass of principal (load designed) components is cubic, provided aerodynamic loads are design drivers.

Upscaling with similarity (and assuming invariant wind input):

1) *Maintains*:
    a) all aerodynamic-related stresses and deflections;
    b) centrifugal stresses in the rotor blades;
    c) aeroelastic stability of the rotor;

  d) natural frequencies of the wind turbine system in proportion to the rotor frequency.
2) *Reduces*:
  a) rotor rotational speed and associated system frequencies.
3) *Increases*:
  a) moments due to self-weight loading as D^4;
  b) tower clearance but less than in proportion to D (as would be expected from similarity) because of the effect of the blade mass moments scaling as D^4.

As has been noted, geometric similarity in design is only an approximation because of boundary-layer effects (Section 4.4.8.3), which introduce fractional exponents into scaling in a fundamental way. The aerofoil boundary-layer effect can be ignored for medium- and large-scale rotors, but the earth's boundary layer is much more significant. The more complex boundary-layer effects usually mitigate the severity of the 'square-cube' law by offering an energy output that may upscale by more than D^2 but always by substantially less than D^3. Moreover, extra energy from an increased hub height wind speed always implies some associated level of increase in loads.

Thus, upscaling is fundamentally adverse from the standpoint of the effect on wind turbine system specific mass[7] (and cost). Upscaling is therefore justified only when decrease in specific operation and maintenance (O&M) costs and or specific BoP costs more than offsets the increase. Regardless of how O&M and BoP costs may vary with scale, even should they fall to zero which is obviously quite unrealistic, the implication of the essentially cubic upscaling of major wind turbine components is that there will be an economic limit at some scale. This looks to precede any final technology feasibility limit, say, on the size of a cantilevered beam that can be engineered with state-of-the-art blade materials.

4.5 Analysis of Commercial Data

The analysis of commercial data on turbine dimensions, component mass, and so on is fraught with difficulties.[8] Designs are in general for different wind conditions, in the context of different regulatory requirements, to different design standards and embody different design philosophies. It is all the more remarkable that some consistent trends can be observed.

In an overview of present large-scale wind technology, it is valuable to review how salient parameters like mass, cost, cost per rated kilowatt, cost per swept area of rotor, and so on, scale with size of turbines. Relevant data is available from wind turbine manufacturers who produce technical data sheets on their designs and who also provide information to sources such as *Windkraftanlagen Markt* [7] or *Windenergie* [10] where turbine data is collated and published. Manufacturer's data sheets are however frequently revised and the published sources are not free from errors. Moreover, data is rarely

---

7 'Specific' mass or cost here refers to mass or cost per unit of installed capacity (say, kilowatts or megawatts).
8 Data and text portions in this section have appeared in similar or broadly similar form in various issues of the EU publication *Wind Energy the Facts*. This is not plagiarism as this author was the source of the text in these documents.

reported in an exactly common basis (especially costs and masses) with classifications of components and systems varying among the available sources.

Scaling trends need to be interpreted with great care. Data indiscriminately lumped together may suggest spurious trends or at least provide only superficial descriptions rather than insight into basic issues like the inherent specific cost (cost per kilowatt or cost per kilowatt hour) trend with upscaling. Some of the main issues are as follows:

- *Geometric similarity*: With strict geometric similarity, volume, mass and cost of items will tend to scale as the cube of any characteristic dimension. Very small turbines (say <30 kW output power rating) are generally too dissimilar to the larger turbines for valid interpretation of inherent scaling rules if all sizes are grouped together.
- *Parametric similarity*: Designs basically similar in concept (e.g. three-bladed, pitch regulated with glass epoxy blades and tubular tower) may have significantly different choice of key parameters. Tip speed is a key parameter that very directly influences the tower top mass and cost of a wind turbine. Different ratios of power rating or tower height to diameter will also clearly influence mass and cost. These influences can sometimes be effectively considered by normalisation processes allowing more data sets to be grouped together.
- *Duty similarity*: Machine designs, mass and cost are influenced by the class of design site, that is, the severity of the design wind conditions.
- *Stage of development*: The latest and largest wind turbines are at the most advanced state of knowledge of the manufacturers with ever-increasing emphasis on cost and mass reduction affecting design. This can obscure scaling trends that would appear if all sizes were at the same stage of technical maturity. Viewing an axis of increasing diameter is similar to looking along an axis of time because, over the history of development of modern wind technology, wind turbine size has generally increased.

There are also many other factors which complicate scaling comparisons such as manufacturers' preferences for electric or hydraulic systems, or for simple heavy structures versus more lightweight optimised structures, and so on. Finally, in moving beyond technical issues to costs – and a major motive in addressing the technicalities of scaling is to get insight into how they will influence costs of large wind turbines – a number of non-technical factors are added (exchange rates, labour cost variations globally, marketing ploys, and so on).

4.5.1 Blade Mass Scaling

Some analysis is first presented of comparatively old data on blade mass trends. A relevant part of the story enlightened by the older data is how technology changes with time and how this may obscure inherent scaling trends.

A power-law fit to the data as a whole (Figure 4.4) yields an exponent of 2.29. Thus, blade mass would appear to scale with increasing blade size by a power law that is significantly less than cubic. This is principally because of the following:

- The large blades of 40 m length or more are very substantial structures and promote ongoing development efforts to reduce mass and cost.
- The largest blades are the more recent and at a more advanced stage of manufacturing technology.

Figure 4.4 Blade mass trends based on blade manufacturers' data.

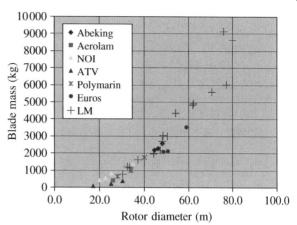

There is further clarification of blade mass trends in Figure 4.4, where blade mass data (mass of a three-blade set) from manufacturers' data sheets is presented. It is apparent (Figure 4.4) that the small carbon blades of Atout Vent (ATV) are very light for their size, as would be expected with carbon-fibre-reinforced plastic (CFRP) as the sole structural material. A wood-epoxy blade of NEG Micon Aerolaminates[9] (Aerolam in the legend of Figure 4.4) is also, as expected, quite light for its size.

Table 4.1 corroborates the comments explaining why blade mass apparently increases as less than cubic. Trend line equations (not shown in Figure 4.4 to avoid complication) in the form of best-fit power-law curves were determined for the data of each manufacturer. In the case of ATV, who has had a consistent technology for small blades, and Euros, who were then comparatively recently established as blade manufacturers with little development time behind them, the power law exponents are close to the predicted cubic relationship. The other manufacturers, LM, NOI[10] and Polymarin, have been trading for a long time, their technology has gone through major developments and, consistent with that, the power-law exponents appear to be less than cubic.

It should also be noted that given the small data set and extent of scatter in it, power-law exponents cannot be determined with any significant statistical rigour. The

Table 4.1 Curve fit exponents of blade mass versus diameter.

NOI	2.05
ATV	3.07
Polymarin	2.76
Euros	2.92
LM	2.30

---

9 NEG Micon were absorbed by Vestas, who now own the wood-epoxy blade technology.
10 It is logical that the lowest exponent of all applies to NOI since their blade range includes small Aerpac blades originally manufactured with polyester resin in a wet lay-up process, whereas all recent designs of large blades are based on resin infusion using epoxy resin. Some of the manufacturers mentioned are no longer trading or have been subject to further acquisitions within the wind turbine industry.

case for a fundamentally cubic scaling of blade mass is not affirmed statistically by the limited data available. It is rather that there is an underlying logic that the scaling should be cubic and that when the data is analysed with care taking account of technology changes, and so on, there is nothing to contradict cubic scaling.

Near-cubic scaling is generally apparent when a consistent set of designs at a similar state of development are evaluated. The WindPACT study on the scaling of wind turbine blades [11] involves some detailed case studies of large blade designs and broadly confirms this conclusion.

This historical picture is underlined by recent analysis of blade mass scaling trends by the Centre for Renewable Energy Sources (CRES) [12] within the UPWIND project. Characteristic cubic curves (Figure 4.5) have been developed for a variety of different blade materials systems and manufacturing methods. It is very clear that the commercial blade designs have been subject to ongoing development as blades have become larger.

There is about 30 years of development between the oldest blade technology (glass chop strand mat with polyester resin manufactured by hand lay-up) and the latest, and Figure 4.5 underlines the huge relative blade mass reduction that has been achieved. It also shows how speculative it may be to estimate blade mass for very large turbines of 10 and 20 MW rating with diameters up to 250 m. Commercial blade mass data, represented by the points embedded in Figure 4.5, have a trend line close to a square law as is confirmed in Figure 4.6, which is a more detailed view of data for rotor diameters >40 m. Note, however, a suggestion from Figure 4.6 that the largest machines are struggling to avoid blade mass increase beyond the square law trend.

Since blade mass scaling is fundamentally cubic, the scaling to a lesser power apparent in Figure 4.6 is principally because blades of 40 m length or more are very substantial structures and promote ongoing development efforts, as evidenced in Figure 4.5,

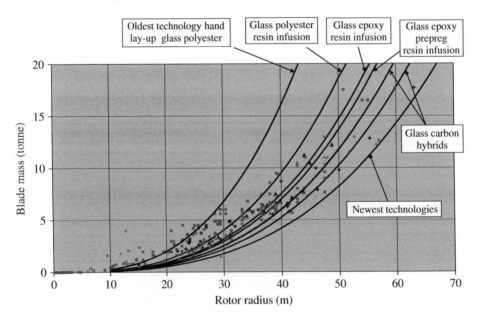

Figure 4.5 Blade mass scaling related to blade technologies.

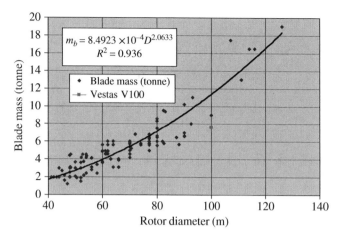

Figure 4.6 Blade mass trends.

to reduce mass and cost. The largest blades are therefore the most recent and at the most advanced stage of manufacturing technology. Both the impact of blade mass on blade cost (direct proportionality) and the impact of blade mass on hub fatigue loading are strong incentives to minimise blade mass as wind turbines get ever larger. Vestas blade technology with carbon prepregs leads to very lightweight blades, as is evident considering the point for the V100 in Figure 4.6.

It was noted that self-weight bending, being the product of the force due to blade weight (scaling at least[11] cubically with diameter) and a moment arm proportional to diameter, will scale at least as the fourth power of diameter.

This implies the following:

- At some very large scale, self-weight bending loads will drive design.
- Should this occur, then blade mass will scale to a higher power than the cube.

Further to the argument presented for fundamentally cubic scaling of blades, the extensive study [11] within the WindPACT programme supported by the US Department of Energy and administered by the National Renewable Energy Laboratory (NREL) also confirms the view that basic mass scaling of blades (at a fixed state of technology development) will be cubic.

4.5.2 Shaft Mass Scaling

At the wind energy conference in Husum, Germany, September 2001, Enercon displayed a set of main rotor shafts for the range of sizes of their direct drive wind turbines. The mass of each shaft was provided on plaques beside each exhibited shaft and is recorded in the table attached to Figure 4.7. Figure 4.7 shows a cubic curve fitted by the method of least squares that describes the data very well. The shafts are simple solid steel forgings; and Figure 4.7 is convincing evidence that, when there are no factors like change of

---

11 The phrase 'at least' is used in this text because extra blade material above a cubic increase in mass will be required to upscale such a design to meet the self-weight loading and, if the mass increases more than cubically, then the mass moment will increase as more than the fourth power.

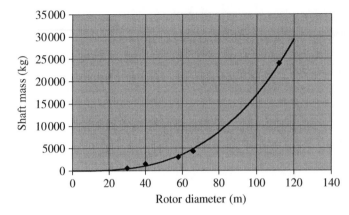

Wind turbine design	Shaft mass (kg)
E30	550
E40	1 500
E58	3 000
E66	4 300
E112	24 000

Figure 4.7 Enercon rotor shafts.

materials, change of design concept or technical developments intervening, scaling of components designed by rotor torque or bending moments is cubic.

Wind turbine design	Shaft mass (kg)
E30	550
E40	1 500
E58	3 000
E66	4 300
E112	24 000

4.5.3 Scaling of Nacelle Mass and Tower Top Mass

Figure 4.8 shows nacelle mass (mass above the yaw bearing excluding the rotor blades and hub). The trend line for large turbines of 80 m diameter or greater is cubic. If all

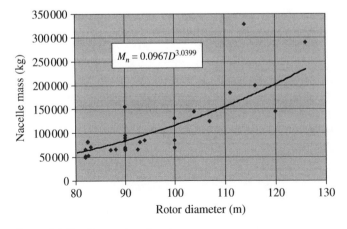

$$M_n = 0.0967D^{3.0399}$$

Figure 4.8 Nacelle mass trends.

data from turbines >20 m diameter is considered, the trend line exponent is found to be around 1.8. Thus, overall the data can be viewed as containing two distinct groups, older smaller scale designs where technology improvements had helped restrain increasing mass with increasing scale and the latest designs above 80 m diameter where technology development is ongoing but now presently failing to avoid cubic scaling.

In Figure 4.9, the trend of torque with diameter, which would be exactly cubic with strict similarity of scaling, is approaching a cubic relationship with a scaling exponent of ~2.8. The departure from a cubic relationship is almost entirely accounted for by the trend for design tip speed to rise with increasing diameter (see Section 4.7).

Figure 4.10 shows the essentially linear trend in nacelle mass versus wind turbine torque rating.

Figures 4.8–4.10 are evidently interconnected and all consistent with broadly cubic upscaling of very large wind turbines.

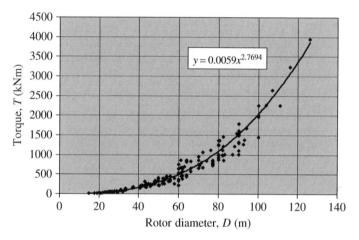

Figure 4.9 Torque trend.

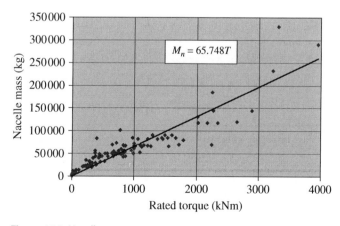

Figure 4.10 Nacelle mass versus torque.

4.5.4 Tower Top Mass

Considering the rotor and nacelle assembly, that is, the complete tower top mass (Figure 4.11), the scaling trend is similar to the rotor around a square law for all wind turbines above 40 m rotor diameter. If only the more recent large turbines above 80 m diameter are considered, unsurprisingly the exponent rises to around 2.8.

4.5.5 Tower Scaling

4.5.5.1 Height versus Diameter

Hub height clearly increases with rotor diameter (Figure 4.12) and, on average, is approximately equal to rotor diameter. The data exhibits great scatter for three main reasons:

- At small scale, tower heights are relatively higher in order to access wind flow unobstructed by obstacles (trees, buildings, etc.). Also, small turbines operate in a region of the earth's boundary layer nearest the ground where wind shear is most severe and there is greatest benefit from additional height.
- To suit different wind classes, it is common for a range of tower heights to be offered with each turbine model.
- The largest turbines are for offshore. There is less severe wind shear there and therefore less benefit from higher towers. Moreover, the higher the tower, the lower is the fundamental frequency of the system and this may make the response to wave loading more severe and add cost to foundations.

In summary, the data needs to be considered in three categories: small scale, larger turbines for land use and very large offshore turbines.

This will not, however, sort the data in a way that significantly reduces scatter. Great scatter will remain due to the impact of wind class. It can be seen, for example, from Figure 4.12 that at 70 m diameter, tower height ranges from just above 50 m to nearly 120 m. Alternatively, over a range of designs from 50 m diameter (800 kW machine on low wind speed site) to the largest offshore machines at 126 m diameter, a tower height of 80 m is employed.

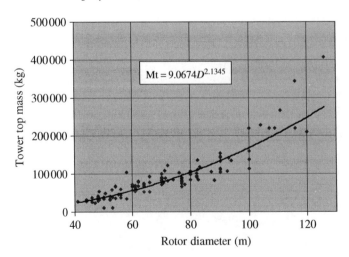

$$Mt = 9.0674D^{2.1345}$$

Figure 4.11 Tower top mass.

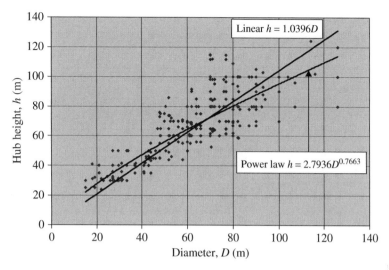

Figure 4.12 Hub height trends.

4.5.5.2 Mass versus Diameter

The scatter in hub height is consequential in producing great scatter in tower mass data (Figure 4.13). The trend line and interpretation of the intrinsic variation of tower mass with diameter under conditions of similarity (fixed design site conditions, fixed state of technology and material systems) cannot be effected without some method of normalising the results taking account of different tower heights and masses at any given diameter.

Concrete and lattice tower are excluded from all discussion in this section. The scaling is not expected to be different in principle, but there is a lack of data by comparison with tubular steel towers. The actual masses of such towers will however be greatly different, concrete towers approach being around an order of magnitude greater in mass than that

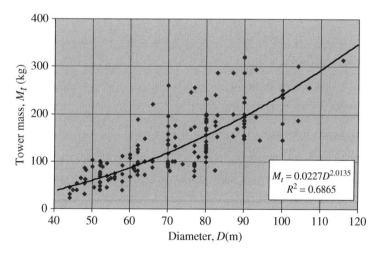

Figure 4.13 Tower mass trends.

of tubular steel ones. Lattice towers are intrinsically lighter and stiffer than tubular steel towers. However, they have other design issues, for example, maintaining bolt torques and accurate alignments of tower members.

4.5.5.3 Normalised Mass versus Diameter

Tower designs vary considerably and are very much adapted to site conditions. For a given turbine model, there may be as many as five tower height options, usually reflecting different wind classes as determined by mean wind speed and wind turbulence conditions. Thus, a vertical array of data points – all at the same rotor diameter – commonly reflects the fact that a series of towers of different heights are associated with only a few different turbine models. In order to rationalise the tower and wind turbine system mass and cost data in reducing data scatter and revealing trends, it is useful to have a means of normalising tower mass and cost.

There is much scatter in tower mass and cost data for a wide variety of reasons. Different design codes, different manufacturer's methods and commercial factors, and so on, apply. There is therefore little point in attempting a very sophisticated normalisation process.

The following assumptions are proposed:

- Assume constant tower base bending stress for all tower heights.
- Assume constant aerodynamic design thrust acting at tower top for any given rotor diameter.
- Tower base thickness varies in proportion to tower base radius as tower height varies.

With the following notation:

Σ	Tower base bending stress
H	Tower height
M	Tower base bending moment $= Fh$
I	Section modulus of a thin cylinder
R	Tower base radius
T	Tower base thickness
Y	Distance to neutral axis $= R$
F	Tower top aerodynamic thrust
M_t	Mass of tower

$$\sigma = \frac{My}{I}$$

$$\sigma = \frac{Fh}{\pi R^2 t}$$

$$M_t = 2\pi \rho R t h$$

Assume F constant, σ constant and $t \propto R$ as h varies. Then:

$$\sigma \propto \frac{h}{R^3} \quad t \propto h^{1/3} \quad M_t \propto h^{5/3}$$

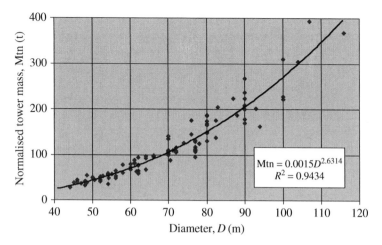

Figure 4.14 Normalised tower mass trends.

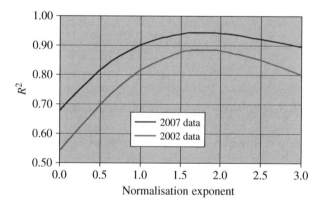

Figure 4.15 Verification of normalisation exponent.

As tower height h is, in general, proportional to diameter (Figure 4.12), the tower masses are normalised by the factor, $(D/h)^{5/3} h$. It is apparent that this normalisation reduces scatter significantly (compare Figures 4.13 and 4.14).

There is still large vertical spread of data at similar diameters, but a considerable degree of order has been produced from the initial chaos of Figure 4.13. As a test of the optimisation of the normalisation, the exponent predicted as 5/3 by the analysis presented was varied. The value of zero for the exponent corresponds to the data correlation without any normalisation factor applied.

It is apparent (Figure 4.15) that the R^2 value (square of correlation coefficient of a best-fit power-law equation as in Figure 4.14) is near maximum for exponents in a range between 1.6 and about 1.8 and, in particular, for the theoretically determined exponent of $5/3 = 1.667$. The two curves are results of this analysis from data available in 2002 (when the method was first conceived) and then as later repeated with a larger data set of 2007.

This corroborates the present normalisation analysis since it shows that no empirically chosen exponent can normalise the data significantly better than the theoretically determined value of 5/3. It is also encouraging that trends in the normalised data of the new larger data set and the older smaller data set are similar and that, with a larger data set, the correlation improves. It should be noted that the normalisation as far as it is developed here is a means of reducing scatter in the data and revealing trends. Normalised tower mass values correspond to an average hub height for scaling similarity rather than the actual hub height. What the process shows is that, due to the reduction of scatter by the normalisation procedure, the exponent of 2.6 from Figure 4.14 is more likely to reflect the tower scaling trend than the lower value of 2.0 from Figure 4.13.

Although the normalisation is very successful in reducing data scatter, there is still a degree of vertical scatter in data at several specific rotor diameters. It is suspected that this is due to the effect of wind class on design not being accounted for lack of sufficient information.

The value of the exponent at 2.6 (Figure 4.14) is more nearly approaching the exponent of 3 than would be expected by simple scaling logic. In fact, taking account of wind shear increasing the hub height wind speed with increasing turbine scale, an exponent >3 would be expected. However, mainly due to a compensating effect of many wind turbines around 70 m diameter having large diameter rotors for low wind speed sites, present rating trends indicate that power rating and rotor thrust increase as square of diameter as if wind shear were not accounted for. Thus, if that rating relationship is preserved, a tower mass scaling exponent of around 3 can be expected.

4.5.6 Gearbox Scaling

Figure 4.16 is based on data of Winergy and others as published on the Internet. It broadly confirms the expected direct relationship between gearbox mass and torque. Differential gearboxes of Bosch and others are increasingly used in large wind turbine designs, and in some cases are significantly lighter. Although the input torque and first

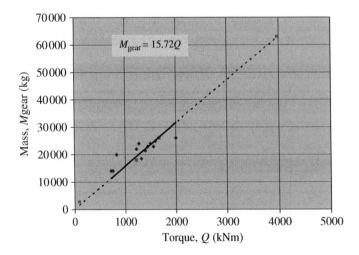

Figure 4.16 Gearbox mass trends.

stage of gearing dominate mass and cost of a gearbox, the primary purpose of a gearbox is to provide a gear ratio and mass; and cost will have some sensitivity to this, which is not reflected in the data as presented in Figure 4.16.

4.6 Upscaling of VAWTs

Horizontal-axis and vertical-axis wind turbine (VAWT) design are compared in Chapter 13. Although large-scale VAWT designs have never been commercially established, some see future potential in this technology. The effects of scale on the blades and central column of a Darrieus rotor wind turbine have been studied by Malcolm [13], and this is referenced in a review by L. Schienbein Associates [14]. An extract from that review observes:

> …assuming similar construction and materials, their mass (and likely cost) was proportional to the cube of the rotor size diameter. This also held for the cost of the major drive train costs (gearbox and couplings) so that it was likely that the total manufactured cost would increase with the third power of diameter. This conclusion is probably optimistic since large components are normally less available and require special manufacturing techniques. It is also less likely that large components will enjoy the cost benefits of mass production. The other major costs are for transportation, site preparation and infrastructure. These are likely to increase at least with the square of the diameter.
>
> The energy capture, however, is proportional to the swept area and to the cube of the wind velocity. In a region with 'normal' vertical wind shear, this will result in the total energy increasing with the diameter to the power of 2.4. There is, therefore, an overall negative benefit of increased size. The conclusion concerning the negative benefit of size is given some support if the total mass (of rotor and guy cables) per swept area is compared to the swept area for actual and planned machines.

Limited data (Table 4.2) for the total mass of VAWTs of the Darrieus type (including rotor and guy cables or rigid frame) shows a large increase in mass to swept area ratio

Table 4.2 Mass trends of Darrieus-type VAWTs.

Turbine	Swept area (m^2)	Total mass (kg)	Mass/area (kg/m^2)
FloWind 17 m	241	7 254	30.1
FloWind 19 m	315	10 962	34.8
Adecon 19 m	316	9 100	28.8
Indal 6400	495	17 770	35.9
Eole	4000	300 000	75.0

between swept areas up to 500 m² and the Eole wind turbine at 4000 m². This suggests upscaling to a significantly higher power than as square of diameter.

4.7 Rated Tip Speed

Rated tip speed is defined here as the maximum steady-state tip speed of a wind turbine in normal operation. As was indicated, for strict similarity of design, tip speed should be invariant with scale. What are the trends observed?

Figure 4.17 shows a gradual drift towards higher tip speeds with increasing rotor diameter, all the more so because the latest very large wind turbines are intended for offshore sites where acoustic noise may not limit tip speed. Note that the Gamma 60, a two-bladed wind turbine with a remarkably high tip speed of 138 m/s, and the Lagerwey LW 80 are excluded from the fitted trend line as quite untypical of general design routes. Among recent designs, the Xiangtan Electric Manufacturing Corporation Ltd (XEMC) Darwind 5 MW wind turbine has a particularly high tip speed of 108 m/s.

The great scatter of data points in Figure 4.17 and extremely poor correlation (Table 4.3) may make the interpretation of a rising tip speed questionable. However, this interpretation is consolidated with reference to torque trends (Figure 4.9) and power trends (Figure 9.3), each of which have much better correlated data.

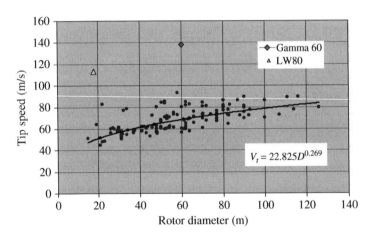

Figure 4.17 Tip speed trends.

Table 4.3 Tip speed/torque/power trends.

Variable	Exponent	Reference	R^2
Tip speed	0.27	Figure 4.17	0.41
Torque	2.77	Figure 4.9	0.96
Power	2.05	Figure 9.3	0.96

For a rated power, P, rated torque, T, diameter, D and the rated tip speed V_t will have the dimensions:

$$[V_t] \equiv \left[\frac{PD}{T} \right]$$

Thus, from Table 4.3, an exponent for tip speed of $(2.05 + 1 - 2.77) = 0.28$ is inferred, which is similar to the value as determined directly from the tip speed trend line data of Figure 4.17.

4.8 Upscaling of Loads

The similarity rules of Section 4.4, in general, give straightforward predictions for the upscaling of wind turbine loads. To what extent are these predictions upheld when real data is considered?

Real data in this context means loads on which commercial wind turbines have been designed. Over the years, GL Garrad Hassan has conducted a great many sets of load calculations for commercial manufacturers, mainly for purposes of design certification. Individual data points are not presented in what follows in order to preserve the confidentiality of each manufacturer's information. Each data set is therefore characterised by a trend line equation, an R^2 value and the number of data points. The load calculations are, in general, according to established standards for certification loads calculations such as International Electrotechnical Commission (IEC) [9]. The data have not been filtered except to exclude stall-regulated designs and consider only pitch-regulated designs. Thus, different site classes are all lumped together; and in the case of extreme loads, the load cases associated with each extreme data point may differ.

An example of the trends for the extreme value of the blade root out-of-plane bending moment (designated M_y in standard GL coordinates) is presented in Figure 4.18,

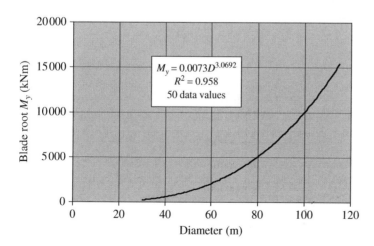

Figure 4.18 Extreme blade root out-of-plane bending moment.

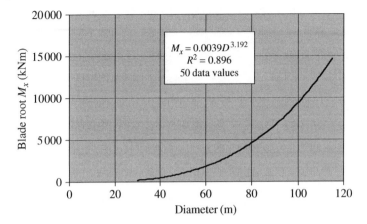

Figure 4.19 Extreme blade root in-plane bending moment.

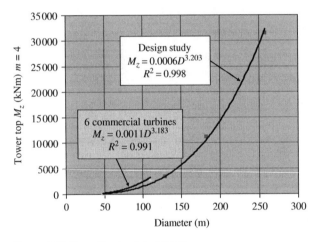

Figure 4.20 Upscaling loads to 20 MW.

indicating a nearly cubic trend for that data set. A similar result is obtained for the in-plane blade bending moment M_x (Figure 4.19).

As turbines increase in size, the effect of wind turbulence changes both in terms of the rotor size in relation to turbulence length scales and on account of the variation of wind turbulence with height within the earth's boundary layer. To explore this, turbine designs were created scaling with similarity from 5 to 10 MW and 20 MW scale.

Figure 4.20 shows the yaw torque on the tower, M_z. Also included are six commercial designs, obviously of smaller scale, which appear to lie on a steeper but approximately cubic curve. There is no indication of a strong effect of turbulence on bending moment scaling up to 20 MW. Nevertheless, from the similarity of the three designs considered, strictly cubic scaling would be expected if turbulence had no effect and the exponent is a little higher than 3.

Even where there has been data from 50 or more wind turbine designs, there are so many variables differentiating the designs and load case situations that there is little statistical certainty that any of the trends are clearly established. As yet, however, there is little to contradict the simplistic scaling of most bending moments as cubic with diameter and forces as square of diameter.

4.9 Violating Similarity

In the UPWIND project, Sieros and Chaviaropoulos [15] evaluate a hollow tapered tower as a typical wind turbine structural component designed by bending stress. They prove that, in upscaling when self-weight loading is significant, no design optimisation route can avoid an increase in mass greater than cubic and conclude as follows;...

> This indicates that, for a given technology level, upscaling always results in an unfavourable weight increase. Although these conclusions are 'exact' for the examined tower structure only, there is no obvious reason why they should not apply for other wind turbine components that can be similarly modelled as beam-like structures, like blades, low speed shaft etc.

The cost of the major components of a wind turbine is generally related to a mass that upscales, as has been discussed. However, some components have more of a fixed cost largely independent of scale. This often applies to intelligence-related functions – a bigger turbine does not need a bigger brain (control computer) but could easily afford one if it is of any benefit. A very large turbine can more readily justify the cost of expensive spectacles (e.g. light detection and ranging (LIDAR)) to see the wind better and use in turbine control for more effective alleviation of system loads. Where weight has knock-on design implications beyond the cost impact on the component itself, more expensive materials, say more carbon in blades, may optimise overall cost, for example, by reducing hub cost and weight. This does not, however, make the upscaled system intrinsically more competitive than a smaller one.

A discrete change in available component sizes is normal in engineering and disturbs similarity in continuous upscaling. Among an almost infinite list of examples are the following:

- Standard metric bolt sizes (ISO 898-1, 1999, Sections 3–7 and Appendix A),
- Generator frame sizes,
- Components related to choice of generating voltage level.

Practical limits have been applied on crane lifts to given heights, on steel plate sizes that can be rolled, on the largest diameter of foundation piles, on castings of various types. These limits reflect worldwide engineering standard practice and not what is ultimately physically possible. The growth of a large-scale wind industry has already shifted some of these limits beyond what had been established industrial practice.

The lack of a smooth continuum in the scaling of components will often distort apparent scaling trends (see discussion in Section 9.7 on hub scaling). The practicality of what is available as standard must definitely be considered at a detailed stage in realistic engineering design, but should not colour what is physically possible (M13 bolts can be made if they are vitally required); nor should such step change effects be built into generic scaling rules. This is clarified in the discussion in Section 9.7 regarding scaling impact on cost.

4.10 Cost Models

The commercial data of Section 4.5 has shown how inherent scaling trends may be confused with general technology progress, especially if the technology trend in a component or system is systematically towards larger scale as time progresses. Sørensen *et al.* [16] proposed a methodology in which such issues can be formally separated. A theoretical framework for risk-based optimal design of large wind turbines is developed in which three levels of formulations are considered:

1) A risk/reliability-based formulation,
2) A deterministic, code-based formulation, and
3) A crude deterministic formulation.

Specifically, a general cost model is formulated in which the impact of technology change with time is separated from the intrinsic scaling that is expected from similarity. The basic formulation, as extracted from Sørensen *et al.* [16], is

$$C(sf, T)_{comp} = C(1, T_0)_{comp} \frac{c(sf, T)}{c(1, T_0)} \frac{m(sf, T)}{m(1, T_0)}$$

$$= C(1, T_0)_{comp} \frac{c(sf, T)}{c(1, T_0)} \cdot \frac{m(sf, T)}{m(1, T)} \cdot \frac{m(1, T)}{m(1, T_0)}$$

$$= C(1, T_0)_{comp} \frac{c(sf, T)}{c(1, T_0)} \cdot sf^{\alpha_{comp}(T)} \cdot r(T) \tag{4.2}$$

The notation of Equation 4.2 is

$C(sf, T)_{comp}$ – cost of component with technology T when upscaled with scale factor sf

$m(sf, T)$ – mass of component with technology T when upscaled with scale factor sf

$c(sf, T)$ – cost of component per mass unit with technology T when upscaled with scale factor sf

$\alpha_{comp}(T)$ – scaling factor for component, based on similarity.

The cost can be broken down in the following factors:

$C(1, T_0)_{comp}$ – cost of component with technology T_0 when not upscaled

$c(sf, T)/c(1, T_0)$ – 'technology improvement' in cost per mass unit from technology T_0 to T and upscaling with scale factor sf. This accounts for changes in cost per mass unit due to technology-based changes in materials, manufacturing process.

$sf^{\alpha_{comp}(T)}$ – upscaling of mass with technology T based on similarity

$r(T)$ – technology improvement (decrease) of mass from technology T_0 to T can, for example, be estimated by evaluating the effect of technology improvement on mass with same size of the component(s).

The practical problem in implementing such methodology in cost models is usually a lack of enough good data to clearly separate effects such as technology progress from inherent scaling trends. This is especially because data sets such as wind turbine designs, even at one point in time, embody such a variety of technologies in each component and in overall system design. Nevertheless, it is extremely valuable to formulate the scaling model in this way whatever the limitations in available data. Thus, similarity scaling in a cost model is embodied as a core element. Therefore, any factors that modify scaling beyond similarity rules are recognised as such and a causal assumption that they will necessarily apply at all scales can be avoided.

4.11   Scaling Conclusions

Simple scaling rules are often regarded as unhelpful and too simplistic because of the complexity and dissimilarities in the scaling of real engineering systems. This is far from being the case. They may be of limited direct application because similarity is so often violated, but they form the only solid foundation for a broad understanding of scaling behaviour.

Consideration of basic physics dictates that scaling rules under conditions of similarity will only have integral exponents or rational factional exponents when these integral relationships are inverted. Empirical scaling relationships are full of arbitrary fractional scaling exponents. The only physical justification for such exponents arises when non-linear relationships exist. This arises commonly in viscous flows, for example. Thus, if a gearbox (at fixed ratio) does not scale exactly as torque, the reason should lie in the nonlinear behaviour of the oil and in associated heat transfer relationships. This nonlinearity is physically profound and the fractal relationships arise exactly as do the Reynolds number effects on aerofoils from boundary layers and their interconnection with turbulence.

A ground rule that emerges is that no predictive upscaling model should be justified by empirical data without an underlying, possibly quite simplistic, physical model that can account for the general trend characteristics in that data. For reliable cost modelling purposes, considering a typical scaling term with rotor diameter, D, of the type aD^n, the rule should be that basic physics and engineering considerations determine the scaling exponent, n, whilst the coefficient, a, is a function that varies with materials, design style, manufacturing process technology or any of the factors that can be wrapped into a technology learning curve. Moreover, a polynomial relationship $\sum_i a_i D^{n(i)}$ describing the scaling of component cost is commonly realistic considering that some cost components are typically volumetric, whilst others such as painting relate to surface area. Clearly, such a relationship will lose extrapolative validity if it is approximated by a single power-law term.

In many instances, technology advances are proposed which 'defeat the square-cube law'. If there is a genuine benefit of scale, the acid test is not only that the new technology reduces cost or weight in relation to expectations from similarity scaling but *also that it has benefits that are unique to large size and cannot equally be applied at smaller scale.* This latter requirement is often forgotten and it is comparatively rarely met.[12]

---

12 Occasionally, smaller wind turbine designs in the hundreds of kilowatt range are revisited due to new markets appearing for them. The approach has generally been to upgrade old proven designs rather than undertake the much longer development programme to completely refresh the design with latest technology benefits.

With so much emphasis on upscaling of designs in the wind industry, scaling rules are presently of critical importance in estimating mass, cost and feasibility of larger designs. Cost models with empirically based scaling may have reasonable accuracy for interpolative purposes, but the underlying formulae are restricted to a scale range and restricted to a specific period in time when certain technologies and technological solutions are prevalent, that is, they date. Cost models that are physically based on similarity rules with other effects functionally separated can be developed with a methodology and underlying structure that need not change although the input information undoubtedly will.

References

1 Harrison, R. and Jenkins, G. (1994) Cost Modelling of Horizontal Axis Wind Turbines. Energy Technology Support Unit (ETSU) W/34/00170/Rep, University of Sunderland, Sunderland.

2 Fingersh, L., Hand, M. and Laxson, A. (2006) Wind Turbine Design Cost and Scaling Model, National Technical Information Service. Technical Report NREL/TP-500-40566.

3 Stiesdal, H. and Kruse, H. (2004) *The Growth in Turbine Rating – A Case Study*, Bonus Energy A/S, Brande

4 EWEA (2009) *Wind Energy the Facts. Part 1 Technology*, Earthscan. ISBN: 978-1-84407-710-6.

5 http://www.clipperwind.com/pr_091609.html (accessed January 2011).

6 http://nextbigfuture.com/2010/05/american-superconductor-sea-titan-10-mw.html (accessed January 2011).

7 *Windkraftanlagen Markt*. Issues from 1997 to 2007, Sun Media GmbH.

8 Batchelor, G.K. (1982) *The Theory of Homogeneous Turbulence*, Cambridge Science Classics Series, Cambridge University Press. ISBN: 0 521 041171

9 IEC International Standard 61400-1. (2005) *Wind Turbines – Part 1: Design Requirements*, 3rd edn, International Electrotechnical Commission, Geneva.

10 Windenergie (1999) (and other years). Bundesverband WindEnergie e.V. ISBN: 3-9806657-0-4.

11 Griffin, D.A. (2001) WindPACT Turbine Design Scaling Studies Technical Area 1 – Composite Blades for 80 to 120 m Rotor. Subcontractor Report NREL/SR-500-29492, by Global Energy Concepts.

12 Chaviaropoulos, T., Lekou, D. and Sieros, G. (2011) *Cost Models Applied to Blades*. Presentation for Work Package 1B4 Upscaling (ensure that this is public by mid 2011).

13 Malcolm, D.J. (1990) *Feasibility of Large Scale Wind Turbines*. 6th CanWEA Wind Energy Conference, October 1990.

14 L. Schienbein Associates (1992) *Darrieus Wind Turbine Assessment Study*, Renewable Energy Systems Ltd.

15 Sieros, G. and Chaviaropoulos, T. (2010) *Aspects of Upscaling Beyond Similarity*. TORQUE 2010 Conference, Heraklion, Crete, June 28–30, 2010.

16 Sørensen, J.D., Chaviaropoulos, T., Jamieson, P. *et al.* (2009) *Setting the Frame for Up-Scaled Off-Shore Wind Turbines*. ICOSSAR Conference, September 2009.

5

Wind Energy Conversion Concepts

Figure 5.1 shows a top-level concept map for wind energy converters. The thick bold path defines the route of options that can be considered mainstream. There is no hope of being comprehensive or rigorous in such categorisation, but the main options for wind energy conversion concepts are illustrated.

The dominance of concepts based on the rotor rather than alternatives based on translating or oscillating aerofoils was explained in Section 1.6. Systems based on translating or oscillating aerofoils have dimensions which are commensurate with source flow area. Thus, they are much more demanding in the required volume of energy extraction materials in relation to energy source area and likely to be much less cost-effective than a rotor.

Non-rotor concepts including the electrostatic wind energy converter (Section 8.6.1) and piezoelectric wind energy converter (Section 15.10) and a variety of systems including airborne ones such as kites (Section 8.4) are based on translating aerofoils. The oscillating aerofoil is demonstrated in laboratory flutter engines but has substantially been rejected for wind energy conversion. Three oscillating aerofoil converter concepts were evaluated by South and Mitchell [1]:

- The galloping characteristic of a stretched cable enhanced by various aerodynamic shapes with energy being extracted from linear oscillatory motion at a cable end;
- An oscillating vane concept involving a vertical blade that is flexible or flexibly mounted;
- An aerofoil oscillating up and down normal to the wind on a vertical track with energy being extracted from work done by the lifting force.

They concluded that these options had no advantages over conventional rotor solutions and had the basic disadvantage of most non-rotor concepts, as discussed in Section 1.6.

Considering designs based on the rotor concept of Figure 5.1, there are a number of possible rotor axis configurations, although the horizontal axis is mainstream and the vertical axis (see Chapter 13) is the only other option that has been much considered. Rotors principally have rigid or semi-rigid blades designed as powerful lifting surfaces, although a few more primitive designs such as the Savonius rotor are substantially, but not exclusively, drag devices. Some older rotor designs, among them the famous wind pumps on the Lasithi Plain in Crete [2], have used sailcloth blades.

The essence of modern rotors is aerofoil sections designed to have high lift-to-drag ratio. Most commonly, such sections derive their function entirely from their invariant

Innovation in Wind Turbine Design, Second Edition. Peter Jamieson.
© 2018 John Wiley & Sons Ltd. Published 2018 by John Wiley & Sons Ltd.

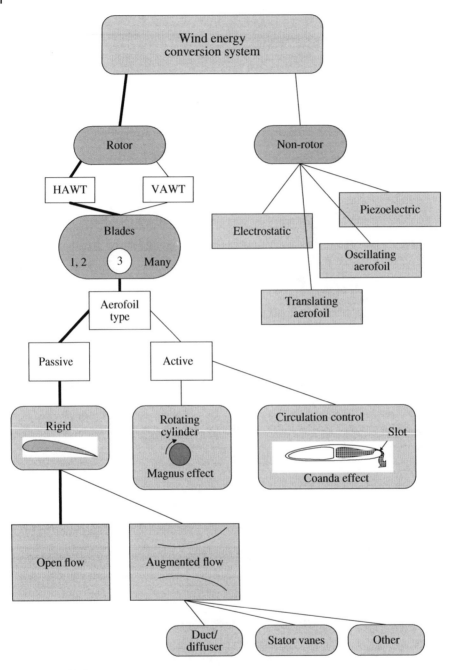

Figure 5.1 Wind energy converter concepts.

shape. However, as shown in Figure 5.1 and discussed in Section 1.2, lifting or non-lifting bodies (cylinder) can generate lift actively using air jets to control the boundary layer or through rotation of the section. As a means to produce lift, this is clearly much more troublesome to engineer than the conventional aerofoil section but has the considerable virtue that the lift can be regulated, switched off or even reversed in direction.

A large number of wind turbine designs have sought to augment flow using a wide range of systems or devices. These include duct and diffuser concepts discussed in Chapters 1 and 18 and some ingenious designs to exploit strong vortices to create areas of concentrated flow. An early notable investigation of innovative concepts took place at a Solar Energy Research Institute (SERI) conference in Colorado (1979). Many of the non-rotor concepts and unusual augmenter concepts (using rotors) [3–6], were first explored then. A particularly ambitious kind of concentrator [4, 7] involved an attempt to create an artificial tornado. As yet, none of these designs has made inroads into the main commercial markets for large wind turbines.

Kitegen [8] proposed a very interesting system which is also a minor challenge to the taxonomy presented here! However, it can reasonably be regarded as a vertical-axis, multi-bladed rotor with each 'blade' comprising an aerodynamically inactive tether length with an active (kite) aerofoil section at its extremity.

Beyond the basic choices for the type of energy conversion system are many options for mechanical and electrical power conversion systems, some of which are reviewed in Chapter 6 on drive train design.

Small wind turbines are not specifically considered here and would present many more design options, especially in operational concepts. For example, control and safety options for small wind turbines include yawing[1] often passively actuated, perhaps through spring-hinged tail vane(s) or by an offset in the line of action of the rotor thrust from the rotor shaft axis which will create a yawing moment that overcomes a restoring spring in strong winds.

Although only the mainstream route of a rotor with passive aerofoils, without augmentation and in a horizontal-axis three-bladed configuration, will receive much further attention, it is worth noting the variety of other systems that have been tried. This both adds to understanding of the chosen preferred rotor system and also provides awareness of ideas which, although they have not been successfully realised commercially, may in some cases have future potential.

References

1 South, P. and Mitchell, R. (1983) Oscillating Wind Energy Conversion Systems. SERI Report WPA NO. 171-83, US DOE.
2 http://www.travel-to-crete.com/place.php?place_id=84 (accessed April 2011).
3 Whitford, D.H. and Minardi, J.E. (1979) *Utility-Sized Wind-Powered Electric Plants Based on the Madaras Rotor Concept*. Wind Energy Innovative Systems Conference Proceedings, Colorado, May 1979.

---

1 Yaw control was used on the Gamma 60 wind turbine, but this use on a large wind turbine is quite exceptional.

4 Yen, J.T., DeCarlo, J. and Zywan, W. (1979) *Recent Developments of the Tornado-Type Wind Energy System*. Wind Energy Innovative Systems Conference Proceedings, Colorado, May 1979.

5 Sforza, P.M. and Stasi, W.J. (1979) *Field Testing the Vortex Augmentor Concept*. Wind Energy Innovative Systems Conference Proceedings, Colorado, May 1979.

6 Lissaman, P.B.S., Hibbs, B., Walker, S.N. and Zambrano, T.G. (1979) *A Definitive Generic Study of Augmented Horizontal Axis Wind Energy Systems*. Wind Energy Innovative Systems Conference Proceedings, Colorado, May 1979.

7 Haers, F. and Dick, E. (1983) Performance analysis of the tornado wind energy concentrator system. *Wind Engineering*, 7 (4), 223.

8 http://kitegen.com/ (accessed April 2011).

6

Drive-Train Design

6.1 Introduction

Drive-train design covers the range of hardware that communicates power between the aerodynamic rotor and electrical output terminals. The aim is to look particularly at the implications of new concepts for weight, cost, efficiency and energy capture. Rather than simply conduct ad hoc investigations of a number of drive-train types, a possible approach to optimisation of drive-train design is also discussed. With an almost limitless number of alternative system components, electric, hydraulic, magnetic, and so on, 'optimisation' of design is more than ambitious. Nevertheless, much insight can be gained in the attempt. There are many issues around the mechanical and structural design, bearing types and bearing arrangements, cast and fabricated nacelle support structures, and so on, but these are largely left to detailed design discussions beyond the scope of this book. Around the start of the new millennium, a major study of drive-train options [1, 2], funded by the US DoE within the WindPACT programme managed by National Renewable Energy Laboratory (NREL) had led to the manufacture of new drive-train systems and laboratory testing at full scale.

Drive-train design remains one of the most active areas of innovation. Many new wind turbine designs continue to appear in the megawatt range using permanent magnet generators (PMGs), some direct drive (no gearbox) and others with one, two or three stages of gearing. Many innovative drive systems have fallen by the wayside, often as much related to the challenges in commercialisation as to deficiencies in technology.

6.2 Definitions

After the rotor, the next key area of development interest in a wind turbine system is the power conversion system or drive train that produces grid standard electricity from the shaft power developed by the rotor. 'Drive train'[1] is a widely used term, often referring to all the critical equipment between the aerodynamic rotor and the point of connection to the main power cables from the wind turbine.

---

1 'Power train' may be a better term as the power transmission system generally includes electronic components and not only the mechanically 'driven' gearbox and generator. However, the term, 'drive train' is more generally in use.

Innovation in Wind Turbine Design, Second Edition. Peter Jamieson.
© 2018 John Wiley & Sons Ltd. Published 2018 by John Wiley & Sons Ltd.

There is scope for confusion in the terms used to describe generic and proprietary drive-train types. General terminology relating to drive-train systems is adopted as follows:

- *Conventional*: gearbox and high speed generator with few pole pairs
- *Direct drive*: any drive train without a gearbox
- *Hybrid*: any drive train with a gearbox but with unconventional features such as multi-pole (say, more than eight) generator[2] or differential drives
- *Multi-generator*: any drive train with more than one generator
- *Multibrid*: only the specific single stage gearbox and PMG combination licensed by Aerodyn.

The term 'hybrid' is not widely used in the industry with any agreed definition and is introduced here especially as a generic name for a variety of systems that have fewer stages of gearing than are required for a conventional high-speed generator and consequently employ a medium-speed multi-pole generator.

6.3 Objectives of Drive-Train Innovation

The main motives for investigating drive-train technology relate to the following:

- Improving reliability, for example, in the direct-drive generator concept by eliminating components like the gearbox;
- Improving efficiency and in that connection, part load efficiency is especially relevant;
- Reducing cost – by searching for the most cost-effective system and in some cases by radical concept changes, in other instances by more integrated design.

The discussion will return to these topics, reliability, efficiency, cost and cost of energy impacts following a general review of drive-train options

6.4 Drive-Train Technology Maps

Two maps are now presented, comprising:

- A top-level map (Figure 6.1) summarising drive-train options;
- A more detailed map (Figure 6.2) showing the options related to power flow from the rotor aerodynamic input up to the point of connection to the electric network.

The general map of Figure 6.1 serves as an introduction, showing the mainstream design route to the doubly fed induction generator (DFIG) following the bold thick lines. A few manufacturers are cited as having designs exemplifying particular technologies. As is evident in Table 6.2, there are manufacturers other than Multibrid and Gamesa involved in hybrid drive-train designs.

The DFIG has been a mainstream choice offering sufficient variable speed capability to realise most of the benefits in energy capture and operational flexibility. However, power quality requirements in the European market are driving design

---

2 The use of 'hybrid' as a term for the whole drive train is not to be confused with the hybrid generator, a type of electrical generator that has both magnets like a PMG and excited windings.

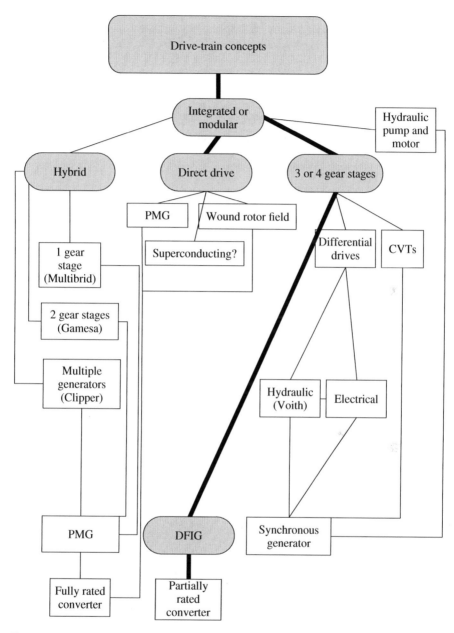

Figure 6.1 Top-level drive-train technology map.

towards other systems with generally wider speed range capability involving fully rated power converters. Superconducting generators promise large mass reductions in very high capacity systems, perhaps around 10 MW rating, but there are many technical challenges especially for the long life, low maintenance duty of a wind turbine system. The need to maintain cryogenic systems and. in particular, the time to cool down and restore operation after a stoppage is one concern.

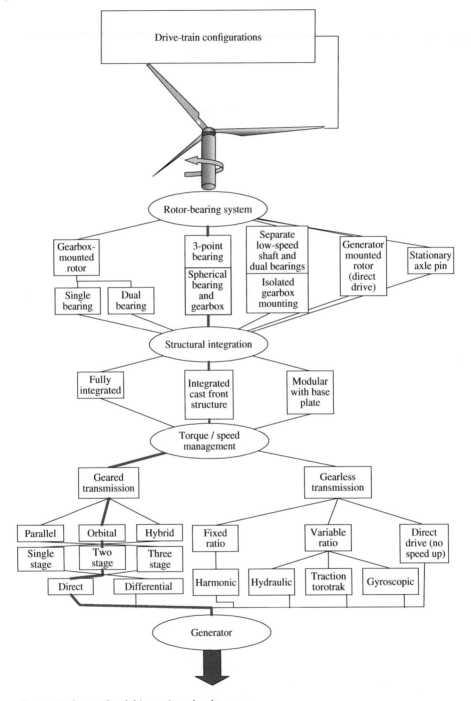

Figure 6.2 System-level drive-train technology map.

Direct-drive systems are increasingly popular in new designs. The excited synchronous generator with wound field rotor design is a solution long established in the market by Enercon, a leading manufacturer. However, with the availability of comparatively cheap high-strength neodymium magnets, WindPACT [1, 2] and other recent studies favour permanent magnet designs and the current interest in direct-drive developments is predominantly with PMG designs. In parallel with the development of direct-drive, hybrid systems have evolved such as Multibrid (developed from the ideas of Böhmeke [3]) which have one or two stages of gearing rather than the three or four demanded by a conventional four- or six-pole high-speed generator at multi-megawatt scale. These innovative gear systems are integrated with one or more medium-speed generators, again most usually PMGs.

Variable speed can, in principle, be realised by mechanical and fluid mechanical rather than electrical systems. In general, such systems are categorised as continuously variable transmissions (CVTs). Systems such as Torotrak [4] are based on traction and another evaluated by DNV GL [5] employs gyroscopic torque transmission with variable ratio resulting from control of speed and/or angular position of gyroscopes. The Torotrak system is established and used in a few specialised non-wind applications, but a previous study of DNV GL (as the former Garrad Hassan) suggested that there would be a considerable challenge in adapting the technology to megawatt-scale wind turbine systems with adequate fatigue endurance. A gyroscopic transmission system (reviewed in Chapter 21) has some attractive features but awaits prototype development.

A hydraulic transmission system developed by Artemis Intelligent Power [6], with potential for wind, wave and automobile applications (see Section 6.9), has advanced towards commercialisation following the acquisition of Artemis by Mitsubishi Heavy Industries, Ltd (MHI) in December 2010. MHI has installed a 7 MW wind turbine, the world's largest floating system, at Fukushima in July 2015. Chapdrive [7], who also were developing a hydraulic transmission concept, is however no longer in business. The power density of hydraulics suggests that it may be particularly promising for multi-megawatt offshore wind turbines in achieving compact lightweight nacelle systems.

The drive-train arrangements and the options for management of torque and speed in each system can be further classified as in Figure 6.2. The issues of whether rotor and gearbox or direct-drive generator bearings can be integrated, the desirability of isolating the gear and generator shafts from bending loads, the balance in physical size between gearbox and generator and how their individual requirements for structural stiffness and support integrate in the whole load path taking rotor loads through the nacelle structure into the tower top all figure in the determination of an appropriate system design.

Various arrangements have been considered for the main rotor bearing system. Initially, a main shaft directly connected to the hub and rotor, rotating within a pair of bearings was the most common arrangement. More recently several alternatives have been tried. These include direct mounting of the rotor on a single slew ring bearing (as in the Vestas V90) and setting the rotor hub on bearings that rotate around a fixed axle pin.

The wind industry initially favoured a modular design approach in which gearbox, bearings, generator, and so on, were all separate components of more or less standard supply from experienced component manufacturers who, subject to a valid specification being provided, took design responsibility for these components. Increasingly more individual designs, Vestas V90 single bearing, GE Wind's and Samsung's designs with high-speed PMG, WinWind, Multibrid and others have appeared.

The harmonic drive mentioned in Figure 6.2 is an elegant concentric drive which was used for pitch drives of a 750 kW wind turbine developed by Markham around 1994 in the United Kingdom but is not presently available in large enough units for the pitch drives to main transmission system of megawatt class wind turbines.

A quite different approach to speed or torque control is to use a differential gear stage at the low torque end of the drive train, that is, an epicyclic stage in which the input speed is proportional to a linear combination of the output speeds, and each output torque is individually proportional to the input torque. In practice, the concept often assumes one constant speed output and one variable speed. In principle, either an electric, hydraulic or mechanical variable speed device may be connected to the variable speed shaft, or between the two output shafts. In some implementations, this concept is essentially a mechanical analogue of the wound rotor induction generator, with the differential high-speed stage doing the job of the slip rings in decoupling shaft speed from synchronous speed of field rotation. Examples of differential drive trains are discussed in Section 6.8.

The drive-train map of Figure 6.2 continues into diverse options for generator technology, power electronics and network connection. Electrical power conversion options depend strongly on generator speed.

For the 'high-speed' generator option which has been dominant, there are four common sub-choices:

- Fixed speed with an induction generator allowing a small percentage slip;
- Variable slip – essentially the same but with a substantially increased slip range;
- Limited variable speed, that is the popular DFIG, with an alternative brushless configuration which has not proved popular;
- Full-range variable speed.

For the 'medium-speed' alternative, the generator diameter will be larger, and it is generally found that induction machines are less attractive at low speeds and large diameters. Therefore, unlike the 'high-speed' alternative, only synchronous machines, especially PMGs, are considered. It is notable that Vestas has favoured a medium-speed PMG generator [8] for the offshore flagship V164 design.

For the 'low-speed' alternative (i.e. direct-drive), switched-reluctance generator designs have been explored but appear uncompetitive. This leaves synchronous machines, again predominantly PMGs, for the medium-speed alternative. For direct drive, the specialised nature of the application throws up some additional design decisions. The converter may comprise back-to-back insulated gate bipolar transistor (IGBT) units or machine side converter components may be integrated with the stator coils.

6.5 Direct Drive

Gearboxes have been controversial for some time in wind technology. Statistics collected by the wind industry have suggested that the most frequent turbine system failures are electrical, but that gearbox failures are the most consequential, Tavner *et al.* [9]. Gearbox technology remains an integral part of most commercial wind turbine designs, yet reliability concerns and the argument of a simpler drive train with fewer components have fuelled interest in direct-drive technology. Initially, reliability stats did not always uphold the logic that the simpler direct-drive system should be more reliable, but this was probably due to many direct-drive machines being new and needing time to eliminate teething problems and become more mature technology.

In new multi-megawatt designs, wind turbine manufacturers such as Siemens and GE are moving away from gearbox turbines towards permanent magnet direct-drive turbines. The share of global wind turbine installations accounted for by direct-drive turbines increased from around 16% in 2006 to 26% in 2013 and now (2017) has increased further with Goldwind becoming since 2015 the world market leader ahead of Vestas in turbine supply. The PMG has become the main choice among a variety of options in the design of the direct-drive turbine (Table 6.1), although there is no necessity for direct-drive and PMG technology to be combined.

Enercon (as in, e.g. the E126 with a 12-m-diameter generator) was by far the earliest supplier of direct-drive turbines in commercial supply. Their design involves a rotor with windings which can be magnetised by excitation current rather than using permanent magnets. Wound field rotor designs were initially favoured in the 1990s when high-strength neodymium-iron-boron magnets were much more expensive, whilst designs involving a PMG with ferrite magnets would have been very heavy. Compared to a PMG, the wound field rotor (rotor with a separate set of electrical windings) gives extra control of the rotor magnetic field including the ability to switch it off. However, it requires a power supply to the windings involving slip rings and, generally, a direct-drive generator with a wound rotor appears to be heavier than a PMG.

Table 6.1 Direct-drive wind turbines.

	Diameter (m)	Rated power (MW)	Comment
Siemens	154	8.00	PMG
Haliade (Alstom)	150	6.00	PMG (Alstom and GE partnership)
Darwind XE128	128	5.00	PMG
Enercon E126	126	7.58	Wound rotor
Goldwind GW 121/2500	121	2.50	PMG most prevalent PMG design
GE 4.1–113	113	4.10	PMG (upgrade of Scanwind SW3500)
Koncar K104-2.5	104	2.50	Wound rotor
Leitwind LTW101	101	3.00	PMG
Northern Power 2.3/93	93	2.30	PMG
Scanwind SW3500	91	3.50	PMG (Scanwind aquired by GE energy in 2009)
Mtorres TWT 1.65-82	82	1.65	Multi-pole synchronous generator
Dongfang DF70-1500	70	1.50	PMG (drive train provided by the Switch)
Unison U57	57	0.75	PMG Korean wind turbine design
Northwind 100	20	0.10	PMG

Although Enercon had dominated the world market share of direct-drive turbines having pursued that technology at a much earlier stage than others, the Chinese company, Goldwind, became the world market leader in terms of installed capacity and is now the leading supplier of direct-drive wind turbines. Their direct-drive PMG technology was developed after having acquired rights to the technology of the German company, Vensys, in 2008.

Direct drive appeals in eliminating the gearbox as a cost and maintenance concern. However, direct-drive solutions are not obviously lighter or cheaper than the conventional geared drive train. A low-speed direct-drive generator needs to be of comparatively large diameter to achieve an adequate speed at the air gap and the large diameter implies substantial additional structural mass beyond the minimal essential mass related to electromagnetic function. Much attention has therefore been directed to concepts for greater structural efficiency to reduce mass and cost of direct-drive systems [10, 11]. For example, the fundamental idea of the NewGen direct-drive generator [11] is to avoid the stiffness demand associated with a load path through the central main shaft to the periphery of the generator rotor. Instead, generator rotor stiffness demand is much reduced by positioning large-diameter bearings adjacent to the air gap.

In direct-drive designs, with so much structure forming a major part of the mass and cost of the generator, integration of the complete drive train, including bearing arrangements and connection to the main nacelle structure, are major considerations. The bearing solutions generally relate to a choice whether the electrical rotor is external or internal to the stator.

The most usual arrangement is of an internal rotor surrounded by the stator. In all designs, the main issue is whether to employ a single large slewing ring-type bearing (as typically used in pitch and yaw systems) or twin bearings which provide moment resistance more readily and can be of smaller diameter with lower friction. The large slew ring-bearing solution is adopted in the Vensys 100, Leitwind and XEMC designs. All the PMG concepts avoid slip rings by having the permanent magnets on the rotor. In general, more space is required to accommodate the stator windings of a PMG than the permanent magnets and so having the rotor external to the stator and thereby using internal space for the windings in designs such as the Vensys 100 may, in principle, lead to a more compact design with less overall generator diameter for any given air gap diameter. Although there is now major integration in many designs with the electrical and aerodynamic rotors having a common bearing system, usually the hub and wind turbine rotor is overhung beyond the generator system. Exceptionally, the Chinese company, New Unite has a system integrating generator and rotor hub; and the German consultancy, Innowind, has also been developing a design of this type.

Effective cooling, especially to protect the generator windings and maintain the fatigue life of the generator, is an issue and often a significant cost with all designs.

On the one hand, there is obvious merit in protecting generator components from a humid, corrosive and possibly erosive environment as in designs such as New Unite. On the other hand, open hub designs are well suited to passive air cooling.

There has been continuing interest in superconducting (conducting with negligible resistance and losses) technology for direct-drive generators. Superconductivity, first discovered in 1911 by H Onnes, was observed in mercury which superconducts when cooled to 4.2 K. Although scientifically interesting, the discovery had no practical value at the time. However, in 1986, the discovery of superconductivity at temperatures above that of liquid nitrogen (77 K) gave rise to the possibility of achieving practical use of the technology in the development of so-called high-temperature superconductors (HTSs).

The concept is to use HTS wire in the rotor field windings of a synchronous generator. This wire may carry ~10 times the current of a comparable copper wire with the possibility of efficiency gains and mass reduction in a large generator.

Siemens AG in 2005 recorded the successful initial operation of the world's first (4 MW) generator for marine technology, based on HTS designed and manufactured by European Advanced Superconductors (EAS). American Superconductors own Windtec technology, which is exploited by Sinovel. They also provide converters and fault ride through equipment to the wind industry. They claim to have designs for wind turbine systems such as Sea Titan (Section 4.1) that will halve losses and reduce generator weight to one-third. Superconducting technology was to some extent evaluated in the Innwind. EU project [12]. However, results (see Table 15.2) thus far suggest it is significantly less promising for cost of energy reduction than, for example, the pseudo direct-drive concept based on magnetic gearing (Chapter 25). This can be considered in part as due to lower overall efficiency gains and additional system complexity such as in maintaining cryogenic conditions in lifetime operation. Nevertheless, the development of superconducting technology for wind turbines continues. Within the project EcoSwing (Energy Cost Optimisation using Superconducting Wind Generators) of the EU Horizon 2020 programme, Envision Energy aims to demonstrate a 3.6 MW superconducting generator on a wind turbine. A weight saving of more than 40% compared to conventional direct-drive generators is claimed. Following laboratory testing, the EcoSwing generator is planned to operate for over 1 year on a large-scale modern wind turbine in Denmark.

6.6 Hybrid Systems

Hybrid drive trains involve medium-speed multi-pole generators which are almost invariably permanent magnets. Some examples are listed in Table 6.2. PMGs dominate in new designs of drive train, both direct drive and hybrid, and have even been used in the conventional drive train with a three-stage gearbox. In the 100-m diameter, 2.5 MW, GE 2.5xl wind turbine, a high-speed PMG and fully rated converter are employed. This provides a high-efficiency drive train and the capability to operate on 50 or 60 Hz networks without any hardware changes.

The hybrid drive train can facilitate more compact nacelle arrangements with the generator and gearbox better matched in size. Notably, Vestas in the design of the V164 turbine, which holds the world production record (2014) over a 24-h period [13], chose a hybrid drive train with issues around modularity, assembly and projected IRR (internal rate of return on investment), all considered favourable [14] for the hybrid as compared to direct drive or their more conventional design with a four-speed gearbox as in the V112-3.0 MW.

Quite a few companies have ventured either new hybrid drive-train components or new hybrid wind turbine systems. Irrespective of the merits of their technology, it is

Table 6.2 Hybrid drive trains.

Diameter (m)	Rated power (MW)		Comment
164	8.00	Vestas V 164	Hybrid favoured as most cost-effective
132	5.00	Gamesa 5.0 MW	Two-stage gearbox and PMG
135	5.00	Adwen AD 5-135	1 : 10 gear ratio (formerly Multibrid M5000)
90–100	3.00	WinWinD	Two stage + PMG (no longer trading)
—	3.00 and 7.00	Moventas Fusion Drive	Moventas gear and Switch PMG
89–100	2.50	Clipper Windpower Liberty	Has four PMGs (no longer trading)
62–70	1.25	Innovative Wind Power – Falcon	Two stage + PMG (no longer trading)

proving hard to compete in the market for all but the most established, while the more successful may be absorbed by other companies. Thus, WinWinD, Clipper and Innovative Wind Power are no longer trading, while the Switch technology is proving successful but is now owned by Yasakawa.

6.7 Geared Systems – the Planetary Gearbox

Much of the motivation for direct-drive turbines is for greater simplicity and reliability in drive trains. Gearboxes nevertheless remain a mainstay in wind turbine drive trains. Their design is a subject in its own right and long pre-dates their application in modern wind turbine systems. In a few severe cases, however, gearbox problems have threatened the viability of some wind turbine manufacturers. The following extract of 2005, referring to events in 1999, is from an article in Windpower Monthly magazine [15].

> Gearboxes have been failing in wind turbines since the early 1990s. Barely a turbine make has escaped. Six years ago the problem reached epidemic proportions, culminating in a massive series failure of gearboxes in NEG Micon machines. At the time, the NEG Micon brand was the most sold wind turbine in the world. The disaster brought the company to its knees as it struggled to retrofit well over one thousand machines … bigger turbines, however, are proving to be far from immune to gearbox problems, as Windpower Monthly reports in its November issue. What is different today, is that any damage is discovered and repaired well before it results in total gearbox collapse.

Some analysis of the planetary gearbox follows. The terms planetary, epicyclic and, sometimes, orbital all convey the idea of gear wheels revolving around a central

Subscripts notation: a – annulus, p – planet, s – sun, c – carrier

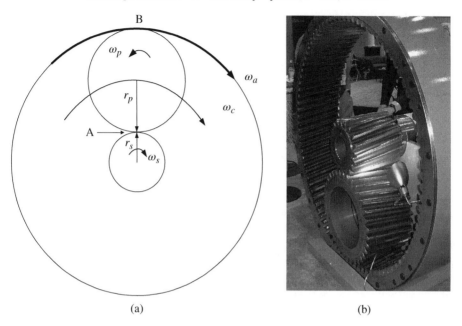

(a)   (b)

Figure 6.3 (a,b) Motion in a planetary gearbox.

gear or shaft like planets round a sun. This type of gear train design is not only important because it can realise transmission ratios effectively with reduced weight and well-balanced loading but because it forms the basis of various forms of differential drives, to be discussed in Section 6.8 that are of interest in new drive-train developments for wind turbines.

Without the surrounding ring gear represented by the outer circle in Figure 6.3, the two inner circles (gears) might represent a so-called parallel stage of gearing where the axes of rotation are parallel but not co-linear. The gear system illustrated by the circles of Figure 6.3(a) and the components of Figure 6.3(b) is intended to represent a planetary or epicyclic gearbox.

It will be apparent from Figure 6.3(b) that the gear teeth are helical and cut at an angle to the face of the gear. When two teeth on a helical gear system engage, the contact starts at one end of the tooth and gradually spreads as the gears rotate, until the two teeth are in full engagement. This gradual engagement makes helical gears operate much more smoothly and quietly than spur gears. It is predominant in wind turbine gear boxes and preferable for many applications, although the angled contact introduces thrust loads which the bearings must accommodate.

As distinct from a parallel stage of gearing in which the gear axes are parallel but not collinear, the planetary or epicyclic gearbox maintains a coaxial transmission between input and output. Consider a single stage of planetary gearing. Referring to Figure 6.3, the essential components are as follows:

- The outer gear ring or annulus, in general, rotating at angular velocity, ω_a, which, in a gearbox designed to provide a fixed ratio, is typically zero with the annulus ring gear being integral with the gearbox outer casing.

- The innermost gear or sun with radius, r_s rotating at angular velocity, ω_s.
- The planets (typically three; although, for simplicity, only one is shown in Figure 6.3) which rotate with angular velocity, ω_p, about their own centres which in turn collectively rotate about the sun with angular velocity, ω_c.
- The planet carrier which is a ring rotating at angular velocity, ω_c, providing a rigid connection of the centres of the planets but with bearings allowing the planets to rotate freely about their centres.

In a typical wind turbine gear train arrangement with a fixed ratio planetary gearbox, the planet carrier is driven by the wind turbine rotor and this will effect a gear ratio, resulting in the sun as output being driven at a higher angular speed. Some elementary analysis of the system allows the gear ratio and torques to be determined as a function of geometric parameters.

However, although the simpler system with fixed annulus and fixed gear ratio is most commonly employed, a primary motive in providing an analysis of the planetary gearbox here is to consider the more general case where all the main elements, annulus, planets, planet carrier and sun are rotating relative to the external fixed reference frame of the gearbox casing or wind turbine nacelle. In this case, by applying an appropriate reaction torque to one of the rotating elements (e.g. annulus), the output (e.g. the sun) can be maintained at constant speed even with speed and torque variation of the input (e.g. planet carrier). This is the concept of a differential drive which provides a variable ratio gearbox with constant speed output. The required control of reaction torque may be provided by an electric servo drive or hydraulically.

At A (Figure 6.3)

$$\omega_c r_s + \omega_p r_p = \omega_s r_s \tag{6.1}$$

$$\begin{pmatrix} \text{Velocity due to} \\ \text{carrier motion} \\ \text{at A} \end{pmatrix} \begin{pmatrix} \text{Velocity of} \\ \text{planet rim} \\ \text{at A} \end{pmatrix} \begin{pmatrix} \text{Velocity of} \\ \text{sun rim} \\ \text{at A} \end{pmatrix}$$

At B (Figure 6.3)

$$\omega_c(r_s + 2r_p) - \omega_p r_p = \omega_a(r_s + 2r_p) \tag{6.2}$$

$$\begin{pmatrix} \text{Velocity due to} \\ \text{carrier motion} \\ \text{at B} \end{pmatrix} \begin{pmatrix} \text{Velocity of} \\ \text{planet rim} \\ \text{at B} \end{pmatrix} \begin{pmatrix} \text{Velocity of} \\ \text{annulus at B} \end{pmatrix}$$

Adding (6.1) and (6.2), $\omega_c(r_s + r_a) = \omega_s r_s + \omega_a r_a$; hence,

$$\omega_c = \left(\frac{\omega_a r_a + \omega_s r_s}{r_a + r_s} \right) \tag{6.3}$$

In particular, from Equation 6.3 with $\omega_a = 0$, a fixed gear ratio between the (output) sun and (input) carrier is provided as

$$\frac{\omega_s}{\omega_c} = \left(\frac{r_a + r_s}{r_s} \right)$$

Continuing with analysis of the general case when $\omega_a \neq 0$, note that the planets are wholly internal to the system. Thus, by Newton's third law that action and reaction are

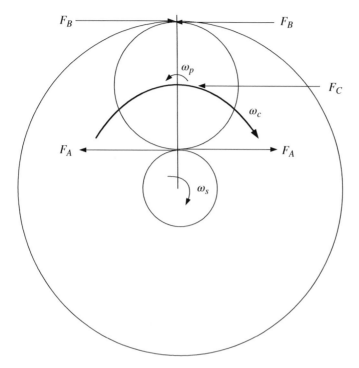

Figure 6.4 Forces in a planetary gearbox.

equal and opposite, their torques will cancel. Referring to Figure 6.4 and considering only external torques T_s, T_c, T_a, associated with the reaction forces;

$$T_s + T_c + T_a = 0 \quad \text{(force and torque balance} \equiv \text{zero net torque)} \tag{6.4}$$

and

$$T_s\omega_s + T_c\omega_c + T_a\omega_a = 0 \quad \text{(power balance} \equiv \text{energy conservation)} \tag{6.5}$$

Eliminating T_a from Equations 6.4 and 6.5;

$$\frac{T_s}{T_c} = \frac{\omega_a - \omega_c}{\omega_s \omega_a}$$

$$\omega_c = \frac{\omega_a r_a + \omega_s r_s}{r_a + r_s}$$

Denoting $E = \dfrac{r_a}{r_s}$,

$$\omega_c = \frac{\omega_a E + \omega_s}{E + 1}$$

Hence

$$\frac{T_s}{T_c} = \frac{\omega_a(E + 1) - \omega_a E - \omega_s}{(E + 1)(\omega_s - \omega_a)}$$

$$= \frac{-1}{E + 1}$$

Consider now an operation with variable input speed and torque but with a requirement for fixed output speed. Assume the planet carrier is the input, the sun the output and the question posed is how the speed and torque of the annulus should vary to maintain fixed speed of the output (sun). If ω_s is required to be maintained at a fixed value as ω_c varies, the speed of the annulus is determined from Equation 6.3 as

$$\omega_a = \frac{\omega_c(E+1) - \omega_s}{E}$$

With ω_s fixed, ω_a is linearly related to input speed ω_c and from Equation 6.4:

$$T_a = -(T_s + T_c) = -T_c \left(\frac{T_s}{T_c} + 1\right) = -T_c \left(\frac{E}{E+1}\right) \tag{6.6}$$

Thus, the torque applied to the annulus is simply proportional to input torque and this defines the requirement for fixed speed output from variable speed, variable torque input.

In the preceding analyses, the gearbox is considered to operate ideally without any inefficiency resulting in energy loss. Since the gear contact transmission enforces speed relationships, these cannot change and the effect of losses is a reduction of output torque. The losses, as measured by the difference in input and output power, can be associated with small frictional torque reactions within the gears and bearings dissipating some of the input energy as heat.

6.8 Drive Trains with Differential Drive

In general, solutions that harness the benefits of a variable speed wind turbine rotor produce grid standard electricity via two main routes:

1) Fixed speed ratio mechanical transmission and hence a variable speed, variable torque generator with additional power conditioning equipment to provide grid-compliant electrical output;
2) Differential drive providing a fixed-speed output speed that will suit a synchronous generator.

The second option of a differential drive is now discussed with reference to wind turbine drive trains. The analysis in Section 6.7 demonstrated the principle of using a planetary gear system to provide constant output speed from variable input. Examples of such systems are as follows:

- Wind Energy Group LS1 [16], an early 3 MW prototype of 1982, which used a small electrical machine on the variable speed shaft to provide narrow band variable speed.
- Windflow 500, discussed in the WindPACT1 study [1] as the Henderson system, in which a hydraulic retarder on the variable speed shaft provides variable slip and hence removes torque spikes.
- Voith-WinDrive system [17] in which a hydrodynamic fluid coupling provides the variable speed relationship between the output shafts.
- SET DSgen-set® [18] – a servo-controlled differential planetary gear box accepts the variable speed low-speed shaft input and controls the ring gear to constant speed as output to suit a synchronous generator.

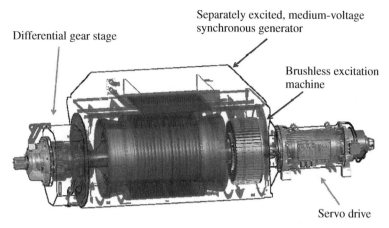

Figure 6.5 SET DSgen-set®.

The Voith WinDrive system is based on differential drive technology long established for electric motor applications and adapted for use with a wind turbine generator. A hydrodynamic torque converter is controlled to provide the required proportional relationship between input torque and its own applied torque.

The SET system (Figure 6.5) and the Voith system do not differ fundamentally. In the SET concept, an electric motor (servo drive) rather than a hydraulic drive receives feedback to regulate the output shaft speed to a constant value.

SET explain their technology as follows;

> This drive technology consists of a differential gear, a three-phase motor with a low-voltage frequency converter as the differential drive, a separately excited synchronous generator, a main gear, and an electro-magnetic eddy current brake. The synchronous generator directly connected to the power supply is driven by the differential's ring gear at constant speed. The planet carrier is connected to the variable-speed main gear by means of an input shaft. The differential's sun gear is connected to the differential drive by means of a drive shaft through the generator. The speed of the differential drive is controlled firstly in order to ensure the speed of the synchronous generator is constant when the speed of the rotor is variable, and secondly to control the torque in the entire drive train. In the event of a temporary power failure (LVRT), the specially developed electro-magnetic eddy current brake is used, which controls the torque in the drive train in milliseconds and significantly reduces loads on the entire WEC.

6.9 Hydraulic Transmission

Hydraulic transmissions are commonplace in heavy industrial applications and have been considered for wind turbines. The high-power density of hydraulics promises, in principle, a drive train of reduced mass. Hydraulics, moreover, can readily provide damping and compliance in the transmission system. However, an early review [19] highlighted critical problems in building a wind turbine transmission system with off-the-shelf commercial hydraulic components:

1) Pumps and motors of the power and torque range required for megawatt wind turbines were not commercially available.
2) Component efficiencies were relatively poor.
3) Design life was generally too short.
4) Reliability was not good enough.

These issues were all too sadly apparent in the experience of an early 3 MW, 50.3-m diameter design of Schachle-Bendix. This design, on account of issue (1), employed 14 hydraulic pumps in the nacelle supplying high-pressure fluid to 18 hydraulic motors at base level [20]. Efficiency and reliability problems were manifest at an early stage of prototype operation on the Devers test site of Southern California Edison near Palm Springs. A later study commissioned the UK Department of Industry constrained to examine only commercially available hydraulic technology, simply confirmed the listed problem issues.

Artemis Intelligent Power [6] in Edinburgh, Scotland, with support from the UK Carbon Trust, developed a hydraulic transmission system for wind turbines. The key components comprising a long-life radial cam pump and a high-speed hydraulic motor are custom designed to address the historical problem issues. The pump (Figure 6.6), rated 1.6 MW, comprises 68 internal modules each with roller and piston assembly driven by the radial cam. It is regulated with very fast control response from electronically controlled poppet valves (key enabling technology initiated at Edinburgh University and developed over 20 years by Artemis). Thus, the pump manages variable speed control of the wind turbine rotor whilst providing a controlled constant pressure output to the hydraulic motor, thereby allowing it to drive a synchronous generator

Figure 6.6 Hydraulic pump of Artemis Intelligent Power. Reproduced with permission of Artemis Intelligent Power Ltd.

Figure 6.7 Schematic of the Artemis 1.6 MW transmission. Reproduced with permission of Artemis Intelligent Power Ltd.

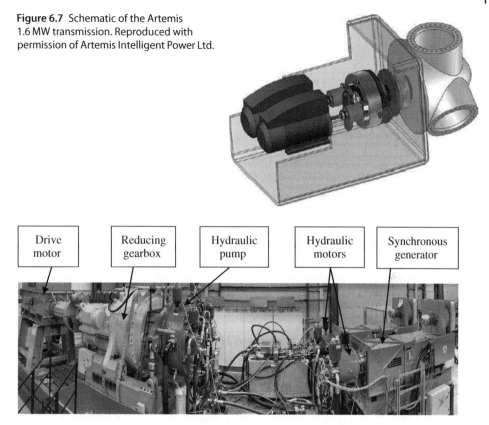

Figure 6.8 Artemis – MHI 1.6 MW drive-train test rig.

at constant speed. The system of 1.6 MW output power rating comprising pump, twin motors and accumulator (Figure 6.7) can replace a gearbox and electrical power converter at a weight of around 6 tonne.

The compact nature of the hydraulic drive train is well illustrated in the conceptual layout of Figure 6.7. Artemis was acquired by MHI in December 2010. Soon after, a full-scale 1.6 MW test rig was built (Figure 6.8). A motor (partially in view at the extreme left of Figure 6.8) drives a speed-reducing gearbox (next major component from the left), thereby simulating the speed and torque levels that would be realistic for a 1.6 MW wind turbine. The gearbox output is closely connected to the hydraulic pump (a short cylinder near the centre of Figure 6.8).

Only for convenience of testing, the twin generators (extreme right of Figure 6.8) and associated hydraulic motors (small cylinders attached to the generator input shafts) are well separated from the pump on the centre left of Figure 6.8. Figure 6.7 indicates how the components would be arranged in a wind turbine drive train.

A speed-increasing gearbox similar in size to the speed-reducing gearbox shown in the test rig of Figure 6.8 and a power converter (not shown) would be required in an equivalent conventional drive-train system. In the Artemis MHI drive train, these are replaced by a hydraulic pump and twin hydraulic motors. These are evidently of quite modest size (Figure 6.8). The speed-reducing gearbox of Figure 6.8 can be considered as

Figure 6.9 Seven megawatt ring-cam pump in MHI factory.

very similar in size and weight to the speed-increasing gearbox that would be employed in a conventional drive train of 1.6 MW rating. The system efficiency was measured at around 93% on full load and therefore about the same as standard drive-train solutions. This is promising technology. It will be critical to maximise efficiency over the operating range, especially as recent systems employing PMG technology are raising the bar especially in terms of part load efficiency.

Further development of the Artemis technology was continued by Mitsubishi Power Systems Europe (MPSE), a subsidiary of MHI. MPSE have since undergone the preliminary testing (Figure 6.9) of their 7 MW-class hydraulic drive train, MWT 167H/7.0, formerly named the SeaAngel.

In 2013, a 2 MW prototype was demonstrated onshore in Yokohama, Japan. Later in 2014, a 7 MW hydraulic drive wind turbine was installed at an onshore test facility at Hunterston on the Ayrshire coast in Scotland. The 7 MW hydraulic drive is now in operation following installation in 2015 on an offshore floating wind turbine located in Fukushima Bay, Japan (Figure 6.10).

6.10 Efficiency of Drive-Train Components

6.10.1 Introduction

It is common enough to hear that a gearbox or generator is 97% or even 98% efficient as if that were the beginning and end of the story. This figure is almost certainly the efficiency at full rated power. It is typically very high and it could be a substantial challenge to push this value upwards by even <1%. However, for a wind turbine, it is the efficiency

Figure 6.10 The 7 MW wind turbine of MHI at Hunterston, with the Fukushima floating system inset.

curve over the whole operational range of drive-train speed and torque that is crucial. In particular, the PMG may not necessarily outperform the DFIG at full load; but at part loads, the advantage in efficiency (Figure 6.11) can be very significant.

The recognition of the importance of part load efficiency has been quite slow to arrive in the wind industry. In the other industries which gave birth to gearboxes, motors and generators, efficiency had generally less intrinsic value. For example, in a process industry, having a low-cost, highly reliable motor may be more important than seeking the very highest efficiency to minimise electricity bills. Where there is significant interest in generator efficiency (e.g. in mainstream large-scale electrical power generation) often the system operates principally at full load and the focus is on efficiency in that state.

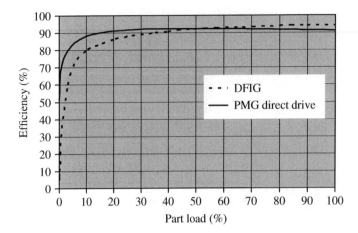

Figure 6.11 Comparison of drive-train efficiency.

There are also many developments aimed at reducing mass and cost of direct-drive generators. PMGs with transverse flux topology are being considered as a possible route to higher torque density. Also, designs with relatively larger air gaps or many topologies which reduce magnetic forces may allow structural mass reduction. However, it is vital to consider such developments in a context (see Section 9.3 on cost of energy issues) that values each percent of efficiency over the load range appropriately and, in general, as much greater in value than each percent of generator capital cost.

The DFIG benefits from reduced losses in the power converter (typically, the converter of a DFIG need only be rated at around one-third of full rated power) but the DFIG system (Figure 6.11, characteristic curve from WindPACT [1]) is disadvantaged by the comparatively high losses of the induction generator at low part loads. However, for the benefits of high part load efficiency of a PMG to be fully realised, it is very important to design for good part load performance in the converter. Strategies such as reduced switching frequency and avoidance of any unnecessary operation of cooling fans may be considered as well as more fundamental design approaches to improving converter part load efficiency.

Against a prevalent trend towards DFIG systems, GE Wind introduced a new wind turbine model, the GE 2.5xl, with a standard gearbox and high-speed PMG. Why did they do this when the DFIG is so well established, when the PMG may initially be more expensive and the power conversion equipment will be more expensive than for a DFIG? Two reasons emerged: the better part load efficiency of the PMG as has been discussed here and, that having a fully rated converter, it provided the ability to supply grids anywhere (whether 50 or 60 Hz) with the same drive-train hardware. Samsung also has developed a design with high-speed PMG.

6.10.2 Efficiency over the Operational Range

The distributions of hours in the year and energy output with respect to wind speed are illustrated (Figure 6.12) for a typical 1.5 MW wind turbine operating at (i) a low wind speed site with annual mean wind speed (AMWS) of 6 m/s and (ii) a good land-based wind speed site with AMWS of 8.5 m/s.

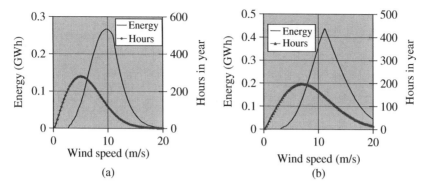

Figure 6.12 Wind speed and energy distributions. (a) 6 m/s AMWS and (b) 8.5 m/s AMWS.

It is apparent (Figure 6.12) that wind speeds around rated (~11 m/s) are among the most important for energy capture. However, it is hard to maximise energy capture here in the dynamic operation of a wind turbine as wind turbulence poses challenges to the control system in managing the transition from below rated wind speeds (no active pitch control) to above rated wind speed when pitch control intervenes to regulate maximum power.

Also, the system is operating near or at full load around rated wind speed and the efficiency of most components is high or maximum there. Thus, it is difficult to achieve much improvement in the performance around rated wind speed of a well-designed, appropriately controlled, variable speed wind turbine.

Considering, however, the typical efficiency characteristic of the most popular drive-train type employing DFIG technology, it is apparent (Figure 6.11) that efficiency reduces substantially in part load operation. It is here that quite large efficiency improvements are possible in principle and alternative drive-train configurations may offer significant energy gains.

Representative part load efficiency characteristics of the major drive-train components are now considered.

6.10.3 Gearbox Efficiency

In a simplified empirical model of gearbox efficiency developed by GL Garrad Hassan, losses (Equation 6.7) are assumed to vary linearly with power throughput. The model comprises a loss varying in proportion to operating power level and a constant loss related to rated power level and the number of stages. N – number of stages

$$L_{gear} = \frac{\{(10/3 + 2N)P_r + 5NP_i\}}{1000} \text{ kW} \tag{6.7}$$

P_r : rated power (constant), kW
P_i : input power (variable), kW

This loss model approximates to 0.8% loss per stage, which is considered typical of gearbox designs. The resultant gearbox efficiencies in terms of part load are provided in Figure 6.13.

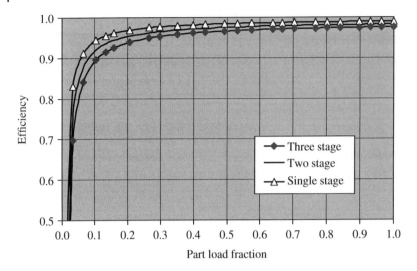

Figure 6.13 Gearbox efficiency characteristics.

6.10.4 Generator Efficiency

The efficiency of the Vensys and Zephyros generators is compared (Figure 6.14) with a generator in the GEC WindPACT study [1]. A problem in evaluating such data is the interconnection of PMG efficiency and machine side converter efficiency. The Vensys system has a diode rectifier on the machine side and IGBT on the network side. The JSW (Zephyros) design uses an integrated gate-commutated thyristor (IGCT)-based power conditioning systems. Among a variety of systems, evaluated in Appendix G of the WindPACT drive-train study [1], the system referred to as 'Single PM diode-IGBT' has been chosen as most similar to Vensys in concept.

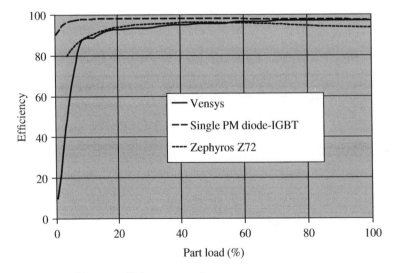

Figure 6.14 Generator efficiency comparisons.

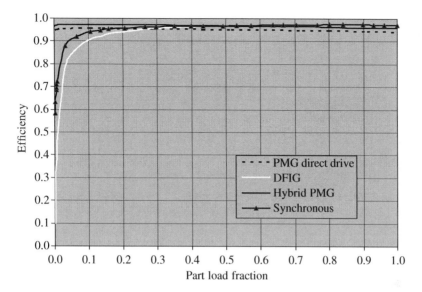

Figure 6.15 Efficiency comparison of generator types.

The test results for Vensys and JSW (Zephyros) appear to agree quite well, whereas the WindPACT1 predictions imply that a much higher performance level can be maintained at part loads. The efficiency of generator types in isolation from the power converter is represented in Figure 6.15.

6.10.5 Converter Efficiency

The efficiency of a converter based on IGBT technology is represented in Figure 6.16. When a converter is used with a DFIG, it is assumed to have one-third full load generator power rating.

6.10.6 Transformer Efficiency

The characteristic of Figure 6.17 has been derived from transformer data applicable to sizes in the megawatt range. Iron losses are taken to vary linearly with power and copper losses as the square of rated power.

6.10.7 Fluid Coupling Efficiency

There are innovative systems [21] that can realise a variable ratio gearbox by mechanical means. This would allow the use of a synchronous generator operating at grid frequency. However, often such a system is mechanically rigid (unless the variable ratio gear box employs some kind of hydraulic system) and both compliance and damping are required in the drive train to avoid unacceptable torsional vibrations. An established solution to this problem is to use a constant volume fluid coupling. Such couplings may typically have around 1% slip and hence thermal loss at full load but will have negligible losses at low part loads.

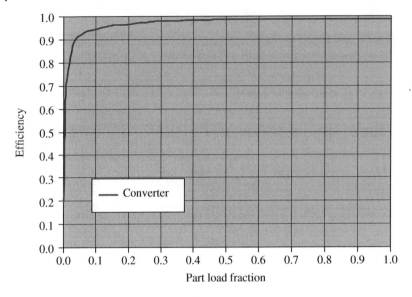

Figure 6.16 Efficiency of IGBT-based converter.

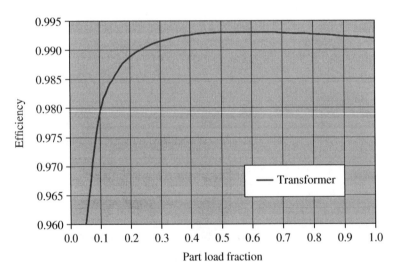

Figure 6.17 Transformer efficiency.

6.11 Drive-Train Dynamics

The drive-train modes of vibration are often damped by the aerodynamic torque which makes the largest contribution to damping, with structural damping being minimal. Surprisingly perhaps, not all rotor vibration modes cause torque variations (see, e.g. the rotor mode discussed in Chapter 24). Aerodynamic damping of edgewise blade vibrations is generally valuable but very light, amounting to no more than a few percent of critical.

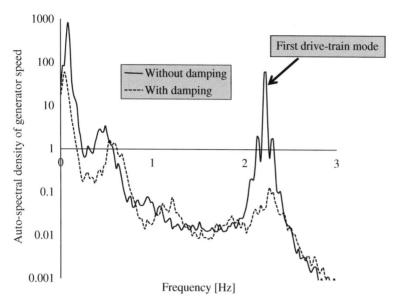

Figure 6.18 Spectrum of generator speed.

The first drive-train mode for a variable speed wind turbine is often highly resonant. The rotor and drive-train components swing against each other in anti-phase relative to their centre of inertia. Not infrequently, the amplitude of the latter can be greater than the amplitude of the former.

A power spectrum of generator rotor speed of a typical 5 MW wind turbine is shown in Figure 6.18. The first drive-train mode appears as a very sharp peak on the spectrum and will do so for any drive-train variable in the absence of active damping. This is normally the lowest frequency dynamic mode present. The wind turbine controller is generally required to actively damp this first drive-train mode, the response with such damping also being illustrated in Figure 6.18.

6.12   The Optimum Drive Train

Given a requirement to produce electricity, the drive train must have a generator, whereas having a gearbox is optional. The following question then arises.

> Without undervaluing the importance of a final, harmonised, fully integrated drive-train design, can the optimum type of generator be determined in isolation without consideration of the rest of the drive train?

In several design studies, GL Garrad Hassan adopted this approach and, on account of the potentially higher reliability of the PMG (having no slip rings or rotor windings) and its potential for high part load efficiency, this was considered the preferred generator solution.

The next stage was for a given rotor diameter and system rated power output to define the operating speed range of the wind turbine rotor. Hence, the input speed

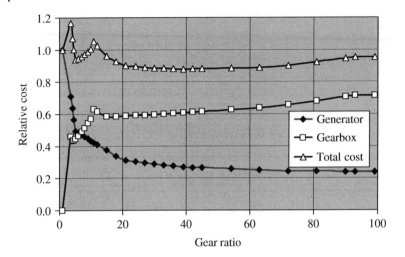

Figure 6.19 Drive-train optimisation.

and torque were determined. With the output power specified and having developed a quite detailed model of PMG design and also models of gearboxes of different types and numbers of stages, a series of designs of gearbox and PMG combinations were evaluated. The gear ratio was varied systematically from 1 (direct-drive PMG) to values around 100 (suiting a high-speed PMG) with the exact upper ratio limit depending on grid frequency and the chosen rated output power. Typical results for a 3 MW rated system are presented in Figure 6.19.

The modelling is least certain at very low gear ratios and at direct drive where generator topology and structural solutions are diverse and can lead to a wide range of results in mass and cost. After review of direct-drive design options, the relative cost of direct drive, taken as unity in Figure 6.19, was found to be 5% more expensive than the conventional solution. The relative 'cost' presented includes capitalisation of losses (accounting for PMG and gearbox efficiency variations with gear ratio) and some quantification of reliability benefits for direct drive in comparison to geared systems. It therefore gives some indication of the overall merit of the drive-train options approximating a cost of energy comparison.

The results are naturally sensitive to details in the models, to costs of key components, to costs of materials (magnets, copper, etc.) and also to financial parameters (discount rates and energy repayment values). Considering the complexity of gearbox and generator design, this type of modelling is very challenging. Nevertheless, it represents a systematic investigation of drive-train optimisation. The results hint that the hybrid designs with one or two stages of gearing, as have been developed by Gamesa, Win-Wind and others, may be the most cost-effective. However, the evaluation of direct-drive design requires considerable investigation of structural solutions for the generator and there is some indication that direct-drive PMG systems may be favoured at the larger sizes – an interesting outcome as there has been some concern that direct drive would upscale less favourably than conventional or hybrid drive trains.

6.13   Innovative Concepts for Power Take-Off

The major trends in current drive-train technology are clearly towards greater use of PMGs including direct-drive designs. What potential is there for more radical innovation? The fundamental assumption in all the drive-train concepts reviewed thus far is that power take-off will be on the axis of the rotor. Examples of alternative concepts that may be considered are as follows:

1) Rimmed rotors with power take-off at the rim;
2) Moving masses within the blades;
3) Secondary rotors on the wind turbine blades.

These solutions are interesting conceptually as they avoid the usual situation with the low-speed, high-torque main shaft being a 'middle-man' in the energy extraction process. In the first option, the rotor rim carries magnets past an arc of stator coils located on a platform. This takes advantage of the high peripheral speed of the rotor leading to a lightweight low-cost stator system, although it is comparatively wasteful in magnets as they interact electromagnetically only over a short sector of arc at any given time. Mechanical rim drive concepts have also been proposed [22] based on traction or friction where the rim engages rollers which turn generators.

The second option was proposed by Garvey [23] and is based on the radial movement of masses internal to the blades controlled in such a way as to absorb energy and provide torque reaction to the main rotor using their self-weight moment. The masses within the blades move only radially in the rotating reference frame of the rotor but, in the nacelle reference frame, they move cyclically in fixed closed loops (Figure 6.20) that are eccentric relative to the rotor shaft axis. This produces a reaction torque as, over a cycle in the eccentric loop, the mass will produce a non-zero average moment about the shaft axis. This has profound logic in the context of upscaling to, say, 20 MW unit capacity. The magnitude of the masses for power take-off relative to the blade mass will decrease with turbine scale as their moment like blade self-weight (see Section 4.4) will scale as R^4, whereas the blades themselves will scale as R^3. Of course, the specific engineering of the power system (perhaps primarily hydraulic or pneumatic) will need much investigation.

The secondary rotor concept (Figure 6.21) also has potential for very large wind turbine systems where the conventional drive-train solutions will be extremely heavy and expensive on account of low shaft speed and high rated torque. This concept involves rotors which are carried by the wind turbine blades of the main rotor. The secondary rotors thereby experience a much higher apparent wind speed than the ambient wind speed and can be of comparatively small diameter, high speed, low torque and low weight. The idea is quite old, possibly preceding patent applications of the 1980s and 1990s Watson [24], St-Germain [25] and Jack [26]. A plausible misconception is that the Betz limit will apply twice to the power take-off. In the limit of low induction on the secondary rotor, all of the (necessarily Betz limited) power of the primary rotor can be extracted [27]. However very low induction would imply slow running large secondary rotors defeating the purpose of having power take-off with small fast rotors and low torque and a compromise will be adopted in realistic design.

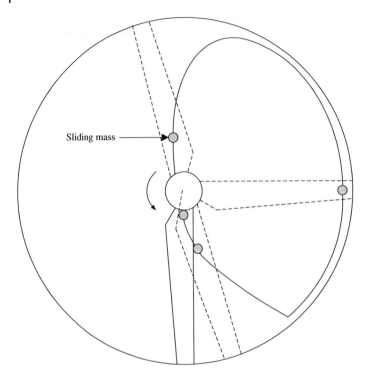

Sliding mass

Figure 6.20 Gravity torque reaction concept.

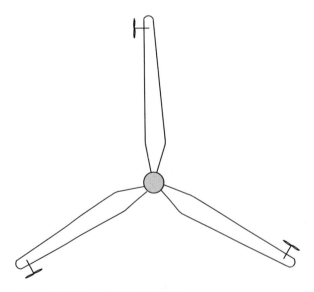

Figure 6.21 Secondary rotor concept.

A brief analysis of the secondary rotor concept follows. It is clearly the thrust (and not power) of the secondary rotor that provides reaction torque to extract power from the primary rotor. It appears that extraction of power from the primary rotor is most efficient when the axial induction of the secondary rotor is small. An interesting trade-off then arises between having larger and therefore more expensive, lightly loaded secondary rotors to improve efficiency and hence reduce cost of the major primary rotor system.

Notation:

Air density	ρ
Primary rotor radius	R
Primary rotor angular speed	ω
Primary rotor blade number	N
Secondary rotor radius	r
Secondary rotor power coefficient	C_p
Secondary rotor power coefficient	C_t
Secondary rotor axial induction	a

The primary rotor produces power P which is as usual subject to the Betz limit. In terms of the thrust reaction T of each of the N secondary rotors and assuming for present convenience that they are mounted at the tip of each blade,

$$P = NTR\omega$$

$$T = 0.5\rho(\omega R)^2 \pi r^2 C_t \text{ neglecting ambient wind speed compared to tip speed}$$

$$P = N 0.5\rho(\omega R)^2 \pi r^2 C_t R\omega$$

$$P = N 0.5\rho\omega^3 R^3 \pi r^2 C_t$$

The power extracted by the secondary rotors is

$$P_e = 0.5\rho N(\omega R)^3 \pi r^2 C_p$$

Thus, $P_e/P = C_p/C_t$

Considering the ideal Betz model:

If

$$\frac{C_p}{C_t} = \frac{4a(1-a)^2}{4a(1-a)} = (1-a).$$

If the secondary rotor is optimised in its own right, then the usual choice of $a = 1/3$ applies and the overall limit is $16/27(1 - (1/3)) = 0.395$. This exceeds Betz squared by a little as $(16/27)^2 = 0.351$.

However, it is much better to trade reduced specific loading, C_t, on the secondary rotors at the cost of making them a little bigger. In a specific design study, $a = 0.2$ was about optimum. Hence, the ratio P_e/P is $(1 - 0.2) = 0.8$ and the overall limit is $(16/27) \times$

$0.8 = 0.474$. The power coefficient of the secondary rotors is reduced to a theoretical limit of $4 \times 0.2\,(1 - 0.2)^2 = 0.512$ and the secondary rotors are somewhat larger and more expensive but this can be a very worthwhile trade-off.

References

1 Poore, R. and Lettenmaier, T. (2002) *Alternative Design Study Report: WindPACT Advanced Wind*. Turbine Drive Train Designs Study, 1 November 2000–28 February 2002.

2 Bywaters, G., John, V., Lynch, J. *et al.* (2001) *Northern Power Systems, Alternative Design*. Study Report, Period of Performance, 12 April 2001 to 31 January 2005.

3 Böhmeke, G. (2004) *Development of the 3 MW Multibrid Wind Turbine*, EWEC, London.

4 Greenwood, C.J. (1990) *Continuously Variable Transmission Traction Drive*, Truck Technology International.

5 Jamieson, P.M. (2004) Evaluation of a GVT System for Wind Turbines. GH Report 3652/GR/01B, June 2004 (internal company report).

6 http://www.artemisip.com/applications/renewable-energy/wind-power/ (accessed March 2017).

7 http://www.chapdrive.com/communications/press-release/chapdrive-secures-nok-86-million-11m-funding-to-commercialize-hydraulic-transmission-for-wind-turbines (accessed April 2011).

8 http://www.mhivestasoffshore.com/innovations/ (accessed 26 September 2017).

9 Tavner, P.J., Xiang, J.P. and Spinato, F. (2007) Reliability analysis for wind turbines. *Wind Energy*, **10** (1), 1–18.

10 Mueller, M.A. (2009) A lightweight low-speed permanent magnet electrical generator for direct-drive wind turbines. *Wind Energy*, **12** (8), 768–780. doi: 10.1002/we.333

11 Engström, S., Hernas, B., Lindgren, S. *et al.* (2006) *Development of NewGen – A New Type of Direct-Drive Generator*. Nordic Wind Power Conference, Espoo, Finland, May 2006.

12 Innwind Project Deliverable 3.13. Fabrication of MgB2 Coils – A Superconducting Generator Pole Demonstrator, http://www.innwind.eu/publications/deliverable-reports (accessed 26 September 2017).

13 http://www.mhivestasoffshore.com/v164-8-0-mw-breaks-world-record-for-wind-energy-production/ (accessed December 2016).

14 http://www.windpowermonthly.com/article/1169347/vestas-v164-drivetrain-choice (accessed December 2016).

15 The Gearbox Challenge (2005) Windpower Monthly, November 2005.

16 Hicks, R.J., Cunliffe, F. and Giger, U. (2004) *Optimised Gearbox for Modern Wind Turbines*. EWEC Conference, London, November 2004.

17 Müller, H., Pöller, M., Basteck, A. *et al.* (2006) *Grid Compatibility of Variable Speed Wind Turbines with Directly Coupled Synchronous Generator and Hydro-Dynamically Controlled Gearbox*. 6th International Workshop on Large-Scale Integration of Wind Power and Transmission Networks for Offshore Wind Farms, October 2006, Delft.

18 http://www.set-solutions.net/products/dsgen-set/as viewed (accessed March 2017).

19 Wilson, R.R., Jamieson, P. and Brown, A. (1981) *A Comparative Study of Wind Power Conversion Systems*. Proceedings of the BWEA Wind Energy Conference, April 1981.

20 Spera, D.A. (ed.) (2009) *Wind Turbine Technology: Fundamental Concepts in Wind Turbine Engineering*, 2nd edn, ASME Press.

21 http://www.iqwind.com (accessed April 2011).

22 Merswolke, P.H. and Meyer, C.F. (Inventors) (2006) Wind turbine with friction drive power take-off on outer rim. US Patent 2006/0275121 A1. Published 7 Dec. 2006.

23 Garvey, S.D. (2010) Structural capacity of the 20 MW wind turbine. *Proceedings of the IMechE, Part A: Journal of Power and Energy*, **224** (A8), 1083–1115. doi: 10.1243/09576509JPE973

24 Watson, W.K. (inventor) (1988) Space frame wind turbine. US Patent 4,735,552, Published 5 April 1988.

25 St-Germain, J. (inventor) (1992) Wind machine with electric generators and secondary rotors located on rotating vertical blades. US Patent 5,151,610, Sept. 1992.

26 Jack, C. (inventor) (1992) Free rotor. World Patent WO/1992/020917, Published 26 Nov. 1992.

27 Madsen, H.A. and Rasmussen, F. (inventors) (2008) Wind turbine having secondary rotors. European Patent Specification EP 1390615 B1, Published 30 April 2008.

7

Offshore Wind Technology

7.1 Design for Offshore

The offshore wind turbine market is growing fast but, at around 3% of the world wind turbine market in 2015, has still some way to go. While 11 GW of offshore wind [1] had been installed in Europe (the main market), cumulative world capacity had reached 433 GW at the end of 2015. With restrictions on land-based exploitation in many areas, offshore is seen as the future and dominates research effort in the wind industry, with the largest wind turbines at the centre of development efforts being designed for offshore applications. There is demand for both innovation to solve the problems associated with the increasing costs and weights of ever larger technology and for conservatism as it is trouble enough to make ever larger units without introducing new technology directions.

Many offshore wind turbines are very much standard, land-based wind turbines that have been 'marinised' with additional anti-corrosion measures and de-humidification capability introduced. Often, the layout of some components differs from typical practice on land to suit access and maintenance strategies. It may be preferred to have all the electrical equipment including transformers at tower top level, but these are hardly design changes that have any major impact on cost of energy (COE). It is also very important that the impact of minor faults is minimised as, whatever the action that may be required, any site visit is expensive and subject to weather constraints that may extend downtime. This may suggest uprating or redundancy (duplication) in various components.

This implied design conservatism is very understandable. In confronting a new and challenging environment (offshore), it is prudent initially to avoid many changes in technology that can be regarded as proved on land.

The turbines may be as little as 30% of lifetime cost for a project in a challenging offshore environment. Thus, innovation and technology development around balance of plant and design for assembly and maintenance can be more important than turbine technology development. In many areas of offshore technology, such as foundation concepts, transportation and erection methods, access solutions, electrical system concepts for turbine interconnection and grid connection, innovation is much needed and will play a very significant role in future developments.

Innovation in Wind Turbine Design, Second Edition. Peter Jamieson.
© 2018 John Wiley & Sons Ltd. Published 2018 by John Wiley & Sons Ltd.

Considering first the turbine itself, what may distinguish its design from the land-based model and what are the opportunities that exist for innovation that is specifically suited to the offshore application? The most obvious answer is to consider higher speed rotors. At least on typical European sites, the design tip speed of land-based wind turbines has been constrained by requirements to limit acoustic noise. Offshore, there is greater freedom from this constraint and higher speed wind turbines may be considered.

7.2 High-Speed Rotor

7.2.1 Design Logic

The logic of a high-speed rotor design is very simple. Since power may be regarded as the product of torque and speed, the higher the speed at which a given level of power is produced, the lower is the torque. It is the input torque that primarily affects weight and cost of major drive-train equipment like gearboxes and generators. This argument has long been evident and some early wind turbine designs of Hutter [2] and a few in the 1980s [3], especially one- or two-bladed designs, chose high design tip speeds. As the European market demand for relatively low acoustic noise emission clarified, such designs fell into decline. Aerodynamic acoustic emission restrictions effectively set a limit on tip speed of around 65 m/s for fixed speed or two speed turbines and in a range from 70 to 80 m/s for fully variable speed designs.

Figure 1.17 shows that too high a design tip speed ratio will imply a degradation in aerodynamic performance and also that the higher the lift-to-drag ratio of the rotor aerofoils, the higher can be the design tip speed ratio without serious drag penalties. Figure 1.18 shows that for lift-to-drag ratios above about 100, typical of current state-of-the-art wind turbine rotors, the reduction in maximum C_p at tip speed ratios above the overall optimum is very gradual. The design tip speed of wind turbines has been generally increasing with time (see Section 4.7). For wind turbines specifically targeted for the offshore market, the increase has been in a range typically from 10% to about 30% maximum [4] from the tip speed levels of the largest land-based wind turbines of each manufacturer.

7.2.2 Speed Limit

There is clearly an ultimate limit on the speed of a high-speed rotor as aerofoil performance degrades when tip speeds approach the transonic region. It is however rather difficult to say what will set an upper limit for practical design. Some key factors that influence this are as follows:

- Leading edge erosion of the aerofoil sections,
- A possible reduction in aerodynamic damping of the very low solidity rotors that may result with a choice of high design tip speed ratio.
- Acoustic noise emission, although, as was already observed, this may not be important for offshore development unless they are very near land.[1]

---

1 Nevertheless, acoustic noise has been raised as an issue for marine life and nothing can be taken for granted about environmental issues without careful investigation.

The polyurethane (PU) tapes presently used in the wind industry would be inadequate for erosion protections of a high-speed rotor. However, advances in helicopter blade erosion protection have come from military operations in desert environments (Kuwait, Iraq, Afghanistan). In the worst cases, blade leading edge protection has had to be replaced daily. This has led to the development[2] of superior PU tapes which at some cost premium may offer adequate protection for wind turbine blade designs with tip speeds around 120 m/s. As erosion protection is a very small part of blade cost (perhaps ~1%), significant cost increase should be quite affordable if major benefits can be demonstrated in a high-speed rotor design. Even so, such improved PU tapes may still be challenged by the lifetime duty of a high-speed offshore wind turbine and metallic erosion protection could be required. It is worth paying a premium for durability in the offshore environment in order to limit or avoid expensive maintenance operations.

7.2.3 Rotor Configurations

Presently, the great majority of large-scale wind turbine designs are in the upwind configuration where the wind passes through the rotor blades before it impinges on the tower. As designs progress. there has been interest in more flexible blades both to economise in blade material and cost and to relieve system loads (discussed further in Chapter 17). However, as blades become more flexible, the problem of providing a safe clearance between blade tip and tower surface becomes more acute.

Methods of improving clearance of an upwind rotor include the following:

- Coning the blades away from the tower (i.e. setting the connection of blade to hub at a small angle out of the rotor plane, shown as γ in Figure 7.1),
- Tilting the shaft axis (as shown by angle β in Figure 7.1),
- Introducing out-of-plane curvature into the manufactured blade shape, tilting the rotor axis,
- Increasing the overhung distance between the rotor plane and the tower centreline.

These options may be combined to a limited extent in various ways. Tilting much beyond 4° or 5° may introduce undesirable cyclic loads. Coning introduces a moment at the blade root from centrifugal forces on the blade elements, which could (in a critical overspeed design case) become designing for the blade root if the cone angles exceed 4° or 5°. Out-of-plane curvature in the blade is a better solution in respect of blade loads and structure. All these options, in principle, reduce power slightly or require marginally longer blades to maintain power. Increasing rotor overhang has no significant effects on rotor loads or performance, but is usually the most expensive option on account of

---

2 Much of the relevant laboratory testing has been done on a highly specialised rotating rig at Wright-Patterson Air Force Base in Ohio. The sample is set on a whirling arm and rotated at suitable high speeds through a precisely metered pattern of rain or sand particles. A company, Hontek, located in Connecticut, was then set up to develop improved erosion protection systems for the US military. Hontek has been working on systems for the V22 Osprey tilt rotor aircraft, which operates with a tip speed of 200 m/s. Over 14 years, Hontek has carried out research contracts to develop protection systems for specific aircraft and has tested thousands of formulations. The rig testing to date runs to hundreds of days. Part of the requirement of the government funding for such work is to aim for products with dual use, that is, military and civilian applications. Therefore, there is active encouragement to apply such products to new applications such as wind turbines.

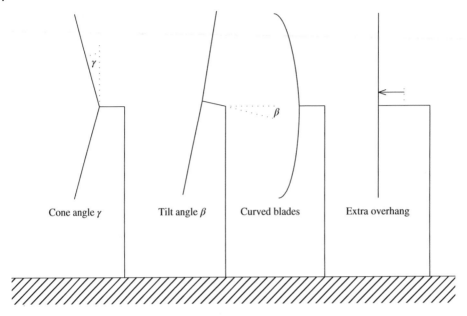

Cone angle γ Tilt angle β Curved blades Extra overhang

Figure 7.1 Rotor blade tip and tower clearance.

extending the main nacelle structure and increasing loads on the nacelle structure or rotor bearings.

For given aerofoil selections, it will be evident that the solidity (or, equivalently, blade chord widths) of an optimum rotor decreases as the square of design tip speed (see Equation 2.1). This implies that a substantial increase in design tip speed from, say, the 75 m/s of current state-of-the-art variable speed wind turbines to 120 m/s will lead to much more slender and flexible rotor blades.

This tendency can be mitigated in several ways, using lower design lift aerofoil sections which would lead to wider chords or eliminating some or all of the constant speed region that typically exists between rated speed and rated power of a variable speed wind turbine (see Figure 2.1). One reason for having the constant speed region (more fully explained in Section 2.2) is that a typical design tip speed of 75 m/s and rated wind speed of 11 m/s would imply a design tip speed ratio of <7 and consequently a rather high solidity blade. Achieving design tip speed at a lower wind speed pushes up the design tip speed ratio of the blade and hence reduces solidity.

However, anything that is done to avoid tower clearance problems that lead to stiffer blades also works against the loading benefits of flexible blades. A two-bladed high-speed rotor may be considered for an upwind high-speed rotor design on the basis that, for a rotor of the same design speed and solidity as a three-bladed rotor, the blade chords are wider and deeper in section width. Thus, given the general similarity in blade materials, aerofoil section types and structural concept, a two-bladed rotor will provide stiffer blades than a three-bladed rotor at any chosen design tip speed and rotor solidity. This is explained in greater detail in Section 11.2.

7.2.4   Design Comparisons

In a number of studies of high-speed rotor design, the approach developed in Chapter 10 was generally followed and high-speed rotor designs were compared with representative (standard speed) baseline designs.

  Initially, attempts were made to develop designs maintaining the usual upwind rotor configuration. It was considered that a high-speed three-bladed rotor was problematic in terms of the added cost that would accrue from providing sufficient blade stiffness and/or introducing the usual design options (Figure 7.1) to provide adequate tower clearance. Two-bladed rotor designs were therefore considered in order to maintain the upwind configuration. However, to bypass the problems and costs associated with safe tower clearance in the upwind configuration, three-bladed downwind rotor designs were also considered.

  It was expected that with reduced rotor solidity, the high-speed rotor designs would experience reduced loading under extreme storm conditions where the rotor was idling. It was also well established that having more flexible blades (as is possible in the downwind configuration) can alleviate fatigue loading – both rotor loads and systems loads. This is illustrated in a design comparison (Figure 7.2) at 5 MW scale where significant fatigue load reductions are evident with the high-speed, flexible-bladed rotor. The distributions shown are of fatigue damage equivalent loads based on the Palgrem–Miner sum [5] for a material fatigue damage characteristic with inverse S-N slope of 10 such as is appropriate for some composite materials. Such loads are not sufficient for design in that rigorous fatigue evaluations must be based on stresses and not on loads, but they are useful relative indicators both of the distribution of fatigue damage with respect to operational wind speed and in the comparisons of different designs.

  More surprising are the widespread reductions in extreme loads (Figure 7.3) as this includes some operational load cases when the wind turbine is in power production mode.

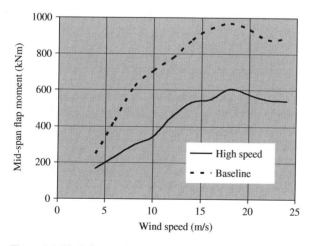

Figure 7.2 Blade fatigue loads comparison at mid span ($m = 10$).

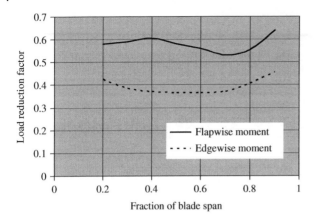

Figure 7.3 Extreme load reductions of a high-speed rotor.

Downwind designs have generally found little favour, particularly because of an effect described as tower shadow. The blades passing behind the tower experience a sharp reduction in incident wind speed due to the tower blocking the flow and highly impulsive loads are transmitted into the system. In addition to loading impacts, there was some concern about infrasound as very large downwind rotors could produce very low frequency sound waves associated with the tower shadow impulse. Later [6], it was argued that infrasound should not be a problem.

It is hard to find justification[3] for a downwind rotor configuration at standard design rotor speeds, when an upwind one can be viably engineered, but the situation is quite different with a high-speed rotor design.

The high speed facilitates a slender lightweight flexible rotor which has an equilibrium position of the deflected rotor blades that is further from the tower. Not only is the rotor lightweight but so also is the drive-train and nacelle structure due to the reduced torque at rated power. This in turn facilitates a more slender tower and these effects combine to mitigate the impact of tower shadow on blade and system loads. Nevertheless, tower shadow still has a significant influence on the design.

In Figure 7.4, the gearbox torque of a high-speed rotor (design rated at 5 MW) is compared with different characteristics in the drive train:

1) With a torsionally rigid drive-train transmission where the signal fluctuates considerably with peaks due to tower shadow and has significant high frequency content,
2) With a soft mounted gearbox when the peaks and higher frequency oscillations are largely filtered out leaving a smooth torque signal.

Various models of tower shadow exist. Simple potential flow models are probably adequate to represent the effect of tower shadow on an upwind rotor, but may underestimate the impact on a downwind rotor. An empirical model by Powels [7] uses a cosine bell-shaped tower wake.

---

3 The downwind configuration may possibly facilitate yaw tracking response in free or partially free yaw operation, but any such benefit depends crucially on rotor and system design details. See also the discussion in Chapter 14. Avoiding the minor cost of an active yaw system is a poor justification for confronting the problems of tower shadow.

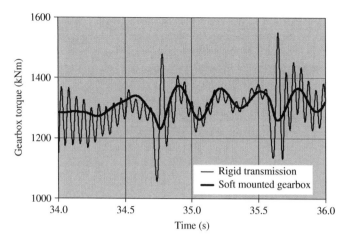

Figure 7.4 Gearbox torque – influence of tower shadow.

Generally, it would appear that in a high-speed downwind rotor design there are issues with tower shadow that need careful attention, but the problems are likely to be manageable without large cost impacts.

Another key issue with the downwind configuration is that if emergency braking is effected by pitching the blades towards fine pitch, which is the usual choice for large wind turbines, then the blade aerofoils are set into negative flow incidence and the lift and thrust on the rotor is reversed drawing the blades towards the tower. This was found to be the only critical load case for safe tower clearance with a high-speed downwind rotor design. However, with flexible blades, even set with 4° downwind coning from the rotor plane, and a natural level of rotor overhang, this load case was still problematic. It could be solved by uprating the rotor parking brake and ensuring it was applied in synchronism with the pitch system. This is effective because slowing rotor speed as the blades are pitched reduces the reverse bending moments which take the blade tips towards the tower.

An alternative solution to the reverse bending problem would be to control power by pitching the blades towards stall. This is unusual, although there have been many designs with so-called active stall control [8] where the pitch system is used only infrequently and the inherent stalling characteristics of the blades are largely the basis of limiting power. In the few studies that have been done, pitching towards stall both with partial span (actively controlled blade tips) [9] and whole blade pitching has been found to be similar overall to pitch in attached flow in fatigue damage implications for the wind turbine system.

Another, more minor, issue with downwind rotor systems as opposed to upwind is that the bending moment on the tower top due to the overhung rotor mass acts in the same direction as that due to aerodynamic thrust.

The design study analyses of capital costs [10] following the methodology of Chapters 9 and 10 suggested that about 15% cost reduction in tower top systems could be realised with a high-speed rotor design targeted at a design tip speed of 120 m/s as compared to a baseline with design tip speed of 75 m/s. This is quite a substantial cost saving especially when all the electrical system costs associated with a single wind turbine unit were included as tower top costs (which is probably appropriate for offshore) and necessarily were considered to be exactly the same as the baseline design for the same rated power output. It should be noted that the cost savings are torque related and are essentially mass savings. This therefore has some knock-on benefit to balance of plant costs in easier transportation and handling. Perhaps, more importantly, the reduced tower top mass is substantially reduced. Although tower top mass as a dead weight load is generally much less important for design than aerodynamic loading, the offshore foundation is designed for stiffness rather than purely for strength in order to avoid natural frequencies close to the centre of the wave spectrum.

The foregoing discussion indicates that the high-speed rotor concept can lead to lower cost turbine technology offshore but, except for the crucial lessening of the environmental constraint, acoustic noise emission, as an enabler for this technology, the high-speed rotor design is not uniquely a design for offshore.

7.3 'Simpler' Offshore Turbines

'Economy' is a universal principle in art, music, literature and engineering, meaning that anything superfluous has been eliminated and that all that remains achieves a maximum of effect with a minimum of resource. 'Simplicity', on the other hand, as a guiding principle is very popular with engineers but can be quite treacherous. Stall-regulated wind turbines, for example, are simpler than pitch-regulated ones as they neither possess a pitch system nor active control. However, the price of such simplicity is that the primary safety of the rotor is not addressed and requires additional engineering. Also, loads on the drive train are appreciably higher as there is no active control to limit them. The design route, termed 'active stall' where blades are pitched into stall to regulate power (albeit with much reduced pitch activity compared to pitching into attached flow) and also pitched for overspeed protection, is a well proven one but, in the introduction of a pitch system, much of the 'simplicity' of stall regulation has been sacrificed.

With large offshore wind farms, high reliability is crucial on account of the additional difficulties for access and maintenance compared with land-based installations. The question has arisen whether simpler, potentially more reliable but probably less efficient, wind turbine units producing variable power that is AC or DC and not fully conditioned to match grid frequency and voltage can be beneficially employed, perhaps eliminating pitch systems and much of the local power electronics at each wind turbine. However, the issue of safety and overspeed protection always challenges any thought of having a rotor of totally fixed geometry. The less local control there is of each wind turbine, the more likely it is to require additional margins on loads adding to weight and cost. Nevertheless, this could be countenanced for a significant benefit in reliability.

In an evaluation of this issue by Siemens [11], it is firstly noted with reference specifically to Siemens wind turbines that the added complexity of the modern design route with variable speed, pitch regulation and an electronic power converter may have cost about 1% in availability (97–98% availability from 2005 onwards). This was contrasted with the older constant speed, passive stall regulation, direct grid connection, fixed power factor correction design route (98–99% availability from 1980 to 1998). It is acknowledged that the difference may be due as much to the greater maturity of the earlier designs as to the complexity of the new designs. This study concludes that central power conversion systems for offshore wind farms have no cost benefit and that the performance and loading benefits of fully independent variable speed much exceeds the potential availability benefit of simpler wind turbine technology (estimated at around 0.5%).

It is argued against the cost benefit of centralised power conversion that power conversion at low voltage level in individual turbines utilises the cheapest components per unit power converted. The case for the estimated 0.5% availability difference associated with central power conversion assumes the mentioned 1% net difference for wind turbine complexity and is based on the following (as quoted from the presentation [11]):

- Experience shows that power conversion accounts for 5–50% of lower availability with the remaining part due to other more complex systems (pitch system, cooling systems, etc.).
- Conservatively assuming 25% share, power conversion has an availability cost of 0.25%.
- Including offshore weather downtime, actual loss may be 0.5%.

Naturally, this preliminary study is far from conclusive on the debate around simpler turbines and centralised wind farm power conversion, but it highlights the need to attempt fully rationalised COE evaluations of such concepts. Work on impacts of control wind turbines in clusters continues. Benefits in reducing costs around electrical and control systems may be considerable [12], but the total impact on reliability, loads and energy capture needs detailed consideration.

7.4 Rating of Offshore Wind Turbines

On land-based sites, the rating of wind turbines, specifically the choice of electrical output power at rated wind speed in relation to rotor diameter or rotor swept area, is well understood. An illustration of the economic issues in optimising rating is presented in Section 10.8. The aim is to balance energy capture (directly proportional to rotor swept area) against drive-train cost (approximately proportional to power rating) taking account of the site wind distribution.

Excepting named turbine designs, the data points of Figure 7.5 relate to land-based machines. For all rotor diameters there is a wide range in power density defined as the ratio of rated power (kilowatt) to rotor swept area (square meter). This reflects site-specific impacts. Low power densities (the Nordex N117 is an extreme example) are equivalent to having relatively larger diameters in relation to rated power as is common on low wind speed sites. Taller towers are also often employed with the larger rotors further augmenting energy capture with COE benefit as the associated system load increases are less severe in a low wind speed site. Opposite characteristics are seen on high wind speed sites. The turbine spends more time in wind speeds above rated wind speed and minimum COE is associated with relatively smaller rotors that restrain loads. Land surface is, in general, more complex and variable than sea surface conditions. Thus, there is a very wide spread in appropriate power density on land to suit a great variety of site conditions. Overall, the average power density of land-based turbines is around $0.4\,\mathrm{kW/m^2}$.

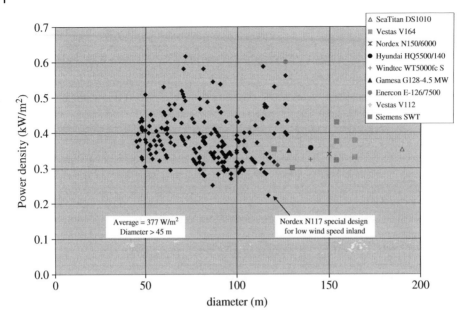

Figure 7.5 Power density characteristics of large modern wind turbines.

Offshore is rather different. Relative to land-based applications, the turbine CAPEX is substantially less of the project CAPEX and hence less again as a proportion of lifetime cost. The rotor systems as a sub-component of the turbine are similarly smaller cost proportions and this would suggest that an optimum minimum COE will arise with lower power density than on land. Initially, offshore cost models pointed clearly in this direction and all the recent very larger offshore turbines have low power densities in spite of generally being designed for site with high mean wind speeds above 8 m/s annual mean at hub height. If enlarging the rotor had no cost impacts other than on rotor and turbine structure, very low power densities would be justified. However, enlarging rotor diameter will increase the impact of wind loading on foundation cost (discussed in Section 7.5) and more recent cost modelling suggests that the optimum power density offshore may more typically be around 370 kW/m^2 rather than as low as say 330 kW/m^2. These average estimates do not of course reflect a considerable variation in what may best suit a particular site. On the other hand, there will be cost associated with providing variety in power density whether through rotor diameter variation or power train rating.

7.5 Foundation and Support Structure Design

7.5.1 Foundation Design Concepts

Design concepts for foundations and support structures fixed to the sea bed are now considered (foundation concepts for floating turbines are considered separately in Section 7.8). The main options for such foundations are as follows:

a) Monopile
b) Gravity base

c) Multi-pile
d) Suction bucket

The preferred solutions are much influenced by seabed conditions and water depth.

a) *The monopile* (essentially a fabricated steel tube hammered into the soil) has been much the preferred option in offshore wind projects. From an industry limit associated with manufacturing and pile driving capability of around 4 m diameter when the offshore market began, there are now projects to develop 10-m diameter piles. EnBW has installed (January, 2017) the first giant monopile at the 288 MW Baltic 2 offshore wind farm in Germany. The company confirmed that Ballast Nedam vessel Svanen has driven the 73.5-m-long, 6.5-m diameter structure more than 10 m into the seabed. Baltic 2, located 32 km north of the island of Rügen, will feature 80 Siemens turbines of 3.6 MW.

b) *Gravity-based foundations* were employed in the Middelgrunden offshore wind farm. Middelgrunden is in the Øresund 3.5 km outside Copenhagen, Denmark. When it was built in 2000, it was the world's largest offshore farm with 20 Bonus (now Siemens) turbines of 2 MW, giving a total capacity of 40 MW and providing about 4% of the electric power for Copenhagen. Jointly 50% owned by Dong Energy and 50% by a cooperative set up for the purpose of establishing the wind farm, Middelgrunden is held up as a leading example of a popular and successful offshore wind project with public involvement. The Baltic is comparatively shallow and freezes much more readily than North Sea areas and ice flows have been known to move large structures such as lighthouses. The bulbous shape of the gravity foundation sketched in Figure 7.6(a) with waterline just below the maximum bulge is intended to encourage ice to ride up on it and break up.

c) *Multi-pile foundations* are used to anchor other subsea structures such as tripods or jackets as in Figure 7.6(b,c). This allows comparatively small diameter piles to be used as foundations for a large subsea structure.

d) *Suction Pile Foundation*, or the suction pile or bucket foundation, is an alternative foundation option avoiding the high stresses induced in pile driving. A structure like a large upturned bucket lowered to the seabed cuts down a small distance. The water trapped inside the bucket is then pumped out, penetrating the foundation to its final position. This is best suited to clays or sands and possibly very soft rocks. Guidance on design calculations is provided by Houlsby and Byrne [13].

7.5.2 Support Structure Design Concepts

The monopile is an integrated foundation and subsea structure solution, certainly very attractive because of its simplicity. A plane tube is not, however, particularly efficient structurally for resisting overturning bending loads with minimal total mass of steel and there may to be tendencies towards over-use and possibly unjustified efforts in the up-scaling of pile technology. Golightly comments [14]:

> The debate is whether the move towards heavier, longer monopiles is desirable for the offshore wind industry. Alternative foundation solutions, including mono and multi-suction caissons, leaner lightweight jackets and tripods and concrete gravity bases may all suffer as a result. Many of these alternative solutions

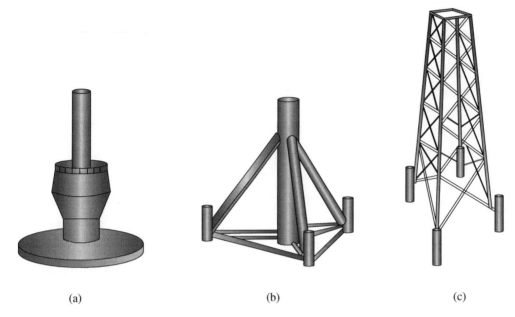

(a)   (b)   (c)

Figure 7.6 (a) Gravity foundation, (b) tripod and (c) jacket.

may be more suitable for the prevailing conditions and also cheaper than giant monopiles.

The jacket, on the other hand, essentially an under- and above-water lattice tower, is much more structurally efficient. The construction of the jacket is somewhat more complex, as shown in Figure 7.6(c), with many small welds. A compromise between these was sought in the tripod of Figure 7.6(b). There are certainly fewer welds but potentially more problematic larger ones.

7.5.3   Loads, Foundations and Costs

In an offshore project, rotor capital cost is proportionally much less as a fraction of life-time cost than on land and, as was discussed in Section 7.4, this may shift optimum economics to lower power densities but only if there is not too much impact on foundation costs

Over the recent years, DNV GL has developed sophisticated engineering-cost models incorporating component design (generators, offshore foundations, etc.) in increasing detail.

Using their Turbine Architect model, a sensitivity study was conducted on how rotor loading affects foundation loading of a generic 7 MW, 160-m diameter wind turbine. This is a one-at-a-time (OAT) sensitivity analysis; reducing selected extreme and fatigue load components individually by 10% in each case, and running the foundation design models to get the effect on the design and cost.

Only the perturbations of load components that yielded significant cost changes are shown (Figure 7.7). It should be noted that these results are valid for a single design case and baseline load envelope. The sensitivities are case-specific and will vary with turbine

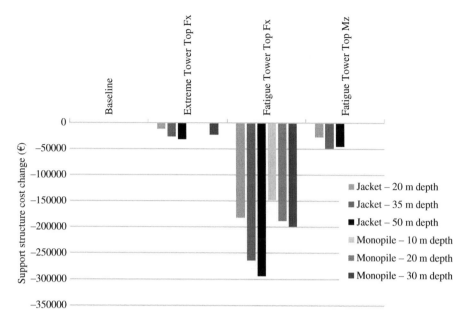

Figure 7.7 Effect of load reduction on support structure cost of a 7 MW wind turbine.

size and load envelope. The fatigue thrust loading, F_x (force in direction parallel to wind) is most influential (Figure 7.7). For jacket structures, the fatigue M_z (torsion-moment about vertical axis) also shows a significant effect due to its designing influence on joint bracing. This effect is absent for monopiles as these tubular structures are very well suited to resisting torsion loads. The cost reduction benefit levels off at greatest water depths, with little change for the monopile between 20 and 30 m and much less change to the jacket between 35 and 50 m as compared with between 20 and 35 m. For a fixed size of turbine, turbine rotor loads, in comparison to self-weight, wave and current loads, can be expected to have reduced influence on support structure design, the greater the water depth and consequently the more extensive the subsea structure.

7.6 Electrical Systems of Offshore Wind Farms

7.6.1 Collection System for an Offshore Wind Farm

The reliability of the system for interconnection of the wind turbines of an offshore wind farm has a strong impact on the reliability of the wind farm. There are several types of electrical collector structures:

- Radial
- Single-side ring design
- Single return with single hub design
- Double-side ring design
- Star design
- Single return with multi-hub design

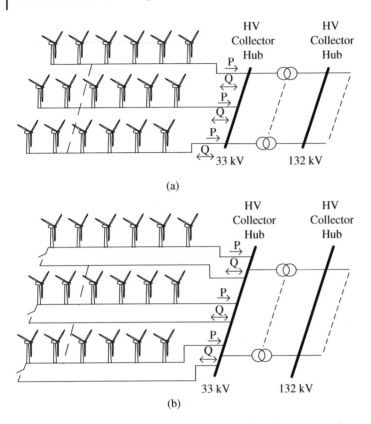

Figure 7.8 (a,b) Radial and single-side ring design of a collector network.

The most popular design for a collector system is the radial design shown in Figure 7.8(a). Each wind turbine is connected to the same feeder in a string, and so a fault in the collector circuit will result in disconnection of all wind turbines. The maximum number of wind turbines on one cable is determined by two factors: capacity of the generator(s) and rating of the subsea cables [15]. Figure 7.8(b) shows a single-side ring design, which provides auxiliary paths and enhanced power reliability in the case of the fault in a radial line.

The single-sided ring design shown in Figure 7.9(a) is based on a radial design with an additional cable connected to the end of each turbine string. The additional cable is of similar current rating as a single string, which results in lower installation costs compared to single-side ring design [16]. In Figure 7.9(b), the double-sided ring design, also known as ring design, is shown. This type of design has an even number of cables. The last wind turbine on one cable is interconnected to an adjacent cable. The size of each cable should be doubled to handle extra loading in case of the one string fault next to the hub. This may increase the capital cost of the wind farm [16]

The star design of Figure 7.10(a) has the advantage of cable length savings. However, more complex switchgear is required in the central wind turbine. The star design offers a better voltage regulation as well as lower power losses between wind turbine link cables.

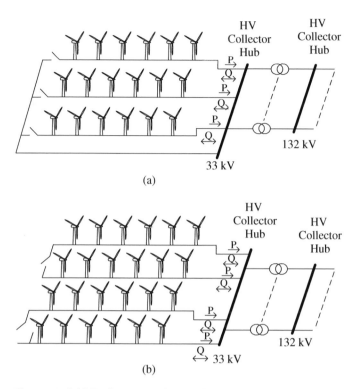

HV Collector Hub

HV Collector Hub

132 kV

33 kV

(a)

HV Collector Hub

HV Collector Hub

132 kV

33 kV

(b)

Figure 7.9 (a,b) Single return with single hub design and double-side ring design.

The most complex design is the single return with multi-hub in Figure 7.10(b). This design offers high levels of security for offshore wind plants, which is an important asset. Due to many redundant paths, power can be generated and delivered even if one hub is lost.

7.6.2 Integration of Offshore Wind Farms into Electrical Networks

Transmission systems have progressed significantly over the past few decades. There are three solutions available on the market for transmitting power from the offshore wind farms: HVAC (high-voltage alternating current), LCC-HVDC (line-commuted converter-high-voltage direct current) and VSC-HVDC (voltage-source commuter-high-voltage direct current). An overview of these key technologies is provided.

7.6.2.1 High-Voltage Alternating Current (HVAC)

Many offshore wind plants are connected via HVAC transmission lines. This technology is well proven and cost-effective for connecting closely located offshore wind farms [17]. The main disadvantage of this connection type is a distance limitation as cables are subjected to a capacitive charging effect which limits the active power transfer [18]. To overcome this limitation onshore, reactive power compensation units are used on one or each end of the cable route. This may not be viable offshore owing to the high cost of offshore platforms. Another major weakness of an HVAC system connected to offshore wind farms is fault propagation. That is, faults on the AC cable side will directly affect

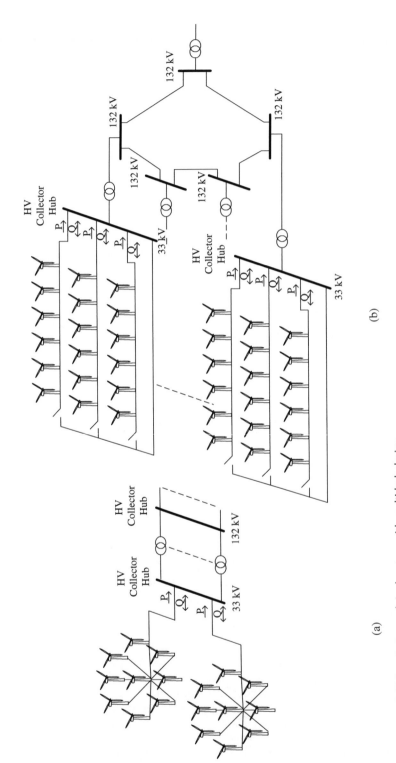

Figure 7.10 (a,b) Star design and single return with multi-hub design.

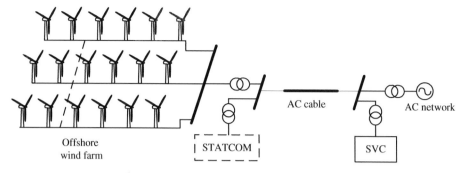

Figure 7.11 HVAC connection for wind farm integration.

wind farms, while faults within the wind farm have direct effects on the AC network. Basic configuration of an HVAC system is shown in Figure 7.11.

7.6.2.2 Current-Source Converter (CSC)

Current-source converter-high-voltage direct current (CSC-HVDC) transmission is a well-established technology, much older than VSC-HVDC, and often been used to interconnect asynchronous AC networks. However, it has not yet been used for the connection of offshore wind farms to onshore grids. Reasons for that include primarily the need for start-up generators or synchronous compensator devices to provide the necessary communication voltage but also the need for reactive power compensation.

Devices such as STATCOM produce large amounts of harmonics and hence filtering equipment is necessary. Conventional HVDC also absorbs reactive power to operate at both ends. It depends on the grid voltage for current commutation within converter valves; it is thus not suitable for connection to weak AC grids. In addition, the footprint is much higher compared to VSC-HVDC, and it cannot provide independent control over active and reactive power [19].

The typical LCC-HVDC layout for connecting the wind farm to the grid comprises an AC collection network, an offshore substation (incubating the converter, a start-up generator or a STATCOM, and filters), DC cables, onshore substation and an AC harmonic filter, as shown in Figure 7.12.

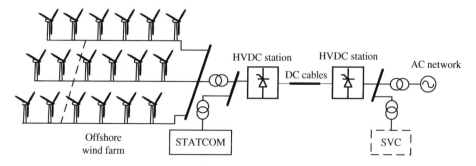

Figure 7.12 Configuration of LCC-HVDC for connecting offshore wind farm.

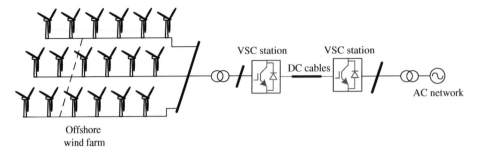

Figure 7.13 Schematic of a typical VSC-HVDC system.

7.6.2.3 Voltage-Source Converter for Offshore Wind Farm Integration

About a decade after the CSC-HVDC was developed and commercially utilised, the appearance of semiconductor switches, such as insulated gate bipolar transistors (IGBTs) and insulated gate-commutated thyristors (IGCTs), has revolutionised the industry and introduced the VSC technology. VSC-HVDC transmission systems have become an area of growing interest due to their benefits over the classical HVDC and conventional HVAC transmission systems. VSC-HVDC was first manufactured and tested by ABB (also known as HVDC-light [20]), by Siemens (known as HVDC-plus), and also by Alstom (commercially known as HVDC MaxSine). This technology utilises IGBT devices and can switch ON and OFF current in a force-commutated manner, as opposed to the line-commutated operating principle of conventional CSC-HVDC. This allows pulsewide width modulation techniques (PWM) as well as multi-level conversion to be used. The general layout of VSC-HVDC is shown in Figure 7.13.

VSC-HVDC can be considered an attractive alternative to HVAC and conventional HVDC as it allows wind farms to meet modern Grid Code requirements. When comparing VSC-HVDC to HVAC, transmission losses are reduced. VSC-HVDC permits asynchronous operations between the wind farm and the AC onshore grid. Moreover, it does not normally require additional reactive power compensation units on the ends of the transmission lines [21–23]. It also has lower footprint compared to traditional HVDC as it does not need harmonic filters and reactive power compensation units. Hence, it requires nearly 30% less space on the offshore platform, which leads to further cost reductions. However, the main drawback of the HVDC technology is the lack of reliable DC protection technology.

7.7 Operations and Maintenance (O&M)

7.7.1 Introduction

Given adequate turbine technology and infrastructure, design for operations and maintenance (O&M) may reasonably be considered the most vital aspect of offshore projects. Turbine reliability clearly impacts on this and so turbine design for O&M is a key part of offshore turbine design. However, the same can be said of all other key project components, foundations, electrical infrastructure, and so on. Design to facilitate maintenance with in-built cranes, for example, is no more important than design to limit or avoid it. Site visits and offshore operations are much more constrained by weather (both wind

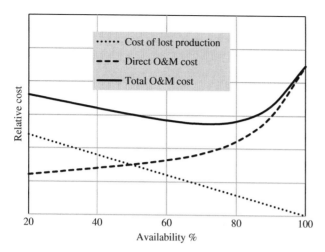

Figure 7.14 Optimising O&M cost.

and wave conditions). This leads to design considerations of greater redundancy in, say, minor electronic components where the cost and repair cost may be slight but the cost impact of downtime and cost of a site visit is always substantial. Simpler systems such as direct drive may be preferred, although that technology may need to mature further to be of proven reliability benefit.

While availability above 97% and acceptable maintenance cost around 1% of lifetime cost per annum is achieved on many competent land-based projects, the challenge for offshore is much more severe. Distance from shore, project scale, water depth, wave and wind climate, turbine reliability and energy prices all affect the optimisation of O&M procedures. The qualitative interaction of key factors influencing O&M cost is indicated in Figure 7.14. In general, the increased difficulty of offshore maintenance may imply that minimum total O&M cost is realised at lower availability than typical of land-based projects. Nevertheless, if O&M lifetime cost is around 30% of total lifetime cost, a somewhat higher availability than that which minimises total O&M cost will minimise COE.

7.7.2 Modelling

Considering all the logistics around size, cost and availability of vessels that may have a role in maintenance, the interaction with weather windows, costs and timescales of repair activities, and so on, it becomes very clear that effective modelling of O&M is of great value in design and management of projects, and extensive modelling has been evolved by many research institutions and industrial parties within the wind industry.

As an example, a very detailed O&M model was developed by a group at the University of Strathclyde [24] with ~75 inputs reflecting climatic conditions, weather constraints, transportation options, maintenance equipment options, wind turbine characteristics, wind farm characteristics and cost factors. Inputs from climate, vessel specifications and fleet configuration, wind farm, turbine and cost are processed. Synthetic climate data is generated. Availability, operability and failure analyses then feed operational simulations. The outputs are wind farm availability and downtime, power production, vessel and failure-specific information and operating expenses (OPEX) costs.

Table 7.1 Distribution of total O&M cost.

	Cost (M€)
Staff costs	24
OEM costs	27
Fixed costs	48
Transport costs	125
Total direct O&M costs	225
Lost revenue	105
Total O&M costs	325

This model has previously been used to determine the most cost-effective approach to allocate O&M resources, which may include helicopter, crew transfer vessels, offshore access vessels and jack-up vessels. It employs a time domain Monte-Carlo simulation approach which includes analysis of environmental conditions (wind speed, wave height, and wave period), operational analysis of transportation systems, investigation of failures (type and frequency) and simulation of repairs. Results from a case study with this model showing the distribution of O&M costs are presented in Table 7.1.

A specific O&M planning strategy was investigated for an offshore wind farm consisting of 150 3.6 MW turbines. The FINO 1 research 10 mast, located 45 km off the coast of Germany, was utilised for the analysis. This mast, located within the German offshore wind development zone, may be considered representative of many current and future offshore wind farms in Europe. Data used is for the period 2004–2012, both years inclusive. In order to cover different variations in the results due to randomisation of the climate parameters and the 14 variables in the Monte-Carlo simulation process, 100 simulations with a period of 5 years were conducted.

7.7.3 Inspection of Wind Turbines

UAV (unmanned aerial vehicle or drone) inspection has been introduced on wind turbines and wind farms. For example, Siemens Wind Power is collaborating with US-based SkySpecs [25] to deploy automated drone technology for onshore and offshore wind turbine inspections.

The following account is synthesised from References [26] and [27]. In the first half of 2016, 250 turbines in the North Sea were inspected, averaging more than 10 per day. At present, the market share among drone deployment, ground-based camera inspection and manual inspection is 7%, 10% and 83%, respectively. Key issues affecting the usefulness of drones in wind turbine inspection are cost, data quality and flying regulations.

- *Cost*: Information about costs is varied. According to Bley, pilot-operated drone inspections in the wind industry could be as little as $300–$500 per turbine, or 20–25% of the cost of manual inspections. However, according to Avitan, the costs are almost the same. Meanwhile, the per-turbine cost for manual inspections can range from $3000 down to €200 ($225) in parts of Eastern Europe. EDF Renewable Energy

Technical Adviser Jon Salmon said. 'A ground-based system has less complexity: no drone to pilot, maintain and recharge batteries.' EDF found that a ground-based camera inspection costs about $300–$500 per turbine, including the technician for turbine operations, blade inspection, reporting time and a written report. Ultimately, if not already, the drone systems will be able to provide lower cost inspections due to the drone's ability to fly around the blade and provide the full inspection without having to reposition the wind turbine rotor and blade for the camera.

- *Data quality*: Regarding data quality, drones can already deliver better data on turbine conditions than can be achieved through a manual inspection. UAVs are able to fly at an optimum distance from the blade, usually from 3 to 10 m, and can circle around it to cover the whole surface area. Data can be automatically meta-tagged and used to build a three-dimensional model of the blade, delivering a level of accuracy and detail that is hard to achieve with manual methods
- *Regulation*: Changes in regulation could also spur the use of drones for wind farm inspections. Drones in the United States are currently required to fly within the visual line of site of the operator, which limits the range of flight to 300–600 m on average, depending on how large the drone is. The U.S. Federal Aviation Administration (FAA) has officially permitted commercial flights of drones weighing <25 kg, flying during daytime at a maximum speed of about 160 km/h and at a maximum altitude of about 120 m above ground level (or, if higher than that, remain within 120 m of a structure), and as long as they remain within the line of sight of the operator. Previously, drones were not allowed to exceed 120 m in altitude. The FAA pilot's license required to fly a commercial drone is replaced with a more easily obtained certificate.

Drones can also provide security surveillance of wind farms and the extension of their role and the introduction of robotic maintenance activities [28] seems a likely future path.

7.8 Offshore Floating Wind Turbines

In 1980 it was widely questioned if wind could ever produce quantities of power that would be significant in relation to a nation's electricity demand and would ever be economic. Later, around 1990, land-based technology was progressing effectively, but it was considered doubtful if offshore technology could ever be economic. By 2000 the offshore market had started and at present (2010) the market is growing, although viability, especially in respect of maintenance costs, has still to be proved. Some quite severe problems were encountered in this first generation of commercial offshore installations, but the industry has nevertheless moved forward. The offshore wind farm at Horns Rev (installed capacity in 2002 of 160 MW using Vestas V80 2 MW wind turbines) at an early stage suffered major problems necessitating large-scale replacement of generators and transformers, yet following that has operated with good availability.

One of the compensations that has improved the prospects of offshore wind is that higher mean wind speeds (also associated with relatively lower extreme winds) are generally experienced offshore compared to land-based areas that could serve a given population area.

Again, following a period of much skepticism about feasibility and costs, there is now a vigorous continuing evolution of wind turbines on floating platforms. Pre-commercial prototype systems are in operation notably at as large a scale as the 7 MW floating wind turbine at Fukushima (see Figure 6.11). The main drivers for floating technology are as follows:

- Access to useful resource areas in deep water, yet often near shore;
- Potential for standard equipment relatively independent of water depth and seabed conditions;
- Easier installation and decommissioning;
- Possibility of system retrieval as a maintenance option.

The main obstacle in the realisation of such technology has been the one given here:

- Development of effective design concepts and demonstration of cost-effective technology, especially in respect of the floater and its mooring system.

Floating offshore concepts are naturally less mature than offshore wind turbines on fixed bed foundations. Perhaps, however, the same pattern that has seen the maturity of land-based systems and the growing credibility of fixed foundation offshore systems will continue in the future to floating wind turbine systems and also possibly to airborne systems (discussed in Chapter 8) that may eventually be deployed offshore and employ floating platforms. Goldman [29] has noted the paradox that this wind turbine evolutionary pattern from land to sea is the opposite of human evolution.

Like many 'new' ideas, floating offshore wind turbine systems are conceptually quite old. Note, for example, Honnef (Figure 15.1) circa 1930 and Heronemus [30] of 1972. However, there is now a context in energy costs and offshore industrial capability which is enabling such concepts to leap from imagination into reality.

StatoilHydro installed (May 2009, Figure 7.15) the world's first full-scale floating wind turbine, Hywind, at a site offshore from Karmøy, north of Stavanger. Their floating system was based on the spar-buoy concept supporting a 2.3 MW Siemens wind turbine at a height of 65 m above sea level. The floatation element has a draft of some 100 m below the sea surface, and is moored to the seabed using three anchor points, a system suitable for water depths ranging from 120 to 700 m.

After proving their technology with that prototype, the Hywind Scotland project presently under construction will establish the world's largest floating offshore wind farm, comprising five Siemens SWT-6.0-154 direct-drive offshore wind turbines. The wind farm, located offshore of Peterhead in Aberdeenshire, Scotland, is estimated to power ~20 000 households when in production. The Hywind Scotland substructures have been constructed by Navantia in Spain. The launching weight of the floating foundations is around 3200 t, including more than 1000 t of high-density concrete. The wind farm was fully commission and formally opened in October 2017 by the Scottish First Minister, Nicola Sturgeon. The third and last rock installation campaign to protect the export cable was in progress in December 2017.

Trident Winds [31] plan an ambitious 650 MW floating wind farm in 800–1000 m water depth 48 km in the Pacific off the coast of Port Estero in California.

Among floating support structures, the spar (as in Hywind designs) is perhaps the simplest solution and possibly the cheapest but requiring relatively deep water depth near the dockside. Barge floaters are also relatively simple structures but generally more affected by wave loading. Floating Power Plant (Figure 7.16) look to the development of hybrid systems where the platform may support wind turbines and incorporate

Figure 7.15 Hywind floating wind turbine system. Reproduced with permission of Øystein Fykse/Alligator.

wave energy devices. The main alternative solution to spars and barges, adopted in the Fukushima floating project and also by Principal Power, the leading US-based exponent of floating systems, is the semi-submersible typically in a tri-floater configuration.

Among the most innovative concepts for offshore floating systems, Aerodyn has unveiled an integrated floating platform and turbine design it claims can reduce offshore wind costs by 40% called SCDnezzy (SCD for super-compact design).

The aim is to combine a very compact nacelle design with a lightweight floating structure that is particularly economic using concrete as a primary structural material. In a phased development programme, a 1 : 36 scale model has been tank tested at LiR in Cork, Ireland, with up-scaling later to commercial systems at 3 MW (see impression of a floating wind farm in Figure 7.17) and 8 MW.

The two-bladed downwind turbine is joined to a concrete platform with three floating cups. The system is designed for onshore construction and towing to the project site. The downwind two-bladed design enables self aligning yaw motion of the system in the sea facilitated by a swivel mooring joint. The platform, which sits at shallow depth under the water, has been designed in a Y-shape, using concrete pre-stressed tubes. A profiled tower is supported by pre-tensioned steel ropes.

Figure 7.16 Floating power plant.

Figure 7.17 Aerodyn 3 MW SCDnezzy.

In June, 2107, Aerodyn announced the further development of a twin rotor SCDnezzy[2] system (Figure 7.18). This system, rated 15 MW, has counter-rotating rotors operating with a blade-phase angle difference of 90° to avoid mutual excitation and claims some economies of scaling (essentially related to the multi-rotor scaling benefits discussed in Chapter 15). Bending moment loading on the cantilevered tubes is reduced and hence

Figure 7.18 Aerodyn 15 MW SCDnezzy[2].

overall structure weight and cost with various tension wires. Clearly, the long-term maintenance of tensioning will be important. Aerodyn have long been involved in the development of efficient, low mass highly integrated drive-train concepts dating from the Multibrid development (discussed in Chapter 6).

The SWAY® system (Figure 7.19) is another floating wind turbine system capable of supporting a 5 MW wind turbine in water depths from 80 m to more than 300 m designed for severe wave climates. SWAY are no longer trading but the design concept, also involving a downwind turbine yawing in the sea, remains interesting.

According to SWAY,

> … the wind pressure on the rotor for a 5 MW turbine is approximately ~60 tons. The tower is stabilised by elongation of the floating tower approximately 100 m under the water surface and by around 2000 tons of ballast in the bottom. A wire bar gives sufficient strength to avoid tower fatigue. Anchoring is secured with a single tension leg between tower and anchor.
>
> The tower finds its equilibrium tilt angle (typically some 5 to 10 degrees) due to the wind pressure on the rotor. In addition the tower is typically only tilting +/−0.5 to 1.0 degrees from this equilibrium tilt angle due to the waves in a storm condition during power production. Through … new … solutions related to active thrust control we have created a new floating foundation system with radical cost savings … for deep water applications.

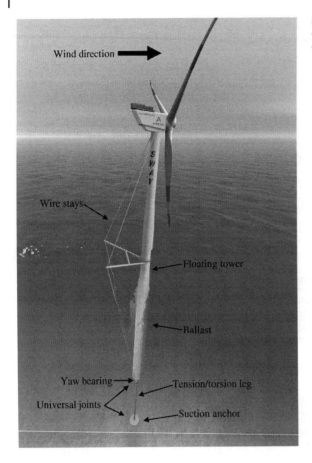

Wind direction

Wire stays

Floating tower

Ballast

Yaw bearing

Tension/torsion leg

Universal joints

Suction anchor

Figure 7.19 SWAY offshore floating wind turbine system. Reproduced with permission of SWAY A/S.

References

1 Ho, A., Mbistrova, A. and Corbetta, G. (2016) The European Offshore Wind Industry – Key Trends and Statistics 2015. EWEA Report, Feb 2016.

2 Hau, E. (2005) *Wind Turbines: Fundamentals, Technologies, Application, Economics*, Springer-Verlag. ISBN: 3540242406

3 Avoloi, S., Arcidiacono, V., Botta, G. *et al.* (1993) *Dynamic Analysis of Gamma-60 Wind Turbine Generator and Control Systems: Design and Experimental Validation.* Proceedings of EWEC, Lubeck-Travemunde, March 1993.

4 Wind Energy the Facts Technology, vol. 1, http://www.ewea.org/fileadmin/ewea_documents/documents/publications/WETF/Facts_Volume_1.pdf (accessed 2009).

5 Miner, M.A. (1945) Cumulative damage in fatigue. *Transactions of the American Society of Mechanical Engineers*, **67**, A159–A164.

6 Jakobsen, J. (2005) Infrasound emission from wind turbines. Danish Environmental Protection Agency. *Journal of Low Frequency Noise, Vibration and Active Control*, **24**, 145–155.

7 Powles, S.R.J. (1983) The effects of tower shadow on the dynamics of a HAWT. *Wind Engineering*, **7** (1), 26–42.

8 Spruce, C.J. (2004) *Power Control of Active Stall Wind Turbines*. Proceedings of EWEC, London, November 2004.

9 Anderson, C.G. and Jamieson, P. (1988) *Mean Load Measurements on the Howden 33 m Wind Turbine*. Proceedings of 10th British Wind Energy Association Conference, London, 1988.

10 Jamieson, P. (2009) *Light Weight, High Speed Rotors for Offshore*. Proceeding of EWEC Offshore, Stockholm 2009.

11 Skjærbæk, P. (2009) *An Idealized Offshore Wind Turbine*. European Offshore Wind Conference, Stockholm, September 2009.

12 Hur, S. and Leithead, W.E. (2012) *Collective Control of a Cluster of Stall Regulated Wind Turbines*. International Conference on Sustainable Power Generation and Supply (SUPERGEN 2012) Hangzhou, China, September 2012, pp. 1–8

13 Houlsby, G.T. and Byrne, B.W. (2005) Calculation Procedures for Installation of Suction Caissons. Report No. OUEL 2268/04. Department of Engineering Science, University of Oxford.

14 Golighly, C. (2014) Tilting of Monopiles Long, Heavy and Stiff; pushed beyond their limits. Technical Note, Ground Engineering, January 2014.

15 Dutta, S. and Overbye T.J. (2011) *A Clustering Based Wind Farm Collector System Cable Layout Design*. Power and Energy Conference at Illinois (PECI), 2011 IEEE, 2011, pp. 1–6.

16 ABB (2014) *National Offshore Wind Energy Grid Interconnection Study Executive Summary*, ABB, Inc.

17 Sathyajith, M. and Philip, G.S. (2011) *Advances in Wind Energy Conversion Technology*, Springer-Verlag, Berlin Heidelberg.

18 Machado, J., Ventim, N.M. and Santos, P.J. (2015) *Economic Limitations of the HVAC Transmission System When Applied to Offshore Wind Farms*. 2015 9th International Conference on Compatibility and Power Electronics (CPE), pp. 69–75.

19 Barnes, M. and Beddard, A. (2012) Voltage source converter HVDC links – the state of the art and issues going forward. *Energy Procedia*, **24**, 108–122.

20 Eriksson, K. (2003) HVDC Light™ and development of Voltage Source Converters, ABB Utilities.

21 Adam, G.P., Ahmed, K.H., Finney, S.J. *et al.* (2013) New breed of network fault-tolerant voltage-source-converter HVDC transmission system. *IEEE Transactions on Power Systems*, **28**, 335–346.

22 Zhang, L., Nee, H.P. and Harnefors, L. (2011) Analysis of stability limitations of a VSC-HVDC link using power-synchronization control. *IEEE Transactions on Power Systems*, **26**, 1326–1337.

23 Korompili, A., Wu, Q. and Zhao, H. (2016) Review of VSC HVDC connection for offshore wind power integration. *Renewable and Sustainable Energy Reviews*, **59**, 1405–1414.

24 http://www.offshorewindindustry.com/news/automated-drone-technology-offshore-turbine (as viewed March 2017).

25 http://www.atsite.dk/index.php/about-us/news/141-new-world-record-16-offshore-wtg-s-inspected-in-one-day (as viewed March 2017).

26 http://analysis.windenergyupdate.com/operations-maintenance/fully-automated-drones-could-double-wind-turbine-inspection-rates (as viewed March 2017).

27 Dong, G.L., Sehoon, O. and Hyoung, I.S. (2016) Maintenance robot for 5-MW off-shore wind turbines and its control. *IEEE/ASME Transactions on Mechatronics*, **21** (5), 2272–2283.

28 Dalgic, Y., Lazakis, I., Dinwoodie, I. *et al.* (2015) Advanced logistics planning for off-shore wind farm operation. *Ocean Engineering*, **101** (1), 211–226.

29 Goldman, P. (2003) *Offshore Wind Energy*. Presentation at Deep Water Technology Workshop, Washington, DC, October, 2003.

30 Heronemus, W. (1972) *The US Energy Crisis: Some Proposed Gentle Solutions. The Congressional Record*, US Government Printing Office, vol. **118** (17), February.

31 Trident Winds website, http://www.tridentwinds.com/projects/ (as viewed February 2017).

8

Future Wind Technology

To set the scene for future technologies a summary is presented of the evolution of present wind technology (Section 8.1) and present trends (Sections 8.2 and 8.3). Following that, among future technologies, airborne technology (Section 8.4), energy storage (Section 8.5) and innovative concepts not based on rotors (Section 8.6) are discussed. Floating offshore technology is now (2017) considered as effectively a present technology with the first commercial wind farm, Hywind Scotland, scheduled in 2017 (see Section 7.8).

8.1 Evolution

How has wind turbine technology evolved since the early 1980s? There has always been a wide variety of designs on the margins of commercial technology; but in the 1980s, the Danish three-bladed, single fixed speed, stall-regulated turbine dominated the market at rated power levels generally <200 kW. Blades were almost invariably of glass-polyester resin manufacture.

In 2002, the focus of attention shifted to technology around and above 1.5 MW rating and by 2006, 2 MW wind turbines were commonly employed in land-based projects. Variable pitch, variable speed designs now (2016) predominate and direct-drive generators have become more prevalent. New blade technology is predominantly involving glass, sometimes carbon and most usually epoxy resin, most usually vacuum infused or sometimes as prepregs.[1] The main trends are as follows:

- Speed variation is mandatory. The solutions range from high slip of a single generator, through dual speed to a variety of variable speed systems with speed range ratios from about 1.5 : 1 to 3 : 1 or more in a few cases.
- Pitch regulation is prevalent on the largest wind turbines and the independently actuated, electric pitch system is the preferred solution.
- An increasing number of manufacturers are developing direct drive systems. Almost all new direct drive designs are based on high-strength permanent magnets.

Dual speed was introduced in stall-regulated wind turbines in the 1980s. Initially, this was achieved with two separate generators and a belt drive connection, but in recent

---

1 Pre-impregnated materials (prepregs) are reinforcement fibres or fabrics into which a pre-catalysed resin system has been impregnated by a machine. The prepreg is presented as a dry sheet; and in manufacture, the resin is infused by heating.

Innovation in Wind Turbine Design, Second Edition. Peter Jamieson.
© 2018 John Wiley & Sons Ltd. Published 2018 by John Wiley & Sons Ltd.

years the solution is almost invariably a single generator with pole switching achieved by dual sets of stator windings. It was then recognised that pitch control in conjunction with fixed speed operation on an induction generator could be problematic in high wind speeds, and Vestas developed the 'OptiSlip' system which allows up to 10% speed variation in wind speeds above rated. This solution continues in the present designs of Suzlon. As noise, power quality and energy capture on low wind speed sites became particularly prominent issues on the German market, variable speed systems became the norm for large wind turbines. In parallel with this and in recognition of the fact that the certification bodies were prepared to accept a rotor with three independently pitching blades as having in effect two rotor brakes (the wind turbine would be held at safe speed even if one blade failed to pitch), independent pitch actuation[2] became popular in large wind turbine designs.

An increasing interest in direct-drive systems is fuelled by perceived benefits:

- Simpler drive train with enhanced reliability including the avoidance of the gearbox as a maintenance centre,
- Improved part load efficiency of the drive train,
- Possible but probably as yet unrealised mass and cost advantage compared to geared transmissions.

With various grid compliance issues becoming more prominent, the case for a fully rated converter (as opposed to partially – around one-third – full load as in the doubly fed induction generator (DFIG) system) has strengthened. It may be argued [1] that the permanent magnet generator (PMG) should be preferred to an induction generator *primarily on the grounds of part load efficiencies*. When the cost of losses is capitalised, there may be a net cost of energy disadvantage in employing DFIG technology in particular and induction generators in general, especially if more stringent grid power quality and fault ride-through requirements add cost to the DFIG system.

Since the 1980s the wind industry has made huge strides, well documented in numerous publications, for example, the European Wind Energy Association (EWEA) [2] and the American Wind Energy Association (AWEA) [3]. In particular, rotor aerodynamic efficiency has been pushed upwards since 1980 when C_p max was typically around 0.44 to values now possibly above 0.50 for the best designs. This is more than 10% improvement relative to the earlier values, a huge gain in an area where it had been supposed that only marginal improvements were likely. This general improvement in turbine performance is alternatively expressed in the report, EWEA [2]:

> The development of electricity production efficiency measured as the annual energy production per square metre of swept rotor area (kWh/m^2) at a specific reference site, has improved significantly in recent years owing to better equipment design.

---

2 The essential requirement is that the pitch actuators are physically independent of each other and cannot experience a common fault. This is distinct from generally older collective pitch systems where a single actuator via linkages moves all three blades in synchronism. In such a system, a single fault can disable all three blades from pitching or cause them all to pitch inappropriately. With independent actuators, the control system may instruct synchronised pitch movement which was the norm when such systems first gained ascendancy, although there is now much interest in independent pitching of blades to alleviate system loads in conjunction with regulating power.

Taking into account the issues of improved equipment efficiency, improved turbine siting and higher hub height, overall production has increased by 2–3% annually over the last 15 years.

Prior to widespread industrial cost increases of the order of 20% in 2004, cost reduction due to market growth is estimated to have been at a learning rate of 10% per annum, according to Morthorst [4]. This means that each doubling of world installed capacity has seen a 10% reduction in cost. The wind industry has now matured considerably due to market expansion to the point where it is the fastest growing sector of the power industry. In technology development, the evidence of a maturing industry is that some major component manufacturers tailor their designs specifically for wind applications and principally supply to the wind industry.[3]

8.2 Present Trends – Consensus in Blade Number and Operational Concept

The early failure of the vertical axis types of design in the market place is attributed mainly to the inherently lower efficiency and added system cost associated with low optimum speed and consequent high torque. Considering the vast potential variety of wind energy conversion systems deriving from the options of Figure 5.1, there is a remarkable

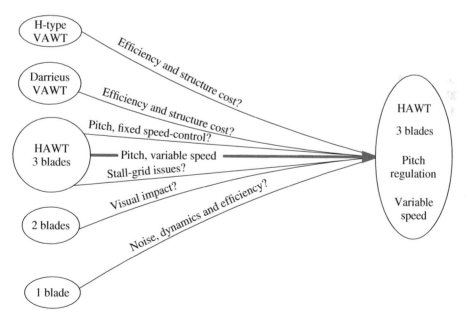

Figure 8.1 Consensus.

---

3 For example, in 1995, in connection with the development of one of the first 1.5 MW wind turbine designs for Tacke Windtechnik, the author was told how various component sub-assembly weights should be limited to accommodate the available crane capacities. However, within 5 years, the German crane industry was developing cranes to suit turbine installation requirements!

consolidation of design towards a consensus around the three-bladed, upwind, variable speed, pitch-regulated design for large-scale electricity-producing wind turbines (Figure 8.1).

8.3 Present Trends – Divergence in Drive-Train Concepts

In contrast to the consensus about basic wind turbine architecture, the technology of the drive train has become much more of a focus for innovation and a variety of alternative configurations with greater or lesser degrees of structural and component integration have arrived. A summary picture of this divergence in drive-train concepts is represented in Figure 8.2. The options described have generally been reviewed in Chapter 6.

8.4 Future Wind Technology – Airborne

8.4.1 Introduction

Airborne turbine concepts had appeared in patent documents for many years, but now such concepts are generating increased interest, with much research being conducted and some prototypes being tested. Such systems continue in the spirit of moving from land into the other elements, water and air. The key attractions are access to new areas of generally superior wind resource with the possibility especially in the case of kite-related

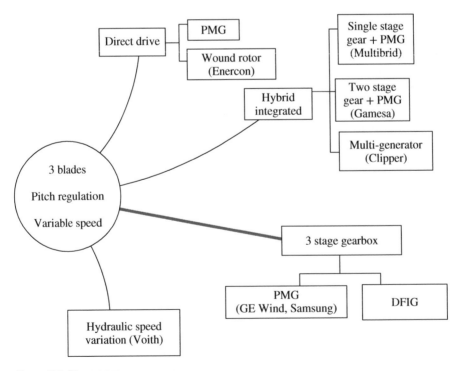

Figure 8.2 Divergence.

systems to capture the energy with surfaces of low mass and cost. A further benefit of many of the airborne systems is the capability to ground the system in anticipation of severe weather conditions. This evades the conflict in design with many conventional wind turbines, where structural demands and cost can be much influenced by conditions in which the system produces no power. There are naturally new risk factors with airborne energy conversion. When a kite is producing substantial power, there is high strain energy in the tether, which, if it fails, can fracture explosively; and crashes of soft kites may be as hazardous, or more so, than crashes of rigid systems because the soft kite may re-launch unpredictably.

There are already a multitude of airborne power concepts challenging any very neat and tidy taxonomy of designs. The systems may, in general, be distinguished by methods of power take-off and by aerodynamic characteristics. Regarding power take-off, some use the cable as mechanical power transmission to a ground-based electrical generator system and this divides into two main categories:

a) The prevalent concept of generating power as the product of cable velocity and cable tension in reeling out the tether as the kite travels essentially downwind but often in complex cross wind trajectories (most typically figure-of-eight or circular paths);
b) A less usual concept, avoiding progressive reel-out of the tether lines, using a tethering system that can function as a means of rotary torque transmission from a kite rotor to a ground-based generator system.

Examples of (a) in the system of Kite Power Solutions (KPS) and of (b) in the Daisy Kite system are discussed in Sections 8.4.2 and 8.4.3, respectively. The other fundamentally different power take-off concept is to generate electricity locally at altitude and use the tether as an electrical conductor. With this type of system, aerodynamic solutions to maintain altitude are broadly categorised as follows:

1) Kites and rigid wings (lifting aerofoils),
2) Buoyant balloons to support or assist support of systems in a few cases rotating to exploit the Magnus effect,
3) Autogyros typically with assisted power take-off and landing.

The aerodynamic and power take-off solutions are often combined in various ways. Table 8.1 lists some of the active companies developing designs or prototypes and summarises the concept choices. Power take-off using the tether extension to rotate a cable drum with a generator predominates, although Makani, Altaeros (with a balloon serving as a duct for a rotor) and a few others generate power at altitude and use the tether (carbon or aluminium in a Kevlar sheath) as an electrical conductor. An excellent review of airborne wind solution concepts with much further detail about the systems of leading companies involved and with over 100 useful references is provided by Cherubini *et al.* [5].

TU Delft has been much involved in researching and promoting airborne wind technology. Delft's system developed in 2010 has a rated power of 20 kW, with the airborne system being a leading edge inflatable (LEI) kite that is controlled by a 'kite control unit', which is a cable robot suspended below the wing. Power take-off is ground, tether reel-out. A start-up company, Kitepower, was launched in 2016 aiming to develop a system to be operational in 2017 of 100 kW rated power with a kite area of 70 m^2 using new technology to enable larger membrane areas. The ground station is

Table 8.1 Airborne wind power developments.

	Rated power (kW)	Airborne system	Power take-off
Makani	600	Eight-rotor powered take-off, autogiro generation at altitude	At altitude, electrically conducting tether
Ampyx	45	Rigid wing glider	Ground, tether reel-out
Enerkite	30	Semi-rigid wing	Ground, tether reel-out
Kitemill	30 (winch rating)	Rigid wing plane with four rotors (vertical take-off and landing)	Ground, tether reel-out
TU Delft	20	Leading edge inflatable kite	Ground, tether reel-out
TwingTec	10	Lightweight rigid wing with vertical axis tip rotors for control of take-off and landing	Ground, tether reel-out, two tethers to dual winches
Omnidea	6.5	Cylindrical balloon, rotated with motors to create Magnus effect lift	Ground, tether reel-out
eWind	12	Complex rigid wing kite/plane	Ground, tether reel-out
SkySails Power	55	Soft kite – para-glider profile	Ground, tether reel-out

presently the largest ground station for the power take-off among tether reel-out system concepts. Project funding is provided by the EU through a 'Fast Track to Innovation' project (H2020-FTIPilot-691173).

SkySails Power has technology established over 10 years based on kites of around 2 MW capacity designed for assisted towing of oil tankers. The company proposes to follow the 55 kW functional model with a 1 MW offshore prototype leading to commercial offshore units rated up to 3.5 MW. SkySails' AWES is based on a foil kite controlled with one tether and a control pod which controls the lengths of kite bridles for steering the kite and changing its angle of attack. Control pod power and communication with the ground station is provided via electric cables embedded in the tether. SkySails has developed a distinctive recovery strategy. Specifically, SkySails uses high-speed winching during reel-in. The kite is then winched directly against the wind without changing the kite angle of attack. Perhaps surprisingly, this kind of recovery phase has proved to be competitive [6].

Many of the kites are similar to glider planes, although a few such as the eWind kite are very distinctive.

The upsurge of interest worldwide in airborne wind energy (AWE) is illustrated in Figure 8.3, showing institutions involved at some level in airborne R&D or prototype development and testing.

Key issues are controllability in the case of kites [7], general problems of maintaining the systems in flight for long periods and reliable power transmission to land/sea level without heavy drag-down from the power cables or tethers. Optimum cyclic kite trajectories to maximise energy capture are complex three-dimensional paths [8]. Just as limits on land-based sites and better resource offshore may justify the extension of wind technology offshore, much is made of the increased wind energy density at higher altitudes

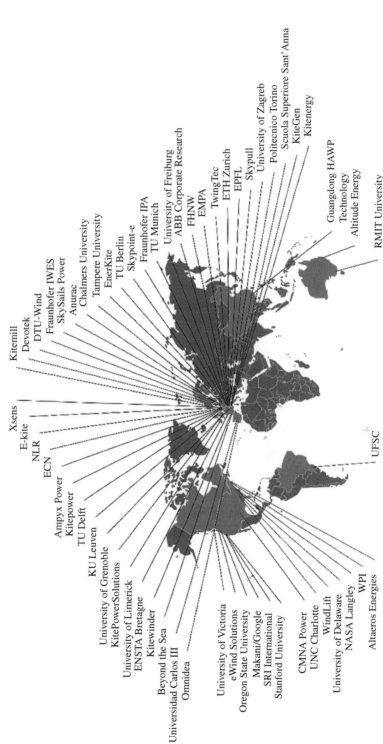

Figure 8.3 Institutions involved in airborne wind energy (2015). (*Source:* Roland Schmehl).

and reduced energy collection area per megawatt of capacity, as compensation for the obvious challenges of airborne technology. Kite systems typically aim to access wind at altitudes from a 100 to 1000 m. North Sea wind speed profiles, from the EU-Project NORSEWInD, for example, indicate clear increases in wind speed with height amounting to between 5% and 10% at 150 m, between 10% and 15% at 250 m and between 15% and 20% around 300 m relative to a reference height of 70 m.

8.4.2   KPS – Cable Tension Power Take-Off

Some systems based on mechanical transmission of power using translation of the tether to turn a cable drum aim to rotate the kite cross wind by control action perhaps employing a pair of kites. KPS, however, has realised a rotating kite with a single asymmetric kite that will naturally rotate in a circle (a helix when the tether is reeling out) transverse to its tether direction without the intervention of active control.

All systems that develop power as the product of reel-out velocity of the cable and cable tension suffer the limitation that the overall kite motion is essentially downwind. In consequence, the higher the cable velocity, the more the relative velocity and associated wind pressure on the kite is reduced. The limiting aerodynamic performance is thus well below the Betz limit. A basic actuator disc analysis (from Professor JMR Graham) follows, which is useful for preliminary design to estimate the aerodynamic performance of such a system. The analysis considers both the ground (earth) reference frame where power take-off occurs and the kite's reference frame being the only one in which the flow can be approximated as steady.

8.4.2.1   Earth Axes

This analysis considers a circular path (as in the power kite concept of KPS) which becomes a helix (Figure 8.4) as the cable reels out.

θ	angle of cable line to the horizontal
U_R	reel-out speed of cable
U_w	wind speed
U_K	kite orbiting speed
U_I	back flow due to wake

The flow is not steady in the earth frame of reference and the kite is now replaced by an actuator disc in axes moving outwards at reel-out speed. Flow is then steady in the kite frame of reference (Figure 8.5).

8.4.2.2   Kite Axes

X	axial force exerted on cable
U_n	relative axial wind speed $= U_w \cos \theta - U_R$
a	wake (inflow) induction factor
A	area of actuator disc swept by kite

Mass flow rate through actuator disc:

$$\dot{m} = \rho A (1 - a) U_n$$

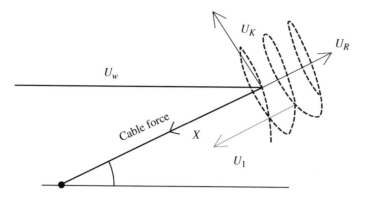

Figure 8.4 Kite motion in earth's reference frame.

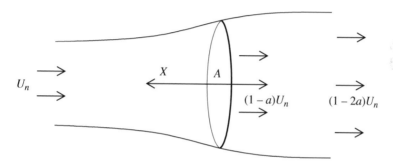

Figure 8.5 Kite in its own reference frame as an actuator disc.

$$X, (\text{momentum lost}) = 2\rho a(1-a)AU_n^2$$

$$X = 2\rho a(1-a)\left(1 - \frac{U_R}{U_w \cos\theta}\right)^2 AU_w^2\cos^2\theta \tag{8.1}$$

$$C_t = \frac{X}{(1/2)\rho AU_w^2} = 4a(1-a)\left(1 - \frac{U_R}{U_w \cos\theta}\right)^2 \cos^2\theta$$

and

$$C_p = \frac{XU_R}{(1/2)\rho AU_w^3} = 4a(1-a)\left(1 - \frac{U_R}{U_w \cos\theta}\right)^2 \left(\frac{U_R}{U_w \cos\theta}\right) \cos^3\theta \tag{8.2}$$

Power P is taken out by force X on the cable. The reduction in C_p compared to a rotor in open flow is essentially due to the cable reel-out velocity reducing the relative wind velocity (the actuator disc relationships can also be derived in earth axes).

8.4.2.3 BEM Application to the Kite as an Aerofoil Section (No Tip Loss Applied)

Consider the kite as if an element of a wind turbine blade (or as several elements for greater resolution) and as viewed in its own reference frame (Figure 8.6).

A_K	kite planform area
Lift	$L = \frac{1}{2}\rho W^2 C_L A_K$
Drag	$D = \frac{1}{2}\rho W^2 C_D A_K$

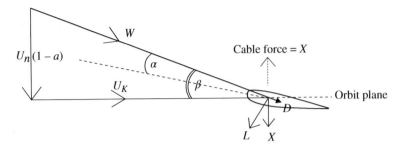

Figure 8.6 Kite as a blade element aerofoil section.

Consider the balance of forces in the 'orbit' plane since there can be no force on the kite in this direction (V_K increases or decreases automatically until this condition is satisfied). Thus,

$$L \sin \beta = D \cos \beta$$

and axially

$$X = L \cos \beta + D \sin \beta \tag{8.3}$$

$$\sin \beta = \frac{(1-a)U_n}{W}, \cos \beta = \frac{U_K}{W}, U_K = (1-a)\frac{C_L}{C_D}U_n$$

$$\Lambda = \frac{(1-a)}{\mu}\left(\cos\theta - \frac{U_R}{U_w}\right) \quad \text{where } \Lambda = \frac{U_K}{U_w} \text{ and } \mu = \frac{C_D}{C_L}$$

Λ may be interpreted as the 'tip speed ratio' of the kite.
From Equations 8.1 and 8.3

$$a = \frac{\sigma C_L(1+\mu^2)^{\frac{3}{2}}}{4\mu^2 + \sigma C_L(1+\mu^2)^{\frac{3}{2}}} \quad \text{where } \sigma = \frac{A_K}{A} \text{ ('Solidity')}$$

Equation 8.2 gives

$$C_{p(max)} = \frac{4}{27}\cos^3\theta \quad \text{when } \frac{U_R}{U_w\cos\theta} = \frac{1}{3} \text{ and } a = \frac{1}{2}\left(a_{earth\ axes} = \frac{U_I}{U_W} = \frac{1}{3}\right)$$

This indicates that the limiting power performance coefficient of the rotating kite system (or any cable reel-out power take-off system) is $<\frac{1}{4}$ of the Betz limit. This C_p definition is however based on the whole of the disc that includes the annulus of actual swept area. The kite flying in a circle crosswind is analogous to a wind turbine blade where only a region of span near the tip is aerodynamically active. With a wind turbine blade, considering the usual blade element momentum (BEM) type of analysis, each blade element would have a local C_p limited by Betz, associated with passage through a specific annular ring of swept area. In an early classic paper, Loyd [9] showed that in relation to its own area, the kite flying crosswind will augment power, compared to direct downwind motion, as $(L/D)^2$, where L/D is the lift-to-drag ratio of the kite. An excellent discussion of power limits for airborne systems by Diehl [10] complements Loyd's earlier work. The 'figure of eight', the circular and other crosswind trajectories greatly magnify the power produced compared to a direct downwind flight. Thus, the kite is quite an effective aerodynamic device in relation to its own area or the area swept in its path if less so in relation to the overall envelope area of its path.

Classic advantages of the kite lie in access to better wind resource at altitude and cheap active surface area (if loads are not too high) associated with design loads that are mainly tension loading. Compared to a conventional wind turbine in continuous rotation, the kite, if not a rigid wing, will have much lower lift-to-drag performance and an intermittent power cycle if based on reel-out power conversion.

The intermittent power cycle of a single kite will usually involve comparatively constant power being developed in the reel-out phase but zero or negative power in the retraction phase. In consequence, the average power is less than the steady maximum power and this would require uprating of power train equipment such as electronic power converters. KPS ameliorates this problem by running two kite systems on one reel so that one is always operating when the other is retracting, thereby moderating net power variations.

KPS argues that for deployment in AWE farms, a circular flight pattern would enable the densest packing and hence the minimum wind farm area. The asymmetric kite design of KPS (Figure 8.7) has been tested at 30 kW scale [11].

Aerodynamic optimisation of the kite's efficiency and stability in turning flight was carried out by Aerotrope (UK) using panel methods and computational fluid dynamics (CFD) software with helical wake modelling in order to capture the quasi-rotor behaviour of this flight mode.

The first stage of aerodynamic optimisation followed a process similar to a classic horizontal-axis wind turbine (HAWT) blade, resulting in a tapered, twisted planform – the first level of asymmetry compared to a standard symmetric kite design. This geometry aimed to maximise energy capture but was not self-stable in circular flight, demanding constant control inputs and hence incurring losses in turning. The

Figure 8.7 Asymmetric kite of Kite Power Solutions and aerodynamic modelling (inset).

second stage of optimisation involved further asymmetric manipulation of the kite's shape to provide a fully balanced condition while circling. In physical testing, this arrangement has proved to be very stable under load in circular flight, that is, during the reel-out phase. Control inputs via the bridles are needed to exit the turn, fly straight or to counter-turn. These manoeuvres do reduce the flying efficiency but are only used during the return phase and so do not have a significant impact on net energy capture.

8.4.3 Daisy Kite – Rotary Power Transmission

It may seem improbable that the tether line could be used for torque transmission. However, this is achieved in the Daisy Kite design of Windswept and Interesting Ltd (Figure 8.8) by creating a tether system of lines spaced apart by light carbon rod structures.

The diagonal tension lines in Figure 8.8(c) provide the equivalent of shear stress bearing capacity that would be necessary in a beam carrying torsion loads. The length of tether that is viable in this form as a torque tube may restrict the operational altitude and associated benefit of accessing higher mean wind speeds more so than with other concepts. On the other hand, the Daisy Kite in operation is a wind turbine approximately fixed in space at a yaw angle, θ to the wind and the limiting power coefficient, in theory at least, is then $C_p = (16/27)\cos^3\theta$, which is four times greater than with the reel-out

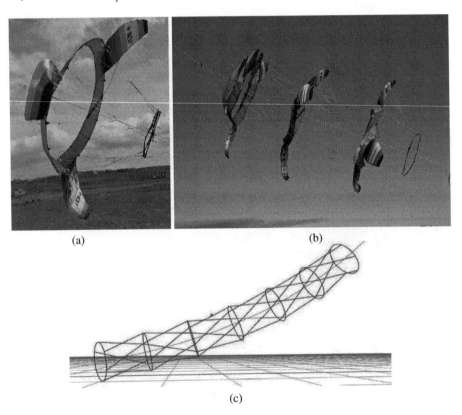

(a)

(b)

(c)

Figure 8.8 The Daisy Kite of Windswept and Interesting Ltd.

concept. Moreover, the land or sea area and areas at altitude associated with a wind farm of such kites may be minimal compared to other concepts. Windswept have cooperated with Benhaïem [12] who has been developing systems similar to the extent of having a kite rotor with soft or rigid blades and with rotary power transmission through the tether system.

Both parties are at an early development stage experimenting with a great variety of arrangements with soft and rigid components and including multi-rotors (Figure 8.8(b)) which may assist in increasing power and communicating torque to base level. A line from rotor centre to the left of the upper figures in Figure 8.8 attaches to a conventional kite (not shown) at a higher level which assists in supporting the weight of the kite-rotor(s) and tether system.

8.4.4 Omnidea – Rotating Cylindrical Balloon as a Lifting Body

Among the great variety of concepts for airborne wind power, Magenn [13] (no longer in business) had proposed a system involving a cylindrical buoyant balloon with fins which enabled it auto rotate in the wind and provide lift via the Magnus effect so as to counteract drag down from wind forces. Power was generated at altitude and transmitted to the ground through the tether system. The Portuguese company, Omnidea, is developing a system with a rotating, cylindrical, helium-filled balloon [14]. However, the company's idea is to power the rotation actively with small electric motors and to use the consequent Magnus lift with cable reel-out power transmission. A high design lift coefficient (Cl ~ 3) enables high cable tension forces to be developed in relation to the balloon frontal area. Powered rotation enables excellent control of cable tension. Consumption is about 500 W for a prototype system nominally rated at 6.5 kW with a cylinder of 16 m length and 2.5 m diameter.

8.4.5 Makani

In the early 1990s, the Oxford-based inventor, Colin Jack, patented AWE ideas [15] based on the autogyro principle where some of the lift of the rotor is used to support its weight. This was prompted by the realisation that a balloon capable of supporting the weight of a wind turbine rotor in power-producing mode requires much more buoyancy than one able to support the dead weight under calm conditions. This arises from the 'drag-down' effect of rotor thrust when the turbine is operating. Jack's ideas anticipated generation at altitude concepts using the tether as an electrical conductor.

Makani uses a carbon fibre tether with an aluminium conductor to transmit power to the ground. With an eight-rotor, 600 kW prototype in operation, Makani may be considered as the current industry leader in AWE. The concept is of a high performance wing which operates in a circular path ~145 m diameter at altitudes in a range from 100 to 400 m. The wing is flown crosswind using the rotors as turbines extracting energy from the relative wind flow. As with many conventional wind turbines, cut-in wind speed is around 4 m/s and rated wind speed at 11.5 m/s.

Multi-rotor systems are in evidence both in the design of drones and in AWE systems. Apart from obvious advantages with multi-rotors of redundancy in the event of one or more rotor failures and enhanced controllability in flight, the same scaling arguments presented in Chapter 15 apply but even more so when minimising mass of the airborne system is at a premium.

8.4.6 Airborne Conclusions

The target applications of airborne systems vary considerably ranging from grid-connected wind farms onshore and offshore to small-scale autonomous or wind-diesel systems. At any given power rating, the airborne system can be much lighter and transportable than standard wind turbines and better suited to deployment in remote sites. This opens up some niche markets such as for mobile military applications or as temporary power supplies for construction activities, and so on. Systems such as TwingTec [16] are designed as standard container size, sometimes with the parked container functional as the ground station.

Overall, airborne wind is preparing to take flight with numerous small-scale prototypes under test, pre-commercial companies in operation and widespread international research in progress. The state of the art is very well summarised in the book *Airborne Wind Energy* [17].

8.5 Future Wind Technology – Energy Storage

8.5.1 Types of Energy Storage

Energy storage concepts are not at all new. Batteries, for example, were developed first by Volta around 1800 and used towards the end of that century in the first electricity wind generation systems discussed in Section 0.4. The development of systems that can store energy over various useful timescale ranges and work as an effective and economic system in conjunction with renewable energy generation is still a challenge which drives ongoing research and innovation. Many renewables are highly controllable over short timescales when the applicable resource is available. An economic combination of effective storage and energy recovery can at the very least add value to renewables and, with sufficient storage capacity, enable renewables to serve as base load.

There are many options for energy storage systems. For example, certain organometallic molecules undergo a reaction upon exposure to light that is reversible [18] with either a catalyst or heat. In some cases, a considerable amount of energy can be stored, as was shown, for example, in the work of Vollhardt in 1996 for fulvalenes. However, the main types of energy storage presently being researched and developed comprise the following:

- Battery storage
- Gas pressure storage
- Compressed air storage
- Flywheel energy storage
- Thermal energy storage

A few examples of each type of system are now discussed.

8.5.2 Battery Storage

GE Wind has introduced battery storage as integral in the power management of some of their wind turbines. According to their website [19]

Over periods of 15-60 minutes, the Predictable Power application smoothes out any short-term wind peaks and valleys, making it more consistent, and more predictable. The application can connect to the converters of both 1 & 2 MW and 3 MW product lines. ... The system is available for new units and can be retrofitted to most existing GE turbines.

Statoil has a new battery storage solution for offshore wind, Batwind, which will be piloted in the Hywind Scotland offshore project (see Section 7.8). Batwind is being developed in cooperation with Scottish universities and suppliers, under a Memorandum of Understanding between Statoil, the Scottish Government, the Offshore Renewable Energy (ORE) Catapult and Scottish Enterprise. Statoil plans to install a 1 MWh lithium-battery-based storage system in late 2018.

8.5.3 Gas Pressure Storage

Gas pressure storage uses a closed circuit of pressurised gas that is subject to pressure and temperature changes via compressors and expanders to allow energy to be extracted in the form of heat. In a recent innovative idea, the power transmission system WIND-TP [20] takes power directly from the main rotor and delivers it to the terminals of an electrical generator. In that sense, it competes directly with several other technical options. Energy taken directly from the wind turbine can be put into storage if not required immediately and recovered later when needed.

WIND-TP uses a gas circulating in a closed circuit to transmit power. A compressor in the nacelle of each wind turbine compresses the gas, raising its pressure and temperature. One expander receives the gas from a group of wind turbines and extracts the work from that gas, causing its pressure and temperature to fall. Gas in the circuit always circulates in the same direction. Energy is put into storage by extracting heat from the gas immediately before it enters the expander and extracting cold from that gas immediately after leaving the expander. Energy is recovered from storage in a reverse process, putting heat back into the gas immediately before the expander and putting cold into the gas immediately after.

8.5.4 Compressed Air Storage

Compressed air energy storage (CAES) plants are similar to pumped-hydro power plants in their applications, output and storage capacity. In the present diabatic method, in place of pumping water from a lower to an upper reservoir during periods of excess power, ambient air is compressed, cooled and stored under pressure typically in an underground cavern, although alternatives including ocean storage in bags in the high-pressure environment at water depth is also being researched [21]. When electricity is required, the pressurised air is heated and expanded in an expansion turbine driving a generator for power production.

In an adiabatic compression to ~70 bar, a large temperature increase would take place and ideally the heat of compression should be stored and recovered. A much higher efficiency of up to 70% may then be achieved if the heat of compression is used to reheat the compressed air during turbine operations avoiding the need to burn natural gas to warm up the decompressed air. An international consortium headed by the German energy company RWE is currently working on this and a pilot plant is scheduled to start operations in 2018.

The only two existing CAES plants in Huntorf, Germany, and in McIntosh, Alabama, in the United States, as well as all the new plants being planned in the foreseeable future are based on the diabatic method. In principle, these plants are essentially just conventional gas turbines, but where the compression of the combustion air is separated from and independent to the actual gas turbine process. In a conventional process, the compression stage normally uses up about two-thirds of the turbine capacity, whereas the CAES turbine – unhindered by the compression work – can generate three times the output for the same natural gas input. This reduces gas consumption and associated carbon dioxide emissions by around 40–60%. The power-to-power efficiency is ~42% without, and 55% with waste heat utilisation.

8.5.5 Flywheel Energy Storage

The concept of storing energy in a rotating disc dates as far back as 2400 BC when rotating wheels were used by Egyptians to handcraft pottery. In the Industrial Revolution of the eighteenth century, metal flywheels were developed and used, for example, in steam engines.

A flywheel can store rotational kinetic energy of an amount $(1/2)I\omega^2$, where I is the mass moment of inertia and ω is the angular speed of rotation. Initially, flywheels focused on concentrating mass (say, cast iron) in the rim, with angular speed limited by stress in the wheel structure, tension in the rim and by windage losses associated with air resistance. Recognising that the speed dependence of stored energy is a square law whereas the mass dependence is only linear, modern high performance flywheels are made of lightweight ultra-strong materials such as carbon fibre rotating at very high speed in a sealed evacuated chamber in order to avoid windage loss. Advanced bearing technology such as magnetic bearings is employed to minimise bearing friction losses.

Beacon Power [26], a leading designer of flywheel storage for the energy sector, has developed just such a flywheel. The rotor assembly spins up to 16 000 rpm enclosed in a sealed vacuum chamber minimising friction using a magnetic system to lift and support the rotor.

8.5.6 Thermal Energy Storage

Siemen's engineer, Till Barmeier, leads a team developing a hot rock bed thermal store system. In this concept, surplus energy is used to generate hot air which heats up pebbles able to withstand temperatures of over 600 °C. To recover stored energy, cold air is pumped through the pebble store and heated to raise high-pressure steam for turbines generating electricity in the conventional way. The main goal of the storage solution is to be able to continuously generate electricity for 2–3 days during energy shortages. Over an area comparable to that required for water reservoirs of pumped storage power stations, the thermal store is estimated to be able to store over 30 times as much energy. Demonstrator plant will be built over the next 2 years with a view to construction of a pilot plant with an output of around 30 MW. As part of a much lower carbon future, with only a heater and heat storage unit such as Siemens is developing, a fossil power plant could then be converted to integrate with renewable energy supply.

8.6 Innovative Energy Conversion Solutions

8.6.1 Electrostatic Generator

Electrostatic generators, also referred to as electro-hydrodynamic (EHD) on account of interaction with wind flow, have been considered for wind energy conversion since the 1970s. A recent example is the EWICON concept under development in the High Voltage Laboratory at TU Delft. Charged droplets are released into the wind and create DC current in collector wires. This system is presently at the laboratory feasibility stage, being tested at milliwatt scale. The considerable attraction is to have a system with few mechanical parts. It is seen as potentially suitable for buildings, having minimal noise and vibration, or for offshore on account of simplicity and potential to be highly reliable.

A key issue with this system is the power demand to charge the droplets. Upscaling to useful power capacities from micro laboratory size is thought to benefit performance significantly but has yet to be demonstrated. According to Djairam *et al.* [22], a conversion of 7% of the wind energy into electrical energy has been measured in tests with improvements suggested that could lead to an efficiency of the EWICON system in the range of 25–30%. To avoid the power consumed in charging feed particles, it may be possible to use natural wind flow to agitate particles in a container which can become electrostatically charged through tribological effects. This was the principle used in one system for an aircraft to detect and measure levels of fine volcanic ash [27].

Figure 8.9 from the EWICON study shows a simulation of the particle flow field in wind of 6 m/s.

A closely related concept of Carmein and White [23] is shown in Figure 8.10 extracted from their US patent application. The same principle of using the wind to move charged particles against an electric field and thereby increase their potential is evident. The company Accio Energy was since formed to develop this concept. In late 2015, the Advanced Research Projects Agency-Energy of the US Department of Energy awarded

Figure 8.9 EWICON concept – charged particle flow simulation. Reproduced with permission of Dhiradj Djairam.

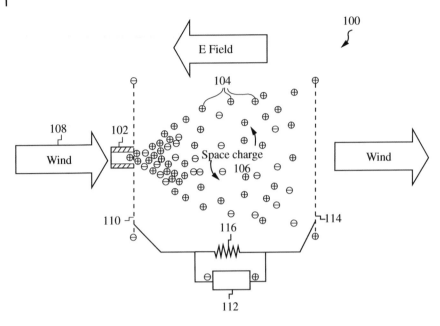

Figure 8.10 Electro hydrodynamic wind energy system.

$4.5 million in funding to support Accio Energy's work [24] in developing utility-scale EHD wind-power generation systems, specifically for the offshore market.

The EHD concept has the obvious attraction of avoiding mechanical moving parts and the key to it being successful is in control of the complex electric fields and hence achieving high enough overall efficiency of power conversion. Accio Energy has apparently made useful progress.

8.6.2 Vibrating Column

Static cylindrical towers (chimneys, cooling towers at power stations, etc.) shed vortices when the wind blows and may vibrate excessively if the vortex shedding frequency is resonant with a low-order natural frequency of the tower. Usually, measures such as adding spiralling strakes round the tower are employed to inhibit such vibration. A small Spanish company [28] has conceived a system comprising a tower which is purposefully designed to exploit vortex shedding and is designed to vibrate in the wind with magnets and coils through linear motion generators directly extracting electricity. The basic feasibility of this concept is not in question [25], but its value cannot really be assessed until the structural demands and cost implications of design for long life endurance in operation with the self-induced fatigue loading are known.

References

1 Jamieson, P. (2008) Getting out of gear and into magnets. *Wind Power Monthly*, November, 61–69.
2 Krohn, S., Morthorst, P.-E. and Awerbuch, S. (2009) The Economics of Wind Energy. EWEA Report.

3 AWEA (2010) American Wind Energy Association 4th Quarter 2009 Market Report, January 2010.

4 Morthorst, P.-E. (2009) *The Economics of Wind Power – Status and Perspectives.* Pre-conference Seminar, EWEC 2009, Marseille, March 2009.

5 Cherubini, A., Papini, A., Vertechy, R. and Fontana, M. (2015) Airborne wind energy systems: a review of the technologies. *Renewable and Sustainable Energy Reviews,* **51**, 1461–1476.

6 Erhard, M. and Strauch, H. (2015) Flight control of tethered kites in autonomous pumping cycles for airborne wind energy. *Control Engineering Practice,* **40**, 13–26.

7 Canale, M., Fagiano, L., Milanese, M. and Ippolito, M. (2007) *KiteGen Project: Control as Key Technology for a Quantum Leap in Wind Energy Generators.* Pre-Proceedings of American Control Conference, New York.

8 Williams, P., Lansdorp, B. and Ockels, W. (2007) *Optimal Cross-Wind Towing and Power Generation with Tethered Kites.* AIAA Guidance, Navigation and Control Conference, Hilton Head, South Carolina, August 2007.

9 Loyd, M.L. (1980) Crosswind kite power. *J. Energy,* **4** (3), 106–111, Article No. 80-4075.

10 Diehl, M. (2013) *Airborne Wind Energy: Basic Concepts and Physical Foundations,* http://homes.esat.kuleuven.be/~highwind/wp-content/uploads/2013/08/Diehl2013a .pdf (as viewed January 2017).

11 Hardy, J.W. (Inventor) (2012) A kite for a system for extracting energy from the wind. European Patent Specification, EP 2748064 B1 and WO 2013/026746, Kite Power Solutions Ltd, filed August 14th 2012.

12 Benhaïem, P. (2013) *Land and Space Used.* Airborne Wind Energy Conference 2013, Book of abstracts, p. 59.

13 Kamini, N.S. and Mohini, D.D. (2012) Magenn air rotor system (Mars). *International Journal of Engineering Research and Applications (IJERA),* **2** (6), 1566–1568. ISSN: 2248-9622.

14 Omnidea (website), http://omnidea.net/site/index.php/research/wind-energy (accessed 26 September 2017).

15 Jack, C.H.B. (Inventor and proprietor) (1992) Free rotor. Patent WO1992020917 A1, published November 1992.

16 TwingTec website. http://twingtec.ch/viewed January 2017.

17 Ahrens, U., Diehl, M. and Schmehl, R. (eds) (2014) *Airborne Wind Energy,* Springer-Verlag, Berlin Heidelberg.

18 Boese, R.J., Cammack, K., Matzger, A.J. *et al.* (1997) Photochemistry of (Fulvalene) tetracarbonyldiruthenium and its derivatives: efficient light energy storage devices. *Journal of the American Chemistry Society,* **119**, 6757–6773. doi: 10.1021/ja9707062

19 GE Renewable Energy GE Wind, https://www.gerenewableenergy.com/wind-energy/ technology/integrated-energy-storage-for-wind-turbines.html (viewed January 2017).

20 WIND-TP, http://nimrod-energy.com/tech_wind-tp.php viewed (January 2017).

21 Garvey S. (inventor). Ocean storage, https://www.theengineer.co.uk/issues/25-april-2011/compressed-air-energy-storage-has-bags-of-potential/ (accessed 26 September 2017).

22 Djairam, D., Hubacz, A.N., Morshuis, P.H.F. *et al.* (2005) *The Development of an Electrostatic Wind Energy Converter (EWICON).* International Conference on Future Power Systems, Amsterdam, ISBN: 90-78205-02-4.

23 Carmein, D. and White, D. (inventors) (2009) US Patent 0, 218,910 A1, Sept. 3, 2009.

24 Accio Energy website, http://www.accioenergy.com/viewed (January 2017).

25 Indiegogo, https://www.indiegogo.com/projects/vortex-bladeless-a-wind-generator-without-blades--3#/ (accessed 26 September 2017).

26 Beacon Power website, http://beaconpower.com/viewed (January 2017).

27 Welland, E.W., Atif, A. and Ganney, I.J. (2014) Sensing system (for volcanic ash). Patent US20140157872 A1, PCT/GB2012/051890, published June 12, 2014.

28 Cajas, J.C., Houzeaux, D.J., Yanez, D.J. and Mónica, M.-T. (2015) SHAPE project Vortex Bladeless: parallel multi-code coupling for fluid-structure interaction in wind energy generation.

Part II

Technology Evaluation

9

Cost of Energy

9.1 The Approach to Cost of Energy

The key cost question about an innovation is

> If an innovation is introduced, will it in mature production realise significant cost benefit?

This recognises, as is evidenced in general learning curves for technologies, that a prototype may be a factor of four times more expensive [1] than the cost that can be realised in established production. Thus, the merit of an innovation needs to be judged in terms of the potential benefit when it is at a similar stage of technological maturity to the standard solution. Before developing the evaluation process in which new technology is compared to a representative conventional system, a key metric in such an evaluation, the cost of energy (COE) is discussed.

COE is not the only metric by which a wind turbine system may be evaluated, but, in evaluating innovation, it is usually the most appropriate one. The cost in COE is the cost of making the turbine, erecting and commissioning it, providing necessary infrastructure at the wind farm site and operating and maintaining the turbine over its design life. The cost of de-commissioning should also be accounted for, although quite often it is not. Energy, more obviously, is the total electrical energy output of the turbine over its design life. A schematic overview of COE components is presented in Figure 9.1.

In the context of evaluating innovation, the objective is not directly to meet a specific COE target, although, within the politics of public funding of innovation, target figures will be set from time to time. Such target figures are transitory, changing with time and varying with market area. Instead, the prime use for COE in evaluating innovation is to determine whether a new system or component (the innovation), essentially on its technical merit, will offer significantly reduced COE. The best approach is then to conduct an evaluation that puts the innovation in a direct comparison with an established state-of-the-art design – the baseline wind turbine of Chapter 10. In view of the costs of implementing design changes and time to market with technology changes, it is important that the projected reduction in COE is significant enough to compensate for the disruption of innovation.

In the financing and operation of wind farm projects, there is often much more emphasis not on the COE but on the market value of energy that sets repayment rates and determines income stream. COE calculations can sometimes conveniently

Innovation in Wind Turbine Design, Second Edition. Peter Jamieson.
© 2018 John Wiley & Sons Ltd. Published 2018 by John Wiley & Sons Ltd.

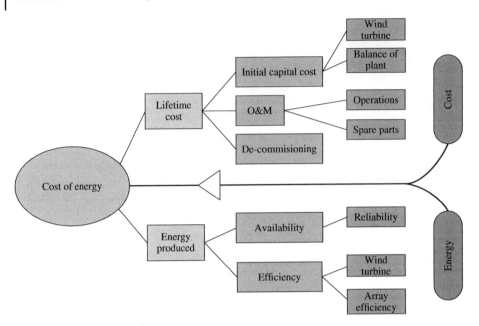

Figure 9.1 Cost of energy overview.

be short circuited considering calculations involving net present value (NPV) and capitalisation of losses based on prevailing repayment prices for energy and relevant discount rates. This can be very useful for evaluating components, say, in comparing types of generators with different capital costs and efficiencies, but it does not put the component fully in context in terms of its significance in relation to lifetime costs. For evaluation of innovation, especially initial top-level evaluation, COE models are more appropriate, the National Renewable Energy Laboratory (NREL) type of financial model [2] being generally very suitable. The International Energy Agency (IEA) had also in the early 1990s developed a standard [3] with similar financial modelling for COE calculation which has useful treatment of uncertainties and sensitivities and their work on wind energy economic issues is ongoing [4].

The cost of wind turbine components is not easily obtained, nor are such costs in any sense stable absolute values. They vary with time and with a huge variety of market influences, for example, costs of raw materials, labour market fluctuations and exchange rates. Moreover, the costs of wind turbine components, at least particularly useful and relevant costs, are generally not freely available. Prices that may be quoted to an individual or consultant may differ substantially among potential suppliers and will rarely reflect the price at which the component may be sold competitively into a project.

The way out of these difficulties is to start from two basic kinds of information. A prime input is cost per kilowatt at which turbines are sold into commercial projects. Data for specific projects may be highly confidential but, from publications of wind energy associations and other sources, representative figures can usually be obtained.

The second vital piece of information is the relative split of costs among turbine components. As it is not linked to specific costs or price data, this sort of information is available for some wind turbine designs or as more generic estimates [5, 6]. Given the

Table 9.1 Representative cost split.

Component	Cost fraction
Blades	0.177
Hub	0.077
Gearbox	0.143
Generator	0.076
Yaw system	0.019
Nacelle cover	0.020
Nacelle structure	0.040
Tower	0.219
Variable speed system	0.073
Pitch system	0.043
Rotor brake	0.006
Couplings	0.003
Shaft	0.041
Other	0.063
Total turbine	1.000

variety of designs, there is no unique list of turbine components and frequently different items are included in categories such as 'hydraulic systems', 'control systems', and so on. Difficulties of categorisation not only have complicated both the attribution of costs within a wind turbine system but also the analysis of failure statistics. Nevertheless, the principal components of a typical state-of-the art wind turbine that will represent the baseline concept discussed previously can reasonably be identified.

Table 9.1 shows such a typical split based on data from the WindPACT study [7]. The cost split and project cost per kilowatt must be centred on an appropriate scale of wind turbine. In recent years. 1.5 and 2 MW turbines have been at the centre of the market place. Such a cost split can not only accurately reflect a specific design but may also be a reasonable approximation for a range of mainstream designs. The aim is not to arrive at refined cost estimates for components but to develop overview perspectives, recognising, for example, that major percentage cost reductions in blades or gearbox will have greater impact than, for example, cost reductions in a high-speed generator or yaw system.

This cost split, the design types sold into projects and obviously the associated cost per kilowatt will change with time. A COE model using this type of input can reflect such changes whilst maintaining a consistent structure to answer the question how an innovative system to be evaluated compares with a state-of-the-art system.

Thus, the input cost and cost split information is never absolute and needs to be refreshed periodically from whatever sources are available to inform it. In the context of evaluating innovation, these ingredients then provide the basis of a cost comparison of a system that may have a number of common components with the baseline but also some other innovative ones. In the case where an innovative wind turbine system is radically different, there may be little in common with the standard concept. In such a case, the cost split is less consequential and some effort must be made to build cost evaluation of the innovative system from the ground upwards.

The cost split approach is never expected to yield high accuracy but then high accuracy scarcely exists in the context of generic costs.

The cost of, for example, a yaw system based on a cost split and cost per kilowatt may be somewhat inaccurate if compared with the real, but usually inaccessible, cost data known to the supplier in a particular project. Suppose the aim is to evaluate whether a new gearbox concept will have much impact on the COE and it is determined that the new concept has no impact on yaw system loads or cost. Then the more important issue with regard to yaw system cost is not absolute accuracy but to ensure that the innovative system (evaluated in mature production) is assigned a yaw system cost value that is generally of the correct order and is exactly the same as that of the chosen state-of-the-art baseline wind turbine.

9.2 Energy: the Power Curve

The energy output of a wind turbine is naturally the result of its power performance over a period. Thus, in order to address energy capture, wind turbine power curve characteristics need to be considered.

The power curve of a large electricity-generating wind turbine has the typical forms of Figure 9.2. A wind turbine usually commences power production operation at a minimum (cut-in) wind speed or when the wind drops below a safe operating maximum wind speed (cut-out). The minimum wind speed is determined by losses at low loads and the capability of the system to produce positive power. This is also influenced by the speed control strategy, with the lowest cut-in wind speeds being achieved by variable speed wind turbines with a wide speed range. Power production is stopped in high wind speeds typically around 25 m/s to protect the machine from excessive loading. In the actual control of startup and shut-down procedure, it is common to consider not only averages of wind speed but also gust levels. Hysteresis is also introduced into the control strategies, making the exact cut-in and cut-out wind speeds a little different from the nominal low-wind and high-wind values to avoid over frequent switching from operational to non-operational states when the wind speed is around the critical levels.

It is apparent (Figure 9.2) that, although power in the wind is rising as the cube of wind speed, the electrical output power reaches a limiting maximum at much lower wind speeds than the cut-out wind speed. In the case of a stall-regulated wind turbine, this limit is maintained passively by the inherent rotor aerodynamic properties relying on the generator and network to provide reaction torque that will hold the wind turbine speed constant. With pitch-regulated wind turbines, the power can be held closely constant at the rated value by adjustment of blade pitch angle. The rated wind speed, the lowest steady-state value at which the pitch-regulated wind turbine can produce rated power, is typically around 11–12 m/s.

Why does the power curve have its characteristic form with cut-in, cut-out and rated power when the input wind power rises cubically over the operational range? The key is the wind distribution which is overlaid in Figure 9.2. This particular distribution is a typical design distribution (Rayleigh distribution) with an annual mean wind speed of 8 m/s. If the turbine produces rated power in a wind of, say, 12 m/s, it could, in principle,

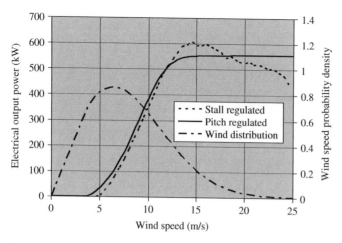

Figure 9.2 Typical power curves of stall- and pitch-regulated wind turbines.

produce $(25/12)^3 = 9$ times that power at a cut-out wind speed of 25 m/s. However, to install a drive train uprated by a factor of 9 would only add a huge cost premium to the system. The part load efficiency of the drive train would be poor in the most frequent wind speed ranges that contribute most to energy output. The system would cost a great deal more (a drive train is typically around 40% of the capital cost of a wind turbine) and will not produce sufficient extra energy considering the comparative infrequency of wind speeds in the tail of the distribution above, say, 15 m/s. This qualitative explanation of the basis of setting the rated power of a wind turbine is supported by an outline quantitative evaluation in Section 10.7 following the methods of COE analysis developed in this chapter.

An old rule of thumb suggests that rated wind speed should be about a factor of 1.5 greater than the site mean wind speed. This sort of relationship is logical on two counts. Power in the wind increases rapidly with wind speed and therefore the most important wind speeds for power production are somewhat above the mean wind speed. Moreover, the higher the mean wind speed of the design site, the more power will be in the higher wind speed ranges and the greater the justification for a higher value of rated power. These issues are equally reflected in having larger diameter rotors in relation to rated power for wind turbines on low wind speed sites. It will be apparent that within an overall trend of rated power varying approximately as square[1] of diameter (Figure 9.3), there are many deviations related to different site conditions and also manufacturers' design decisions and the way in which they have evolved larger wind turbines.

In the context of innovation, it is important to appreciate that the relationship between diameter and rating is fundamentally an economic one – a trade-off between energy output (related to the square of diameter, wind distribution and rated power level) and the costs of the power transmissions system. Whilst these relationships have

---

1 Power as square of diameter, D^2, seems very plausible. However, on land-based sites, because of wind shear effects and increase of mean wind speed with hub height, power varies as $D^{2+3\alpha}$ where α is the wind shear exponent. About 10 years ago, the exponent of diameter in a chart like Figure 9.3 was $\approx$2.4 and recent trends towards larger diameters in relation to rated power of turbines in 1.5–3 MW range have caused the appearance of a square law characteristic.

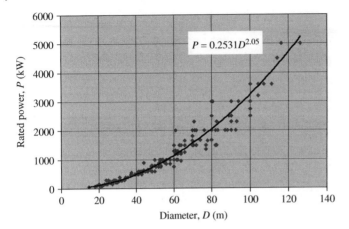

Figure 9.3 Variation of rated power with diameter.

much converged in mainstream design (Figure 9.3), some innovations can completely reopen the issues around this optimisation.

It is usual for manufacturers to publish a power curve characterising their machine and, moreover, a power curve with some comfort margin that will be guaranteed to be met for purposes of commercial trading. It is easy to take the power curve for granted and important to appreciate what underlies it. Terms such as 'wind speed' and 'wind direction' are commonplace and infiltrate daily weather reports. Yet in the three-dimensional turbulent wind field that is real wind, they have no precise meaning unless they are referenced to spatial extents and temporal averaging periods. When this is considered, there is not one power curve but a family of power curves depending on the averaging choices.

Firstly, define an ideal power curve as one that results if there is a steady uniform wind flow for each wind speed and associated power measurement.

The power curve is typically concave (as viewed from the top of the page), in the low wind speed range. The variability of wind is most usually measured by turbulence intensity (TI) defined as the ratio of the standard deviation of the wind speed to the mean wind speed – often expressed as a percentage. In the presence of wind turbulence over a period (the averaging period) where the mean wind speed has a certain value, there will be more power associated with excursions to higher wind speeds than to lower ones, and this has the effect of producing an average power curve value (over the chosen averaging period) that is above the ideal steady state value. An opposite effect will obtain where the power curve is convex. Both effects are evident in Figure 9.4, where an ideal steady-state power curve is compared with the power curve that results in 5%, 10%, 15% and 20% TI.

In general, the more extended the averaging period, the more any curvatures that exist in the instantaneous power curve are apparently reduced in a power curve associated with the chosen averaging period. The comparisons of Figure 9.4 are further complicated by the fact that the measure of TI also depends on averaging period. Preferred averaging periods for power curve measurements are typically around 5–10 min. This is in the so-called spectral gap of the van der Hoven wind spectrum [8], where there is comparatively low energy in the wind spectrum between more energetic short-term wind fluctuations and longer term ones associated with the passage of weather fronts.

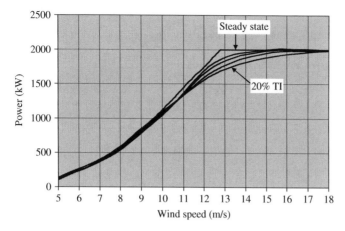

Figure 9.4 Effect of wind turbulence on power performance.

The practical importance of these issues around averaging periods is that the guaranteed power curve must be one based on averaging periods similar to any wind speed measurements used to verify productivity. To be clear, the amount of energy produced by a given wind turbine with given operational strategy under given turbulent wind conditions is unique and in no way depends on averaging periods. However, predictions of power performance and measurements of wind statistics do and this is the crux.

The control system is active in operation around rated wind speed in the operation of a typical pitch-regulated, variable speed wind turbine. Control of the wind turbine is effected by a rapid reaction of the generator to torque fluctuations and a slower reaction of the pitch system to limit speed variations. Typically, the closed-loop pitch control system at rated wind speed operates to maximise energy capture and to avoid excessive power levels. It may also seek to minimise excitation of the support structure. Thus, in principle, in addition to turbulent wind disturbances in combination with the inherent steady-state characteristic of the wind turbine, the control strategy will influence the power produced over a given averaging period. In general, the energy produced by the machine over a given time can be simulated using a model of the wind turbine that accounts for wind turbulence, structural dynamics, drive train dynamics and control system dynamics.

The following simple analysis, considering turbulent wind variations as a Gaussian process, illustrates the effect of turbulence on a power curve at wind speeds around rated.

Denoting:

V_r	Rated wind speed (m/s)
Σ	Standard deviation of wind speed (m/s)
P_r	Rated power
V_c	Cut in wind speed (m/s)
V_S	Shut-down wind speed (m/s)
$P(V)$	Power at wind speed V (kW)

and assuming a normal distribution of wind speed about the mean value, V, the expected value of the power at rated wind speed, $V = V_r$ and $P = P_r$ is

$$E(P_r) = \frac{1}{\sqrt{2\pi\sigma}} \int_{V_c}^{V_s} P(U)e^{\left\{-\frac{(U-V_r)^2}{2\sigma^2}\right\}} dU \qquad (9.1)$$

If the power curve is assumed to be linear approaching rated wind speed and constant above rated wind speed (Figure 9.5), the power, $P(U)$, may be modelled as

$$P(U) = \begin{vmatrix} kU, & U < V_r \\ P_r, & V_r \le U \end{vmatrix} \qquad (9.2)$$

Integration of Equation 9.1, with $P(U)$ defined as in Equation 9.2 leads to a simple result:

$$E(P_r) \cong P_r - \frac{k\sigma}{\sqrt{2\pi}} = P_r - \frac{kIV_r}{\sqrt{2\pi}} \qquad (9.3)$$

where the turbulence intensity, I, is defined as $I = \sigma/V$, so that at rated wind speed, $\sigma = IV_r$.

The interpretation of Equation 9.3 is that, for a typical pitch-regulated wind turbine with an ideal steady-state power curve characteristic such as in Figure 9.5, the apparent power produced at rated wind speed in turbulent wind is reduced in proportion to the intensity of turbulence and the magnitude of the power curve gradient, k, approaching rated wind speed.

The power curves of Figure 9.4 are based on averages from 10-min simulations that include the structural, drive-train and control system dynamics of a specific wind turbine design. Interestingly, the predictions of Equation 9.3 for the power at rated wind speed (Table 9.2) were found to be generally within 4% of the simulation results.

This validation of Equation 9.3 is limited but at least it provides an extremely simple formula illustrating, at least qualitatively, the typical impact of wind turbulence on a power curve at rated wind speed. It is rather surprising that such a simple formula can even get close to a more rigorous prediction. Perhaps, provided that a wind turbine has a control system that effectively tracks the appropriate set point and reasonably well-behaved drive train and structural dynamics, then, irrespective of the detailed nature of the turbine and control system, there is convergence in the power performance characteristics in turbulent wind.

Many issues arise around the predicted and measured performance of wind turbines. Companies such as GL Windtest (now part of DNV GL) have specialised in conducting such measurements as a service to the industry. International standards have been set [9] regarding the quality of instrumentation, extent of data, procedures and site conditions that are acceptable for valid power curve characterisation. Considering the high sensitivity of power to wind speed, it is always critical, whatever the test set-up, that the reference anemometer on which the power performance measurement is based is reliably correlated

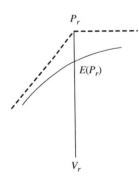

Figure 9.5 Effect of wind turbulence at rated power.

Table 9.2 Comparison of Equation 9.3 with simulation results.

I	Simulation results	Equation 9.3	Deviation (%)
0.00	2000	2000	0.0
0.05	1863	1920	3.1
0.10	1810	1841	1.7
0.15	1742	1762	1.2
0.20	1690	1683	−0.4

with wind speed at the exact location of the wind turbine under test. Ideally, this is done before the wind turbine is erected. In the context of minimising COE, the power curve performance is clearly crucial and aerodynamic design development is now much focused both on increasing aerodynamic efficiency and managing loads more effectively so that larger rotor diameter rotors can be employed to increase energy output.

RES Limited, commencing December 2012, organised a Power Curve Working Group (PCWG) as a broad industry group encompassing developers, consultants, manufacturers, academics and researchers exploring all issues affecting accurate prediction of wind farm power production.

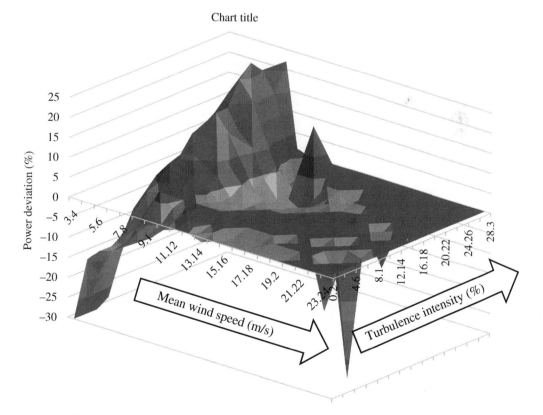

Figure 9.6 Effect of mean wind speed and wind turbulence on deviation from expected power.

Figure 9.6 is derived from data of Senvion presented by Tomas Blodau to the PCWG and at the EWEA conference of 2014. The percentage deviation of power from the expected steady-state performance is illustrated as a function on both mean wind speed and TI. In the wind speed range below rated wind speed, power is rising almost cubically and power gains increase strongly with TI. Above rated wind speed, the control system, regardless of the external conditions, is striving to maintain constant power and the much flatter landscape (darkest area) with generally small deviations reflects that. The large power deficits in very low wind speeds with low turbulence levels may be associated with the cut-in behaviour of wind turbines with less prolonged periods of cut-out when the wind is more turbulent.

9.3   Energy: Efficiency, Reliability, Availability

Energy can be considered the prime value and energy capture the purpose of the wind turbine system. Thus, any increase in energy has a direct proportional effect in reducing COE. Reliability and availability impact directly on energy produced and are therefore are also prime values. In particular, elimination of any component in a wind turbine system always has added value in COE terms beyond the removal of its capital cost as associated issues of reliability and maintenance disappear. Efficiency is also extremely important, directly affecting the greater proportion (most typically two-thirds) of annual energy capture in wind speeds below rated. All other factors within the lifetime costs, machine component costs, wind farm component costs and maintenance costs, have a partial value. Any percentage impact in cost in these components produces a lesser percentage impact on COE.

9.3.1   Efficiency

Efficiency of major drive-train components has been reviewed in Section 6.10 and rotor C_p characteristics have been discussed extensively in Section 1.9. It should be noted that efficiency of the aerodynamic rotor or of drive-train components directly affects power and energy capture only in wind speeds below rated, although it can also affect the value of rated wind speed. Thus, a percentage gain in efficiency is, in general, less than a percentage gain in energy output.

9.3.2   Reliability

Reliability engineering in the wind energy field gathered momentum after major gearbox failures occurred in the late 1990s and after the early offshore turbines showed higher operation and maintenance (O&M) costs and lower availability than expected. Whilst the gearbox failures could be addressed with proper design and quality control, the growth of offshore wind technology with its greater challenges in access and maintenance has focused increasing attention on reliability engineering in wind turbine design. The Dutch Offshore Wind Energy Converter (DOWEC) project was one of the first attempts to include reliability engineering in all design phases, from the concept to detailed design [10]. The revenue losses and the direct O&M costs were included in a life cycle cost model enabling the trade-off between revenue, O&M costs and capital cost to be optimised.

Various initiatives were taken to gather field data on reliability, and wind farm owners have contributed data to large databases [11]. This data was generally very valuable but there were deficiencies; the data grouped together turbines sometimes of different rating and technology with no clear definition of failure given. Nevertheless, the analysis of such publicly available data yielded informative overall trends [12].

Although reliability engineering was receiving increasing attention boosted by offshore wind interests, subsequent failure rates did not always show a trend towards higher reliability. Various explanations are likely. Wind turbine concepts are more complex than in the past and newer technologies are in their early life, lacking the experience to confirm appropriate design margins. In the early 1990s, the mainstream concept was a constant speed-fixed pitch turbine. The manufacturer NEG Micon used a slogan 'Powerful simplicity' to market the turbines. The introduction of variable speed, pitch-regulated wind turbines added complexity in the active pitch system and in the power electronics required to provide fixed frequency electrical output. The latter especially had a negative impact on reliability, which was exacerbated by the upscaling of wind turbines that increased the number components in the inverter. Nevertheless, it is not always a simpler design that affords higher reliability; the typical case is the introduction of the pitch control which has drastically reduced blade failures associated with the higher aerodynamic stresses of a stall-regulated rotor. Failures introduced with the pitch control are compensated for by a higher serviceability of this subsystem.

More recently, wind turbine reliability investigation has seen the active participation of industrial partners in addition to academic institutions, with a huge beneficial effect on data quality, which is of critical importance for effective reliability analysis. In the United Kingdom, the Engineering and Physical Sciences Research Centre (EPSERC) within *Supergen* [13] has funded a major project for the assessment of the wind turbine reliability. The results include wind turbine condition monitoring, gearbox failure evaluation and more general turbine prognosis and reliability-oriented design.

A similar project, but at the European level, is in progress. The Reliawind consortium [14] is a European FP7-funded programme involving major European manufacturers, suppliers, consultants and academic institutions active in the wind industry. Reliawind activities comprise the analysis of field data, reliability prediction, failure mitigation and the development of an integrated tool for fault diagnosis and prognosis.

9.3.3 Availability

The lifetime energy output of a wind turbine clearly depends on the availability to function as intended throughout its design life. Within the complexities of wind farm financial arrangements, there have been many definitions of availability. The renewable energy consultancy, GL Garrad Hassan, considers that a basic definition as follows is most commonly used:

$$\text{Availability}\% = \frac{T_t - D}{T_t}, \quad \text{where } D = [T_g + T_w + T_o]$$

T_t = Total time in period,

D = Total downtime in period,

T_g = Grid downtime in period,

T_w = Time when wind or other ambient conditions are outside specification,

T_o = Other downtime.

The category T_o is intended to cover all excluded events such as

- Scheduled maintenance,
- Force majeure;
- Shutdown instructed by the owner for reasons other than unsafe operation of the wind turbine;
- Owner's risks, for example, vandalism;
- Lightning damage that occurs despite the lightning protection system functioning in accordance with specification,
- *Icing*: events not due to the ineffective operating of anti-icing arrangements;
- *Access limitations*: downtime that arises as a direct result of the maintenance provider's inability to access the site in accordance with defined health and safety procedures;
- *Communications*: downtime arising from the inability to perform remote restarts due to the failure of the communications system.

The abovementioned formula is for individual wind turbine availability. The wind farm availability is then calculated according to the following formula:

$$A(wf) = \frac{1}{N} \sum_{i=1}^{N} A(wt_i)$$

$A(wt_i)$ – individual wind turbine availability in the period,

N – number of wind turbines in the wind farm,

$A(wf)$ – wind farm availability.

9.4 Capital Costs

The capital costs of a wind farm constitute two main items – the ex-works cost of the turbine components in their finished form ready for final system assembly at site and other costs described as 'balance of plant costs'.[2]

Balance of plant costs for a land-based project includes costs associated with assembly at site, erection and commissioning, turbine foundations, access road construction, grid connection, site services such as buildings for site management or maintenance and, possibly, wind farm design, permissions and financing charges associated with the complete establishment of the wind farm.

The cost of the wind turbine support structure (tower) would most usually be regarded as part of the turbine ex-works cost but in some cost analyses, especially for offshore, it may be included in balance of plant.

---

2 'Balance of Station' is the usual terminology in the United States.

Considered at a detailed level, capital costs of wind turbine components depend on costs of raw materials, processes and labour costs which constantly vary with time and place of manufacture. The installed cost includes balance of plant costs which are site-specific and transport costs. Transport costs and also wind turbine ex-works cost may well be affected by whether some extent of local manufacture is established in the regions where wind farm development is taking place. Krohn *et al*. [6], for example, provide data showing the variation of installed cost by country in 2006.

The approach taken in top-level COE analyses aimed at the initial evaluation of innovations is to consider available information on installed cost of turbines of a relevant size or if known, ex-works cost of turbines in current projects. Using relative cost split information such as given in Table 9.1, the costs of components can be inferred.

Capital cost can often be related to mass; and while this is not very useful in making primary estimates of the cost of wind turbine components, it can be useful in assessing the impact of innovations. Thus, saving 20% of mass of some components may equate to 20% cost reduction. This depends, of course, on the impact on process and labour costs. Data is published on costs of materials, for example, steel, including estimated costs of finished product types depending on the complexity of manufacture, for example, cast, fabricated, machined, highly engineered (as in bearings and gears, etc.). Thus, if an innovation reduces design loading, for example, the biggest challenge is then to express this as a mass and cost reduction.

It is interesting that, in spite of the complexity and variability of some industrial processes such as blade manufacture, there is often useful convergence with a fairly reliable average cost per kilogram being established at least for individual manufacturers, sometimes over a number of years.

9.5 Operation and Maintenance

This is a particularly difficult cost item both not only because, as with other wind turbine cost data, much information is held confidential but also because it takes time to build up significant statistics, should they be available. Furthermore, wind technology has been constantly evolving. Thus, the recorded performance of 20-year-old wind turbines has restricted relevance for predicting the performance of current installations. Usually, the innovations to be evaluated are in the turbines themselves so that it is enough to have a lifetime fraction associated with O&M that is appropriate relative to the turbine cost.

If an innovation eliminates a component, a rough estimate of the impact on O&M is to consider the component's importance in the cost split and reduce at least the replacement part's cost fraction in proportion. This impact can be factored accordingly if the innovation does not eliminate a component but reduces its cost. In general, in top-level parametric analyses, it is a matter of approximate judgement how O&M costs may be affected by an innovation.

O&M simulations [15, 16] have been developed as tools to forecast the interaction of reliability, availability, access for maintenance and associated costs so that design and management of offshore wind farms may be optimised.

In a rigorous evaluation of total lifetime costs, decommissioning costs should also be accounted for and/or the possibility of reusing parts of the system such as foundations with replacement wind turbines.

9.6 Overall Cost Split

The lifetime costs of a wind farm project may be regarded as the sum of two components, initial capital costs and continuing O&M costs including, in principle, decommissioning costs at the end of the design life. The initial capital costs, as discussed, are split into two major parts, the wind turbine costs and the balance of plant costs associated with the necessary infrastructure for a wind farm. Offshore project cost evaluation is similar in concept, although the balance of plant items may differ substantially.

Table 9.3 Cost of energy fractions.

			Lifetime costs	
Initial capital costs	Turbine	Rotor	Rotor lock	0.0057
	0.569		Blades	0.1037
0.820			Hub	0.0142
		Nacelle systems	Gearbox	0.0961
			Generator	0.0378
			Rotor brake	0.0085
			Nacelle cover	0.0142
			Nacelle structure	0.0193
			Couplings	0.0057
			Shaft	0.0171
			Yaw system	0.0171
			Bearings	0.0171
		Electrics and control	Pitch system	0.0365
			Variable speed system	0.0551
		Tower		0.0896
		Other		0.0313
	Balance of plant		Roads and civil works	0.0221
	0.251			0.0761
			Electrics and grid connection	0.0073
				0.0365
			Assembly and installation	0.0675
				0.0414
			Transportation	
			Foundations	
			Financial and legal	
O&M	0.180		Labour	0.0792
0.180			Parts	0.0630
			Operation	0.0216
			Equipment	0.0090
			Facilities	0.0072

Table 9.4 Offshore lifetime project cost split.

Wind turbines	0.30
Balance of plant	0.30
O&M	0.33
De-commissioning	0.07

With knowledge of the overall typical split of these major items, the individual project components can then be expressed as fractional lifetime cost values, as in Table 9.3. To relate to overall lifetime costs, changes in the value of money (inflation, etc.) and in the value of hardware (depreciation) need to be considered over the design life. The capital cost *split* is primarily technological and less affected by such issues, although the total capital cost in relation to ongoing lifetime costs (O&M) obviously does depend on the variation of financial parameters over the design life. The data of Table 9.3 is considered typical.

Although this approach is quite crude and represents only typical values, it can give useful initial insights into the likely relative value of components in the project cost chain. For example, rotor blades have a lifetime cost fraction around 0.1. This means that an efficiency gain that directly affects energy output may have up to 10 times the value of a rotor blade cost saving. Conversely, it indicates that using blades that sacrifice more than 10% of annual energy, even if they cost nothing, will increase and not reduce the COE.

Whilst the generic cost fraction data of Table 9.3 is quite detailed, it should be taken only as indicative and not at all as definitively accurate. Such accuracy does not exist as wind turbine concepts, component costs and the classification of subsystems vary a great deal among manufacturers. It should also be noted that, by virtue of the different scaling characteristics of components (Section 4.4), the cost split is intrinsically scale dependent.

Offshore, the cost split changes quite substantially. Costs vary considerably with type of site (e.g. in the North Sea or Baltic) water depth and wave climate. A typical overall lifetime cost split (OptiOWECS ref) may be as in Table 9.4.

9.7 Scaling Impact on Cost

Underlying scaling rules have been extensively discussed in Section 4.4. Working from a baseline at a given scale, it is possible to address how the cost split and costs will change with upscaling.

As an example, consider scaling of hub cost where the hub is assumed to be a casting with cost proportional to mass. The starting point is determination of the scaling rules for the component (hub) based on similarity and underlying physics.

It may be determined that the hub design is primarily driven by fatigue due to aerodynamic bending moments which scale simplistically as the cube of rotor diameter but as a slightly higher power when the impact of absolute length scales of wind turbulence are accounted for. Generic simulation studies examining the scaling of loads might address this and determine, for example, a plausible proportionality to diameter D of the aerodynamic bending moments as $D^{3.1}$.

However, as blades get larger, blade-mass-related moments become proportionately greater in relation to aerodynamic moments and, if blade weight is not restrained by the use of material with ever higher strength-to-weight ratio (mainly carbon considering present technology), then fatigue due to blade mass moments may become a large contribution to the designing loads of the hub or even dominate. Further simulation studies may be undertaken to characterise the fatigue contribution due to blade mass with scale and determine that the hub mass then scales as $D^{(3.1+km_b)}$ where m_b the blade mass, is itself a function of diameter. The 'constant' k may be zero below a certain size of wind turbine or not a constant at all. The point being made at present is that some such underlying physically based model is needed as a starting point.

The next stage is to consider real engineering practice in the design of the component. For example, the hub of a 1 MW scale wind turbine may be thicker than necessary due to minimum casting thickness limitations so that scaling directly with the physical model from mass values for hubs of 1 MW wind turbines would give unrealistically high estimates for upscaled designs. However, even more unrealistic would be simply to fit an empirical diameter dependence to data for hubs in the range 1–5 MW, which may well be less than cubic and upscale from that.

Realistic scaling of the hub then becomes as $f(a, b, c \ldots)D^{(3.1+km_b)}$ where the function $f(a, b, c, \ldots)$ is a function, in principle, of many variables to do with materials, process, production quantities, technology improvement rate, and so on, and thus extremely complex. In practice, it can only be assessed approximately and perhaps reduced to an appropriate constant by consideration of available data coupled with understanding of the design and manufacturing process. It is also worth noting that the 'improvement' that comes to hub design due to the required thickness of the casting exceeding a practical minimum value is not something that should be absorbed in a 'learning' curve. Beyond a certain scale, the effect is not there nor does it improve with time.

A similar issue was evident in some blade designs in smaller size ranges than the blades of the present generation of multi-megawatt machines. They had trailing edge sandwich structures with biaxial glass skins sized to avoid damage from the weight of a man standing on the blade. The skins were therefore relatively too thick and heavy from a stress and cost point of view, at least if viewed as a pointer to what could be achieved in upscaled design.

9.8   Impact of Loads (Site Class)

The loading regime experienced by a wind turbine clearly depends on the wind conditions at the chosen site. These exhibit enormous variation considering different regions of the world and climatic characteristics. For design purposes, particularly in the context of certification of wind turbine designs, various site classes have been defined in standards such as IEC, Germanischer Lloyd (GL) and others. Adopting the site classes of the GL standard of 2003, the defining wind speed characteristics of three site classes to be considered are presented in Table 9.5. The site classes here are as defined for turbulence class A, land-based sites. This sets the level of TI at 18% in a mean wind speed

Table 9.5 Site class wind conditions.

	Annual mean wind speed (m/s)	Extreme 3 s gust with 50-year return period	Turbulence intensity (%)
Class 1_A	10.0	70.0	18
Class 2_A	8.5	59.5	18
Class 3_A	7.5	52.5	18

of 15 m/s. Other less severe turbulence conditions may of course be considered and are classified within the standard.

The prediction of loads on wind turbines is now extremely sophisticated. Most codes such as Bladed of GL Garrad Hassan [17] and Flex 5 developed originally by Øye [18] still rely heavily on blade element momentum (BEM) theory, such as outlined in Section 1.10 with additional modelling of unsteady aerodynamics and coupled with spectral models of wind turbulence and modal or finite element structural models such as the ADAMS code [19] that may include mechanical joints and other nonlinearities. In general, wind turbine BEM or more sophisticated aerodynamic modelling is being coupled with multi-body structural modelling as has been developed in packages such as SIMPACK [20] which evolved from work at the German aerospace centre, Deutschland für Luft und Raumfahrt (DLR). A huge amount of research and development of design tools in this area, validated by full-scale test measurements, has been a focus for national research institutions like Risø, the Energy Research Centre of the Netherlands (ECN) and NREL and also particularly for the commercial consultancy company, GL Garrad Hassan.

The aim of this review is not, however, to discuss loads in any comprehensive way but to find a way into a top-level evaluation of the changes in loading and downstream cost impacts if a wind turbine designed for one set of wind conditions, say Class II, is modified for operation on a Class I site.

Some key issues around wind turbine loads are illustrated in Figure 9.6. The steady-state thrust on a typical pitch-regulated wind turbine rises to a peak at rated wind speed and (assuming the blades are pitched nose into wind in attached flow as is most usual) drops sharply as pitch control is used to regulate power. Typically, above about 25 m/s wind speed, the wind turbine is shut down and parked or put into idling mode for protection from severe wind conditions. In higher wind speeds, loads then vary essentially as the square of the wind speed. Although the blade loads can be vastly reduced at the idling pitch around 85–90°, considering combinations of yawed wind flow from changes in wind direction and also the influence of the random azimuth position of blades in the rotor circle, the blades can still experience high lifting loads approaching a level (solid line of Figure 9.7) as if they had not been pitched. While idling helps greatly to reduce system loads on average, the main benefit of idling is to keep the rotor at a safe speed without requiring a heavy-duty mechanical brake to prevent any rotation of the rotor. Allowing idling also relieves some loads that would arise if the rotor were rigidly parked.

Most of the fatigue loading in the design life experience of a wind turbine arises in operational wind speeds around and above rated wind speed. Extreme loads that

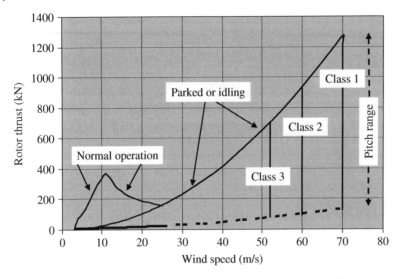

Figure 9.7 Steady-state rotor thrust of a 100-m-diameter 3 MW wind turbine.

may govern the design of some components may occur in extreme events of normal operation (e.g. an extreme wind direction change combined with a gust) but are often much affected by the extreme storm wind speed (typically a 3 s gust with a return period of 50 years). In IEC and GL design standards, this value is up to 70 m/s for a Class 1 site (Table 9.5), although gusts above 80 m/s have been experienced in typhoon climates [21]. The schematic of Figure 9.8 gives a qualitative impression of fatigue and extreme (idling) loads in relation to mean wind speed.

Specific fatigue loading issues arise in offshore sites. Although lower ambient turbulence levels and lower long term extreme wind speeds may be associated with a particular level of annual mean wind speed, machines operating in wakes experience increased TI with reduced turbulence length scale [22]. This may entirely offset any potential benefit from reduced ambient turbulence.

Generally, if lifetime equivalent fatigue loads are in a range from about 25% to 50% (the exact value depending on the material systems employed in a particular component) of the lifetime extreme value, they may dominate design of a component. Thus, as the extreme wind speed for design and the ratio of extreme storm loads to peak operational loads reduces in moving from Class 1 to Class 2 and Class 3, fewer components are designed by extreme loads and more by fatigue.

It is dynamic extreme loads and cumulative fatigue loads mainly from operation in normal wind turbulence that drive the design of most wind turbine components. In general, such loads vary depending on the operational strategy of the wind turbine, its specific structural dynamics and control system dynamics.

Considering a typical wind distribution with mean around 7 m/s (Figure 9.7), and a peak in the energy capture distribution around rated wind speed (see Figure 6.7), it will be evident that much of the energy capture and value in a wind turbine system is associated with wind speeds below rated and most below about 15 m/s. In contrast, most of the design driving loads and consequent cost of components arise from loads in wind

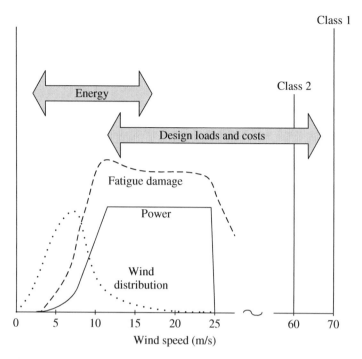

Figure 9.8 Typical load regimes.

speeds above rated and the extreme storm conditions may drive blade loads and other critical components in Class 1 sites. This mismatch between the conditions that provide energy value and those that incur component cost is highlighted in Figure 9.7 and makes the case for more adaptable rotor concepts (Chapter 17).

The primary variables affecting wind turbine capital cost are scale (diameter), design tip speed (affecting drive train torque), hub height (tower), power rating (electrical costs) and the design site wind characteristics. The question therefore often arises whether a turbine is within its design envelope for a particular site which may not exactly fit its certification class. In the context of evaluating innovation, a related issue arises how to compare innovative turbines designed for a different class from an available reference baseline wind turbine or, alternatively, how to convert the baseline to a different site class for comparison purposes.

It is possible to do this at an approximate parametric (spreadsheet) level without a comprehensive recalculation of design loads. Extreme loads (rotor idling) can be scaled as the square of the wind speed. Extreme operational loads are more problematic but unchanged unless related to increased gust levels, for example, annual operating gust when again they may be scaled as wind speed squared. Providing a fatigue load estimation for damage equivalent loads in wind bins over the operational range is available, the data can readily be adjusted to a different mean wind speed accounting for the change in annual hours in each bin and hence in the associated number of rotor cycles. In summary, the same damage equivalent loads associated with any given mean wind speed level are weighted differently according to a change in wind distribution.

Evaluating the cost impact of changes in component loading is challenging. The key stages are as follows:

1) Identify the design driving load(s) for the relevant component of the baseline design.
2) Consider whether, after a change in such load(s), the designing load case and or load type (extreme, fatigue) will change.
3) Assess the impact of the load change on the component mass and cost.

Stage 3 is relatively straightforward for some components. For example, a gearbox of given nominal torque rating and ratio is designed by time at the level of the input torque distribution, and mass and cost reduction can sensibly be related to changes in such loads or, more simply, to a change in nominal rated torque as in the case of a high-speed rotor design (Section 7.2), for example. For other components, the cost impact of load changes may be very difficult to analyse without going to a level of design detail beyond simple parametric analysis.

References

1 Johnson, G.V. (1969) *On Predicting Production Costs and Probable Learning Rates from R&D Investments by S-Curve/Learning Curve Relationships*. National Technical Information Service, October 1969.
2 Fingersh, L., Hand, M. and Laxson, A. (2006) Wind Turbine Design Cost and Scaling Model. Technical Report NREL/TP-500-40566, National Technical Information Service.
3 Tande, J.O. and Hunter, R. (1994) *Recommended Practices for Wind Turbine Testing. Estimation of Cost of Energy from Wind Turbine Systems*, 2nd edn, International Energy Agency, Paris.
4 IEA (2016) *Next-Generation Wind and Solar Power from cost to value*, IEA Publications, 9, rue de la Fédération, 75739 Paris Cedex 15 Layout and Printed in France by IEA, December 2016.
5 http://www.repp.org/articles/static/1/binaries/WindLocator.pdf (accessed June 2011).
6 Krohn, S., Morthorst, P.-E. and Awerbuch, S. (2009) The Economics of Wind Energy. EWEA Report, March 2009.
7 Poore, R. and Lettenmaier, T. (2002) Alternative Design Study Report: WindPACT Advanced Wind Turbine Drive Train Designs Study. 1 November 2000–28 February 2002.
8 van der Hoven, I. (1957) US weather bureau. Power spectrum of horizontal wind speed in the frequency range from 0.0007 to 900 cycles per hour. *Journal of Meteorology*, **14**, 160–164.
9 IEC 61400-12-1 Edition 1.0. (2005) *Wind Turbines – Part 12-1: Power Performance Measurements of Electricity Producing Wind Turbines*, IEC, Geneva.
10 http://www.ecn.nl/units/wind/rdprogramme/aerodynamics/projects/dowec (accessed 11 May 2010).
11 DOWEC team (2013) *Estimation of Turbine Reliability figures within the DOWEC project*. DOWEC Nr. 10048 Issue 4 October 2003.
12 Spinato, F. (2008) The reliability of wind turbines. PhD thesis. Durham University, England.

13 http://www.supergen-wind.org.uk (accessed 11 May 2010).

14 Wilkinson, M., Hendriks, B., Spinato, F. *et al.* (2010) *Methodology and Results of the Reliawind Reliability Field Study*. EWEC 2010, Scientific Track, Oral Presentation.

15 Henderson, A.R., Greedy, L., Spinato, F. and Morgan, C.A. (2009) *Optimising Redundancy of Offshore Electrical Infrastructure Assets by Assessment of Overall Economic Cost*. European Offshore Wind Energy Conference, Stockholm, Sweden, September 2009.

16 Cockerill, T.T., Harrison, R., Kühn, M. and van Bussel, G.J.W. (1998) Opti-OWECS Final Report, *Estimate of Costs of Offshore Wind Energy at European Sites*, vol. **3**, Institute for Wind Energy, Delft University of Technology, Delft, Netherlands.

17 Bossanyi, E. (2010) Bladed Theory Manual. GL Garrad Hassan document, 282/BR/009, issue 20, October 2010.

18 Øye, S. (1999) FLEX 5 User Manual. Lyngby.

19 http://wind.nrel.gov/designcodes/simulators/adams2ad/ (accessed 27 September 2017).

20 http://www.simpack.com/ (accessed 27 September 2017).

21 Lessandro, E.O., Garciano, L. and Takeshi, K. (2010) New reference wind speed for wind turbines in Typhoon-Prone areas in the Philippines. *Journal of Structural Engineering*, **136** (4), 463–467.

22 Thomsen, K. and Sørensen, P. (1999) Fatigue loads for wind turbines operating in wakes. *Journal of Wind Engineering and Industrial Aerodynamics*, **80** (1–2), 121–136.

10

Evaluation Methodology

10.1 Key Evaluation Issues

Vital issues around the evaluation of innovative wind technology are as follows:

- Case for innovation
- Fatal flaw analysis
- Power, torque
- Representative baseline for energy, loads and cost comparisons.

In the first instance, there must be an argument presented with potential benefit as motive for the innovation and not just the ad hoc approach – 'here is a new idea, let's try it'. In gradual stages, the evaluation will address whether the argument is likely to be valid. The first step is to ask if there is anything obviously unsound in the system concepts (see Section 10.2).

Evaluation of any wind turbine system based on the standard concept of an aerodynamic rotor with power take-off from a central shaft considers the following:

1) Power performance which affects the primary value energy output (Section 10.3);
2) Shaft torque which affects drive train weight and cost (Section 10.5).

The power–torque evaluation shines a very clear light on the challenge for vertical-axis wind turbine (VAWT) technology to be competitive (Chapter 13). It also shows in the evaluation example of the Carter wind turbine concept (Section 10.9) how a very simple overview evaluation can clear up confusions about main influences on system mass.

As has been discussed (Section 0.7), in evaluation of an innovative system or standard system with an innovative component, it is valuable to develop a comparable baseline design against which many factors can be compared (energy, reliability, loads, mass, cost, etc.).

10.2 Fatal Flaw Analysis

Whenever a wind turbine innovation is presented, the very first level of examination is to be satisfied that the underlying concept is sound, that the claims do not conflict with the established laws of physics as in some kind of perpetual motion machine. For example, in proposing to mount a wind turbine on a train or any other vehicle, it should be clear that, under conditions where there is no natural wind, whatever energy is extracted can

Innovation in Wind Turbine Design, Second Edition. Peter Jamieson.
© 2018 John Wiley & Sons Ltd. Published 2018 by John Wiley & Sons Ltd.

only be at the expense of the energy providing motive power for the vehicle.[1] However, only very rarely are concepts presented which are patently unsound.

Given a concept that is conceptually sound, the next step is to look for fatal flaws in the way in which it is realised.

Consider, as an example, a blade tip brake in the form of a rotating tip section which is activated by the excess of centrifugal force that will develop when the rotor goes into overspeed, say, due to a loss of electrical load. Typically, the tip section is held in place by a strong spring and only moves outwards under excessive centrifugal force in, say, 20% or 30% overspeed. As the tip section moves outwards, it also turns through about 90° on a spiral track offering high drag to the flow and thereby acting as a powerful aerodynamic brake. The basic concept is perfectly sound and has been much used in stall-regulated wind turbine designs since the 1980s.

However, when the tip is activated and is effective, it will obviously reduce rotor speed. The centrifugal force will correspondingly reduce and the spring may then cause the tip to return to the 0° position. The system may then oscillate through many cycles of overspeed and tip deployment followed by speed reduction and tip retraction. This could lead to a fatigue failure, in which case it may be concluded that the tip brake concept is sound; but there is a fatal flaw in this particular realisation of it. Solutions developed included a lock out device to prevent retraction of the tip but with the penalty that it may require manual intervention to release the lock out, say, when, after clearing a grid fault, normal operation could be resumed. This problem is fully solved in some modern designs which still use such tip brakes [1].

10.3 Power Performance

Provided the overall concept seems physically sound, the evaluation of an innovative wind turbine design focuses initially on power performance. Whatever the merits of the proposed technology, it cannot underperform very much in energy capture compared to a state-of-the-art wind turbine to stand a good chance of being cost-effective. Also, whether any unusual trade-off is being made between, say, energy capture and cost or maintenance, the likely power performance must be quantifiable with reasonable confidence to reach any definite assessment of cost of energy (COE) reduction potential.

10.3.1 The Betz Limit

In general, if a turbine essentially in open flow approaches much nearer the Betz limit than conventional state-of-the-art designs, or exceeds it, there is cause for explanation and this happened following the introduction of a new rotor design of Enercon. An initially sceptical industry was suspicious of claimed rotor power coefficient measurements exceeding 0.5 [2], but gradually became convinced that the power performance measurements were valid. As frequently happens with industry hot topics, Risø DTU sought

---

1 A more subtle version of this occurred when a prospective client suggested extracting energy from the air flows displaced through pressure relief vents when a train passes through a tunnel. Again, the train is doing the work of pressurising the air in the tunnel and the resistance of a wind turbine extracting energy from flow in the vents would simply require the train to do commensurately more work and consume more fuel in its passage through the tunnel.

out analytical explanations of the improved performance. It was considered that the curved tip ends of the Enercon blade, although not acting like a diffuser to induce extra mass flow, nevertheless beneficially inhibited tip loss [3]. At much the same time it was being noted that, in contradiction to standard blade element momentum (BEM) theory (see Section 1.10.4), a reduced or negative induction may exist near the root of a wind turbine blade and it could be surmised that the carefully engineered interface between the large hub cone and blade surfaces of the Enercon may exploit this with benefit. An interesting paper, Johansen *et al.* [4], indicates potential for enhanced performance by improved aerodynamic design in the hub area.

There is a genuine debate about the extent to which the Betz limit is a rigorous ideal limit for energy extraction from a rotor in open flow as discussed in Section 1.5.3. Considering the level of performance that has been measured on wind turbines and the known losses associated with aerofoil drag and the tip effect, it certainly would seem unlikely that the Betz limit is too low as an upper bound for open flow rotor performance. As discussed in Section 1.7, turbines that induce extra mass flow through the rotor plane can legitimately exceed the Betz limit. This may be achieved with a turbine in a diffuser, but other possibilities exist.

For example, high performance is claimed for designs of the French company, Nheolis [5], which has developed several small wind turbines with support from the French National Aerospace Research Centre (ONERA) and the French National Centre for Scientific Research (CNRS). This turbine has scoop-shaped blades; and it is possible that radial flow along the blades expands the wake, acting rather like a diffuser creating a core of low pressure and inducing extra mass flow through the rotor system.

Often, inventors with innovative ideas attract some interest from investors but do not obtain substantial funding in the early stages. With restricted resources, power performance measurements are made which do not fully conform to recommended practice, perhaps on a site complicated by terrain effects or with obstacles in various wind directions all leaving a trail of uncertainties. Usually, this can only be adequately resolved by more rigorous testing at a stage when the wind turbine can be set up with good instrumentation on a better site. Thus, performance evaluation of an innovative design requires understanding the theoretical basis for any unusual claims and, where measurements exist, critical appraisal of the instrumentation and results. There is a further key issue – effective control. A new turbine concept may be sound and power performance measurements perfectly valid, but a poor level of performance is registered because, in wind speeds below rated, the turbine has not been effectively controlled to near its peak C_p.

In order to have a wider critical perspective on the energy changes in wind power extraction, consider now the magnitude of some of the energy-related terms.

10.3.2   The Pressure Difference across a Wind Turbine

It is well known that very small changes in air pressure may be associated with what is perceived as very loud acoustic noise. In the context of wind turbine acoustics, Zhu [6] observes:

> Problems occur in utilizing standard CFD procedures for aeroacoustic problems… due to the extremely small magnitude of waves. For noise levels up

to 100 dB... the magnitude of the fluctuating sound pressure is less than 10^{-4} of the ambient atmospheric pressure.

It may be a little more surprising to note that the energy extraction process of all wind turbines, large or small, is based on pressure differences that are not that much greater. Suppose a wind turbine is operating optimally in an ambient wind speed of 8 m/s, then, from Table 1.4, the pressure difference across the rotor on which the energy extraction depends is

$$\Delta p = \frac{4}{9}\rho U_0^2 = 35\,\text{Pa}$$

This pressure difference is $\sim 3.5 \times 10^{-4}$ of standard atmospheric pressure (101 325 Pa).

10.3.3 Total Energy in the Flow

In the analysis of wind turbines, the usual assumption is that only kinetic energy and pressure energy are relevant. Nevertheless, claims have been made for wind turbine concepts that exploit other energy sources including thermal. Arter Technology in Moscow tested a ducted wind turbine (Figure 10.1) with duct and rotor design rather similar to gas turbine design concepts and consider that thermal energy exchanges play a role in the system performance. The thermodynamic energy equation in the Bernoulli form (dimensionally pressure) can be extended to include such terms. Discounting chemical and magnetic energy sources the equation becomes

$$p + 0.5\rho U^2 + e + \rho gh = \text{constant} \tag{10.1}$$

where e is internal energy per unit volume. In the case of a tidal turbine submerged in relatively shallow water, the gravitational potential term, ρgh is not negligible [7] and the free surface of the water will be depressed in the near wake of the turbine. The Chiral wind turbine [8] based on the Magnus effect makes some claim to exploit energy from the atmospheric gravitational potential term and a few systems have claimed to exploit thermal energy.

The standard wind turbine in open flow may be analysed as in Chapter 1 but including the thermal energy term. Equation (10.1) (ignoring gravitational potential energy) may be re-formulated using γ, the ratio of specific heats of air as

$$\text{Upstream of rotor plane} \quad \frac{\gamma}{\gamma - 1}\frac{p_0}{\rho_0} + \frac{1}{2}U_0^2 = \frac{\gamma}{\gamma - 1}\frac{p_{1u}}{\rho 1} + \frac{1}{2}U_1^2 \tag{10.2}$$

$$\text{Downstream of rotor plane} \quad \frac{\gamma}{\gamma - 1}\frac{p_0}{\rho_2} + \frac{1}{2}U_2^2 = \frac{\gamma}{\gamma - 1}\frac{p_{1d}}{\rho_1} + \frac{1}{2}U_1^2 \tag{10.3}$$

The suffixes 0, 1 and 2 denote, respectively, the upwind source flow, the rotor plane and the far wake locations. p_{1u} is the static pressure at the rotor plane on the upstream side and p_{1d} the corresponding static pressure at the rotor plane on the downstream side.

Then considering an ideal inviscid, adiabatic operational case where the ambient wind speed is 9, 6 m/s at the rotor plane and 3 m/s in the far wake, the temperature change across the rotor plane may be calculated to be $\Delta T = 0.125\,°C$. The change in kinetic energy per kilogram of mass flow is

$$0.5(U_0^2 - U_2^2) = 0.5(9^2 - 3^2) = 36\,\text{J/kg}$$

Figure 10.1 Prototype wind turbine of Arter Technology. Reproduced with permission of Arter Technology.

Denoting the specific heat of air at constant volume, C_V, the change in internal energy is

$$\frac{C_v \Delta T}{\rho} = \frac{714 \times 0.125}{1.225} = 73\,\text{J/kg}$$

It is almost shocking to appreciate that this is twice as much! It is not obvious how any of this can be exploited as wind turbine work considering the extremely low Carnot efficiency of any kind of heat engine operating on a temperature difference of a fraction of a degree. Nevertheless, in the innovative design of Scerbina [9], involving a vertical axis rotor with multi-aerofoil cascades replacing individual blades, it is argued that some of this energy is available as additional work.

10.4 Structure and Essential Mass

In evaluating an engineering structure, it is natural to ask – what scope is there for reducing mass without compromising function? Consider a weight hanging on a string. Neglecting the self-weight of the string, if its length is doubled, its cross section will still equally bear the tension load from the weight but the mass and cost of the string itself will double. This supports intuition that mass, and hence cost of a structure, will depend on the distance over which loads are transmitted.

Structures are most efficient in tension. A cantilevered beam of minimum mass (perhaps 1/16 of the mass of a simple solid beam) can be made with transverse spars issuing at right angles from a main centreline spar and joined at their extremities by pure tension elements such as wires. In bending a beam, one or other of the outer surfaces goes into tension. The idea is to use the wires to resist these tensile forces efficiently at some distance from the neutral axis, thereby creating a resistive moment over the whole beam. Such a beam is complex and often not a good practical solution for specific engineering applications, but it certainly illustrates how mass may be minimised in a structural element.

For the string in tension, its sectional area is determined by the tension force. Its volume and mass is therefore proportional to the product of this force and its length. Garvey [10] generalises this idea of a force–distance product for various types of loading (tension, shear, bending) developing concepts of structural capacity and structural efficiency. Focusing on structures loaded uniaxially, with only one principal stress dominant at most locations, he defines a *force–distance potential* as

$$C = \sum_{k=1}^{m} V_k \overline{\sigma}_k \tag{10.4}$$

where $\overline{\sigma}_k$ is the maximum stress in V_k, a volume in a structure comprising m total volumes. The *duty* of the structure, E, accounting for all actual force–distance transmissions is then defined as

$$E = \sum_{j=1}^{n} F_j d_j \tag{10.5}$$

Garvey goes on to define *utilisation*, η, (which might also be termed structural efficiency) as a non-dimensional measure reflecting how effectively material has been used in a structure to react to the duty loads as

$$\eta = \frac{E}{C} \tag{10.6}$$

Utilisation can be unity in pure tension and Garvey shows it is intrinsically less in structures loaded in shear or bending. He acknowledges simplifications in not distinguishing maximum compressive and tension stresses in the definition of C and complications in the definition of E as some internal forces may cancel. Nevertheless, this makes a start to quantifying the purely qualitative understanding a good engineer has of the interplay of forces, stresses and mass in a structure.

An implication of the force–distance product is that the shorter the path between the incident aerodynamic loading on a wind turbine and the electrical output terminals, the less mass will be required in the system. Garvey outlines an ambitious mechanical-hydraulic concept (see Section 6.13) to extract energy almost at source within the blades. Some time ago, Jack [11] and Watson [12] proposed systems of secondary rotors (also mentioned in Section 6.13). Such secondary rotors could be similar to ram-jets used for auxiliary power in some aircraft. As turbines turning at 20–30 000 rpm, the generators would operate at very low torque and be light and compact, amounting to a streamlined bulge over the surrounding aerofoils rather like modern aircraft engines. With no more than electrical power lines inside the blades passing through the hub and tower system, it is easy to understand that massive saving in drive-train cost and weight would result. The distance between the resultant aerodynamic, edgewise force per blade and generator output is now shortened to the order of a chord width and hence associated with a minimal force–distance product.

Whether very high speed secondary rotors could be practical on a horizontal-axis wind turbine (HAWT) is of course questionable as the generator bearings would need to be fit for long life in an environment of severe asymmetric radial and axial loads, probably amounting to a much more demanding duty than an aircraft even accounting for all manoeuvres in dives and tight turns, and so on. The other main aspect of secondary rotors as high rpm ram-jet-type devices is that acoustic noise would be at quite unacceptable levels for land-based wind farms, possibly similar to jet engine noise. Relatively larger slower running secondary rotors on very large VAWTs (as mentioned in Section 6.13) would not have the same critical problems around bearing duty and noise levels and would still effect large mass savings compared to any system extracting power on the central shaft of a large VAWT.

The force–distance product as related to system mass can be illustrated considering the 20 MW multi-rotor system (MRS) discussed in Chapter 15 in comparison to a single rotor of 20 MW rating. We can determine a radius to associate with the total edgewise force of each rotor blade and then consider the distance to the hub centre and further to the electric terminals as giving a force distance product that will be an indicator of drive-train mass. Considering scaling with similarity, this product, having the same dimensions as torque, will scale cubically leading to the result that the MRS with n rotors will have a reduced force–distance product as $1/\sqrt{n}$ (see Figure 15.2 and surrounding discussion) compared to an equivalent single rotor. In the case of the Innwind, 20 MW, 45-rotor design, the force–distance product is reduced to ~15% of the value of the equivalent single rotor.

10.5   Drive-Train Torque

About 40% of wind turbine cost may typically be associated with conventional drive-train systems (see Table 9.1). For any given power rating and rotor diameter, the level of drive-train torque is determined by the choice of design tip speed. The calculation of rated torque is then elementary, but the impact on weight and cost of many important components is profound. This is very evident considering innovative power take-off options (Section 6.13), high-speed rotors (Section 7.2), the Carter concept (Section 10.9) and HAWT versus VAWT comparisons (Chapter 13).

10.6   Representative Baseline

Consider now the evaluation of a wind turbine system with some degree of innovation. The innovative content may be isolated in a single component such as the gearbox or more widespread. In a typical evaluation of a new concept, a representative baseline wind turbine is established usually as close to the new concept turbine as possible.

This will typically involve matching rotor diameter and generator rating, but it must be noted that some new concepts will optimise differently and logically demand a departure from the usual relationships between rated power and rotor diameter. Thus, it is hard to detail rules for determining a suitable baseline, as often it much depends on the specific innovation and the scope of the evaluation.

The relevant question regarding the merit of an innovation is always, compared to what? The guiding principles for the choice of a baseline system are that the system is

- *state of the art*: it is easy, but of no value (except for irresponsible marketing) to claim an innovation provides 20% improvement relative to a system that is substandard;
- mainstream and therefore of proven viability;
- as good a match to the innovative system as is possible in terms of salient parameters like rotor diameter, rated power, mode of operation, design site class, and so on;
- able to be characterised well in terms of loads, costs, and so on.

In the context of a first analysis of an innovative system, the baseline is a mainstream wind turbine of an appropriate scale which is likely to conform to the typical cost split data such as given in Table 9.1. Depending on the affordable level of initial analysis, loads may need to be estimated or calculated more rigorously using available BEM software according to relevant standards. Detailed data on any real commercial wind turbine is not generally available and the baseline turbine may need to be created as a model in general conformity to such designs. Sometimes suitable reference wind turbines such as the National Renewable Energy Laboratory (NREL) 5 MW [13] are in the public domain, having been created by wind energy institutions essentially for research and development comparison purposes.

10.7   Design Loads Comparison

An ideal comparison of an innovative wind turbine system with a conventional baseline system will, in principle, involve a full re-optimisation of all the main design variables of the new system to minimise its COE. In general, this is too complex and the approach is to keep as much in common between the innovative system and baseline in order to make the comparison as clear as possible. This is best understood considering a specific example.

Suppose a new intelligent control strategy and/or a better blade design can allow a rotor with 5% greater diameter without exceeding the original load envelope of the baseline design. It should first be noted that this can rarely be strictly true across the board. If the new rotor is more productive (there would be no point if it was not), then it is quite likely to attract more edgewise rotor fatigue loading in consequence. However, it may well be that major overturning moments, rotor thrust loads and other system loads are not exceeded compared to baseline.

In the simplest comparison strategy, all the systems downstream of the rotor are kept the same (because they experience no significant load increases) and therefore at the same cost. Rotor blade, possibly pitch system, and/or hub costs are adjusted as appropriate according to load or design changes. Accounting for extra energy produced by the larger rotor will then allow a comparison with baseline on COE. There is a catch, however. If the rotor is made 5% larger in diameter, then the designer must choose between maintaining rotor shaft speed, thereby increasing the design tip speed by 5% (implying increased acoustic noise emission) or maintaining the design tip speed. Maintaining tip speed is the most logically consistent approach, but it will imply a 5% decrease in rotor shaft speed and hence a change in gear ratio and 5% increase in rated drive-train torque levels. The argument continues that if 5% increase in tip speed is all right (and that is what most manufacturers would do as they would not want to modify or overload drive train components), then why did the original design not run at higher speed and lower torque thereby saving cost in the drive train?

Thus, if tip speed is increased so that shaft speed is maintained, it is wrong to attribute all of the resultant COE decrease to improved blade design or more intelligent rotor control. It may be the case, however, if the control strategy is more aggressive in terms of energy capture that some loads will increase requiring, say, a marginal increase in nacelle structure and cost to accommodate them. It may also be that this approach has greater net benefit and can further reduce COE at least viewed from the perspective of new design rather than an expedient adaptation to an existing one. Nevertheless, this is much more complex to evaluate and, depending on resources to be assigned at the design study stage, the simple comparison strategy may be the most appropriate. Even if the result is not optimum and COE benefit is not fully assessed, as has been observed previously, the simple comparison should show (or not show) a large enough benefit to justify the disruption of innovation.

In some cases, especially with larger diameter rotors and also in general, it may be desirable to limit rotor thrust and blade bending loads around rated wind speed. This is effected by introducing a schedule of minor pitch changes ahead of rated wind speed. In reducing thrust, power is also reduced but the trade-off may be beneficial. In the case of an innovative wind turbine being compared with a baseline, such thrust limiting strategy can be a way of keeping loads broadly similar and may facilitate comparisons.

Figure 10.2 shows how the steady-state thrust of the high-speed rotor (Section 7.2) can be restrained not to exceed the level thrust limit of a standard speed baseline design

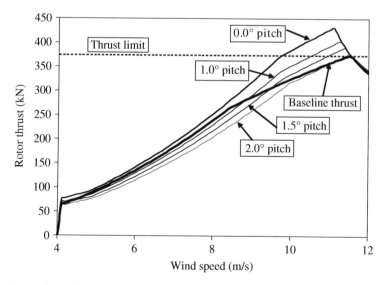

Figure 10.2 Thrust limiting strategy.

(the designs being compared are 3 MW and 100 m in diameter) by progressively pitching blades into wind up to 2° approaching rated wind speed. Although this comparison considers only steady-state loading, peak dynamic loads in normal operation can on average be expected to relate to maximum steady-state loads at rated wind speed.

10.8 Evaluation Example: Optimum Rated Power of a Wind Turbine

Outline COE evaluation methodology can also enlighten generic issues as in the following example.

The reasons for limiting the power of a wind turbine, typically at a wind speed of around 50% greater than the annual mean wind speed, were explained qualitatively in Section 9.2. Let us examine that now in a slightly more quantitative way.

Figure 10.3 illustrates power curves of a 65-m-diameter rotor. A rating of 1 MW at 11 m/s wind speed would be quite typical. However, ratings up to 5 MW are explored. Various assumptions are made about rotor and drive-train efficiency. The rotor is assumed to operate in ideal variable speed mode at a constant maximum C_p of 0.5. At all rated power levels, the generator is assumed to be a doubly fed induction generator (DFIG) type and have efficiency characteristics that depend on part load (Figure 6.16). This has the effect that the higher the power rating, the lower is the part load fraction and the related system efficiency at any given (below rated) wind speed. This is illustrated by a magnified snapshot (Figure 10.4) of a portion of the power curves of Figure 10.2. The curves in Figure 10.3 are not separately identified, but output power is decreasing with increasing drive train rating from 1 to 5 MW.

Based on the cost fraction data of Table 9.3, the variation of lifetime cost with changes in power rating is estimated (Figure 10.5). This cost variation is as a ratio to the cost at a nominal rating of 1 MW. Assumptions are made about which elements of Table 9.3 vary with change of rated power and which are independent of rated power. Power electrics

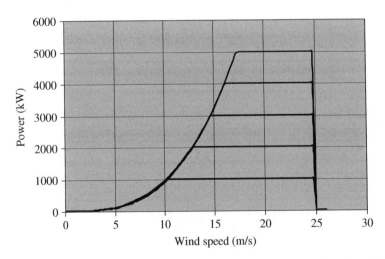

Figure 10.3 Power curves of a 65 m diameter rotor at power ratings from 1 to 5 MW.

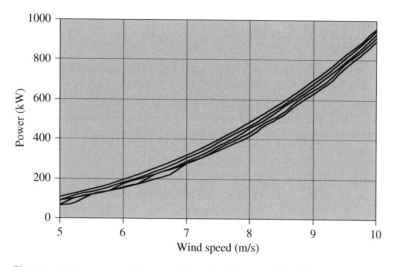

Figure 10.4 Power curves in a selected region below rated wind speed.

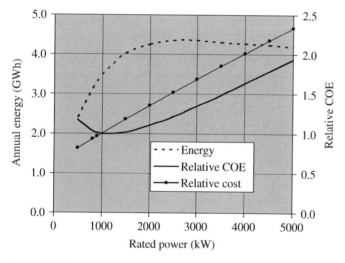

Figure 10.5 Energy and cost of energy.

obviously scales closely as rated power. So does the gearbox cost as its torque rating will increase in proportion to rated power.[2] Increasing power rating also impacts on blade cost due to extra fatigue around rated power levels, but this is considered to scale as less than power and is assumed to vary as the square root of power. Needless to say these assumptions are crude but have some reasoned basis and allow some understanding to be gained where there is not scope to do a much more complete study which would re-evaluate all design loads and cost impacts as power rating is varied.

_____

2 This is true only when power alone is scaled. In scaling with similarity, diameter also increases and the gearbox again scales as torque, but this is now as cube and not as square of diameter.

Overall costs then appear plausibly to vary almost linearly with rated power (Figure 10.5). Energy (based on an assumed annual mean wind speed of 7 m/s) increases with power rating but eventually levels off and even decreases (Figure 10.5) due to the comparatively poor efficiencies associated with low part load operation in the most productive wind speeds. A minimum appears in the COE in the region 1–1.5 MW rating. Whilst this analysis cannot be used as a precision basis to select an optimum power rating, it enlightens all the key issues.

The foregoing analysis can, in principle, be extended to the offshore situation. In particular, the cost of electrical systems offshore (within and beyond the turbines themselves) will be a much larger proportion of lifetime costs than onshore. Thus, much larger rotor diameters in relation to power rating may be justified. This was not happening in practice (although the argument had certainly been noted, e.g. by Snel [14]), until the latest design of Vestas, the V164, 7 MW, was made public (April 2011). A rotor diameter of 164 m at a rated power of 7 MW implies a specific rotor loading of 330 W/m^2, well below the industry average of around 400 W/m^2. Moreover, this is associated with a design annual mean wind speed of 11 m/s and this combination of a particularly large diameter with high annual mean wind speed is exactly the opposite of what would be normal practice on land. This design also has a high tip speed (104 m/s), as has been suggested is appropriate for offshore designs (Chapter 7).

A larger rotor diameter not only impacts on wind turbine costs but may also impact on foundation costs, which are a very significant part of total costs offshore. Thus, detailed modelling of load cost impacts related to offshore foundation concepts (which will vary with seabed conditions and water depth) is needed to determine an optimum rating offshore.

10.9 Evaluation Example: the Carter Wind Turbine and Structural Flexibility

In comparison to mainstream European designs, some lighter weight American designs emerged in the 1980s. This was expressed [15] in such terms as the 'American willow' that bends in the breeze being compared with the more unyielding 'European oak'. Thus, a mystique developed that the principal ingredient in the American lightweight designs of the 1980s and 1990s was structural flexibility. Flexible rotors and towers clearly can provide a system with reduced loads, reduced material demand and hence reduced cost. Indeed, the analysis of a downwind high-speed rotor (Section 7.2) strongly supports this. However, although light weight was often attributed to structural flexibility, a major reason for some of the US designs of the 1980s being lightweight is that they were high tip speed and hence relatively low-torque designs. European designs started conservatively with low design tip speeds, this being partly driven by the need to limit peak power of stall regulated rotors. Subsequently, concern about noise totally inhibited any increase in design tip speed and caused a few European companies that were marketing high tip speed designs to retreat to design tip speeds of about 60–65 m/s and or to develop systems with some degree of variable speed to temper noise generation especially in low wind speed conditions.

An innovative two-bladed, downwind wind turbine design was developed in the early 1980s by Jay Carter. Designs of this type are still being developed by Carter [16]. The

Table 10.1 Design comparison.

	Carter	European	Ratio
Rated power (kW)	300.0	300.0	1.00
Diameter (m)	24.2	29.2	0.83
Tip speed (m/s)	82.4	65.0	1.27
Torque (kNm)	44.1	67.4	0.65
Hub height (m)	47.8	34.0	1.59
Total rotor mass (kg)	1 476	5 000	0.30
Nacelle mass (kg)	3 600	10 000	0.36
Tower system mass (kg)	9 995	20 261	0.49
System mass (kg)	15 110	35 261	0.43

Carter 300 kW design was a plausible candidate for 'American willow' status. The flexible blades allowed a rotation at the blade tip of around 60° in the design extreme wind condition. Data analysis of a Carter 300 wind turbine [17] showed, however, that the alleviation of extreme structural loads by blade and tower flexibility is significantly reduced by dynamic interaction of rotor blades and tower. In addition to having a relatively high design tip speed (82.4 m/s), the Carter 300 had a particularly small rotor diameter (Table 10.1) in relation to its rated power. This is confirmed in Figure 10.6 where general data from a wide range of commercial wind turbines is compared with the Carter design in respect of power density defined as the ratio of rated power to rotor swept area.

The Carter design is so radically different in salient parameters that parallel comparisons are difficult. In Table 10.1, very large mass reductions appear for the Carter 300 compared to a conventional 300 kW machine (not a specific wind turbine design but based on averaged data from European three-bladed wind turbines). These would be

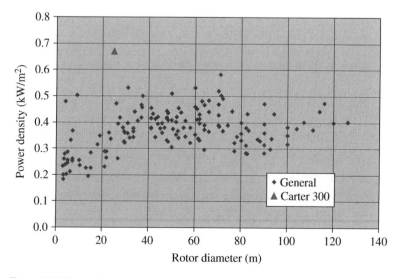

Figure 10.6 Power density.

much less if the comparison were made at the same diameter but would still be significantly favourable for the Carter design. However, a wind turbine of 24.2 m diameter at a typical power density of around $0.4\,kW/m^2$ (see Figure 10.6) would be rated at only 184 kW.

Table 10.1 shows that there is a 35% reduction in rated torque compared to a typical European wind turbine, a 17% reduction in rotor diameter and about 40% increase in hub height. Assuming that power (below rated wind speed) varies as diameter squared and cube of wind speed, and that a wind shear exponent $\alpha = 0.143$ applies in respect of mean wind speed at hub height, then the increase in tower height of the Carter design compensates for power lost in respect of the smaller diameter but not fully. Power below rated is approximately[3] in the ratio:

$$\frac{P_{Carter}}{P_{European}} = \left[\frac{24.2}{29.2}\right]^2 \left[\frac{47.8}{34.0}\right]^{3\times0.143} = 0.79 \tag{10.7}$$

Assume that the small rotor and low top weight of the Carter design along with the concept of supporting the tower with tensioned guy wires allows the torque reduction to be passed through the whole system as a cost and mass reduction. In the spirit of the crude initial evaluations of COE suggested in Chapter 9, assume that the turbine capital cost is 60% of lifetime costs with balance of plant and operation and maintenance (O&M) forming the remaining 40%. The COE benefit of the Carter concept can be compared to the European as baseline at unit COE as

$$\frac{COE_{Carter}}{COE_{European}} = \frac{0.60 \times 0.65 + 0.4}{0.79} = 1 \tag{10.8}$$

This calculation is not intended to be taken seriously – just a pointer that maybe the concept of a small lightweight rotor on a high tower trading cost for energy could just about work. However, it is predicated on the higher tip speed of the Carter design being permissible in terms of acoustic noise emission. If the higher tip speed is acceptable at the design site, then the standard concept turbine can also run at higher tip speed and thereby shed torque, weight and cost. Reworking the calculations of Equations (10.7) and (10.8) for designs with the same tip speed, the torque reduction factor rises from 0.65 to 0.8 (being then associated only with the smaller diameter). The parity for the Carter design associated with the COE ratio of 1 in Equation (10.8) is changed to an 11% penalty.

The lightweight nacelle of the Carter designs facilitates a 'winch-up-and-down' approach to erection and maintenance of the wind turbine which may reduce balance of plant costs. Thus, there may be some maintenance benefit from ground-level access although time and equipment required to lower the wind turbine must be accounted. Transportation may be eased by reduced system weight. Whether site preparations will reduce is unclear considering the need to provide anchorage points for guy wires and a levelled space for access to the wind turbine when winched down. Such factors are not considered in the simple calculations of Equations (10.7) and (10.8).

---

3 The Carter 300 design is further disadvantaged in energy output due to the shape of the power curve, resulting in a very high rated wind speed. However, the factor of 0.84 considers only power below rated wind speed and will only apply to about two-thirds of the energy output. At the crude level of assessment, these influences are assumed to cancel.

The Carter design inspired much government-funded research and development both in the United States and the United Kingdom. In the United Kingdom, the Wind Energy Group in their WEG MS4 design [18] reproduced the blade flexibility of the Carter design, having an internal flex beam. This was very successfully engineered at a much larger scale than any existing Carter designs (600 kW, 41 m diameter), but it proved to be a comparatively complex way of engineering high flexibility (as opposed to simply using higher strength materials with smaller cross sections). Thus, the flexibility and loading benefits were realised in a downwind configuration but probably at increased rather than reduced blade cost. In spite of a successful implementation of flexible system characteristics, the WEG MS4 overall failed to achieve weight reductions comparable to the Carter design mainly for the simplest possible reason. It was designed for a tip speed acceptable in European markets (63 m/s) and therefore had a relatively high torque rating.

The evaluation here has been 'back of the envelope', not at all rigorous and yet very revealing. It is not a matter of reaching definitive conclusions about the quantitative roles of structural flexibility and torque rating in determining system mass. Considering the complex nonlinear dynamics of the Carter design with guyed tower and blades subject to large deflections, the influence of structural flexibility on system mass is not easily evaluated. The main point at issue has been that the mass reduction cannot be credited to structural flexibility alone when, at the same rated power as a standard European design, the Carter design has a much reduced torque rating.

10.10   Evaluation Example: Concept Design Optimisation Study

GL Garrad Hassan is developing wind turbine system cost modelling as part of a new conceptual design tool. The purpose of this tool is to address concept design decisions including optimisation problems about wind turbine configuration considering possible ranges of key design variables and evaluating their impact on COE. The modelling is based on simplified engineering models of all main components, which in turn rely on identification of a few design driving loads and simplified modelling of the loads themselves including load scaling trends such as those illustrated in Section 4.8. Component masses are then estimated and related to cost. This kind of modelling is extremely challenging in aiming to realise simplifications which capture the most important elements of behaviour without introducing unrealistic effects. The model is anchored to engineering design considerations and therefore built from the ground up. Wherever possible the use of empirical trend data is avoided. Nevertheless, some feedback into the cost–mass relationships is essential to calibrate costs and ensure that results are realistic in terms of the approximately known relative costs of components as expressed in cost split data such as is illustrated in Table 9.1.

A study using the conceptual design tool considers an upwind, three-bladed, pitch-regulated, variable speed wind turbine design with typical operation (cut-in wind speed of 3 m/s, shut-down at 25 m/s), glass blades (carbon excluded), drive train with two-stage gearbox with a ratio of 50 and a multi-pole permanent magnet generator (PMG). The drive-train modelling incorporated simplified relationships, some of them derived from the detailed PMG drive-train model discussed in Section 6.12. The impact

Table 10.2 Parameter ranges.

	Minimum	Maximum	Step size
Rating (MW)	6.5	7.5	0.5
Diameter (m)	138	172	1
Design tip speed (m/s)	80	100	5

of parameter variations as indicated in Table 10.2 (in total, 525 wind turbine designs) was considered. The turbine design is targeted at a relatively deepwater offshore project at a location with an annual mean wind speed of 8.9 m/s. Simplified models of offshore balance of plant and O&M were employed, which captured basic effects such as increase in unit turbine size reducing O&M cost per megawatt installed. The wind turbines (considered in the context of large offshore projects of around 500 MW total rated capacity), balance of plant and lifetime O&M costs are approximately equal shares of lifetime cost with variations being introduced according to changes in the parameters of Table 10.2.

The results (Figure 10.7) are in three quite distinct crescent-shaped groups, each associated with one of the three turbine power ratings. The data is classified according to the five tip speed ranges of Table 10.2. It is apparent that the highest tip speeds which provide the lowest rated torque and most economic drive train generally lead to the lowest COE. This result is non-trivial. Especially with upwind designs where blade flexibility must be limited, higher tip speeds lead to blades that are more slender and in some cases more highly stressed and hence heavier and more expensive. On this relatively high wind speed site, the highest power rating is obviously associated with greatest annual energy production. It is also associated with minimum COE and this arises from benefits to balance of plant and O&M in having fewer turbines per megawatt of installed capacity. The

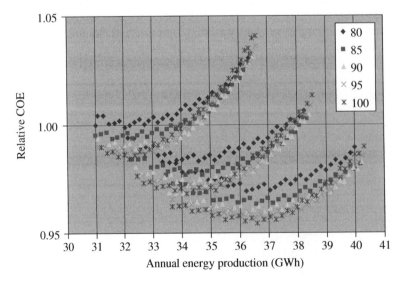

Figure 10.7 Cost of energy evaluation.

COE is expressed in relative terms and it can be seen that all the parameter variations are encompassed by a range of ~5% relative to unit COE, which is arbitrarily based on the first design evaluated with rated power of 6.5 MW, diameter of 138 m and tip speed of 80 m/s.

The overall minimum COE corresponds to a parameter set with a rating of 7.5 MW, diameter of 151 m and tip speed of 100 m/s. The models are still at a preliminary stage of development and these results are not at all definitive, but they certainly illustrate an approach to rationalising conceptual design decisions using cost modelling based on engineering models of components which account for the impact of changes in design loads.

10.11   Evaluation Example: Ducted Turbine Design Overview

Consider the following 'back of the envelope' evaluation (Figure 10.8) comparing a ducted turbine and a bare turbine with diameter equal to maximum duct diameter. Extreme loads, drive-train torque and power performance/energy capture are considered.

10.11.1   Extreme Loads

The tower base bending moment in an extreme storm when the rotors will be parked or idling will be proportional to the total thrust on the rotor system if the rotor hub heights are the same and also affected by wind loading on the support structure. Consider first the rotor loading excluding the support tower but including the duct loading in the case of the ducted rotor. Let the radius of the rotor in the ducted turbine be a fraction, k, of the radius, R, of the bare turbine,

The thrust on the bare turbine is

$$T_{bt} = 0.5\rho U_0^2 \pi R^2 C_t$$

The thrust coefficient when the rotor is parked or idling may be approximated as

$$C_t = \sigma C_{D1}$$

where σ is rotor solidity and C_{D1} is an average drag coefficient representative of the rotor aerofoils. For the ducted turbine the total thrust, T_{dt} is the sum of T_d, thrust on the duct

Figure 10.8 Comparison of ducted and bare turbine systems.

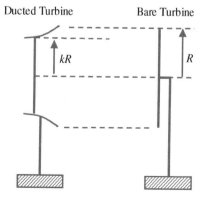

Ducted Turbine   Bare Turbine

kR R

and T_r, thrust on the rotor. Evaluating the thrust on the duct as the product of wind pressure, projected area and a drag coefficient, C_{D2}, T_{dt} is determined as

$$T_{dt} = 0.5 \rho U_0^2 \pi R^2 \{(1-k^2)C_{D2} + k^2 \sigma C_{D1}\}$$

The ratio of total thrust on the ducted turbine to that of the bare turbine is then

$$T_{ratio} = \frac{T_{bt}}{T_{dt}} = (1-k^2)\frac{C_{D2}}{\sigma C_{D1}} + k^2 \tag{10.9}$$

Trying some reasonable values for the terms in Equation (10.9), such as $k = 0.8$, $C_{D2} = 1.5$, $\sigma = 0.05$, $C_{D1} = 1.2$ determines $T_{ratio} = 9.64$.

This shows that the thrust on the ducted turbine system and base bending moment, if the rotor centre is at the same elevation, could be ~ 10 times that of the bare turbine. Although this is a very simplified calculation, it indicates that there is probably a very substantial increase in extreme storm loads on a ducted turbine system and inevitable cost implications. Depending on site class and specifics of particular designs, the extreme load in a storm gust may not be designed for the tower of a conventional 'bare rotor' wind turbine and fatigue loading in normal operation may dominate. However, it is clear that if this extreme load is magnified to the extent it may be in a ducted turbine arrangement, then this could be the dominant load case for the support structure of a ducted turbine. An important point to be taken from this is that such calculations should be undertaken at an early stage in design and inform economic analyses that may be used to justify investment.

10.11.2 Drive-Train Torque

Suppose now that the ducted turbine can realise a flow augmentation factor ~ 2 when optimally loaded. This represents a high level of performance for a ducted turbine which, however, appears to be realised considering power measurements of some of the designs of Kyushu University discussed in Chapter 18. The conventional wind turbine (large, three-bladed, variable speed, variable pitch) may have rotor blades optimally designed for a tip speed ratio of 9 with compromise on speed control limiting maximum tip speed to 80 m/s and performance at rated wind speed of, say, 11 m/s, implying a tip speed ratio of 7.3 at rated power. If the duct induction is doubling the local wind speed through the rotor, then preserving that local tip speed ratio with a tip speed limit of 160 m/s would be rather excessive. However, Takahashi *et al.* [19] have demonstrated substantial attenuation of tip noise in a duct related to rapid cancellation of the tip vortex by interference with induced vortices. So, perhaps, a maximum tip speed of 120 m/s could be considered for the ducted turbine. With $k = 0.8$, as in the illustrative values of the preceding extreme loads comparison, the rated rotor speed of the ducted rotor would then be a factor, $120/(80 \times 0.8) = 1.875$, higher than the bare rotor and almost halving weight and cost of drive train.

10.11.3 Energy Capture

If the flow augmentation of 2 in the optimally loaded state is realised, then the upstream energy source area is effectively doubled. In terms of power, this is equivalent to the power of a bare rotor with an increase in diameter of $\sqrt{2}$. Thus, for power production estimates, in the comparison with $k = 0.8$, the ducted rotor may be considered as equivalent in power performance to a bare rotor of diameter $0.8 \times \sqrt{2} = 1.13$. Rated wind speed (upstream external wind speed) of the ducted rotor, U_{rd}, compared to the bare rotor of unit diameter with an assumed rated wind speed of U_r m/s is then estimated from the equation; $U_{rd}^3 \times 1.13^2 = 11^3 \times 1^2$ giving $U_{rd} = 10.1$ m/s. Now if the ducted turbine is equivalent to a bare rotor of diameter 1.13 greater than the bare rotor of duct exit area considered in this comparison, it will produce more energy below rated power by a factor approximately as $1.13^2 = 1.28$ and if the turbine spends two-thirds of its operating life below rated, maybe 18% more energy overall than the rotor of area equal to the duct exit area. Note that under the assumptions given, the local wind flow through the ducted turbine at rated upstream wind speed ~ 10 m/s is ~ 20 m/s and the *kinetic energy* through the rotor plane is augmented by a factor of 8. However, as has been discussed at length in Chapter 1, kinetic energy cannot be extracted by the rotor. The power depends on the pressure difference developed across the duct and is proportional to the mass flow rate which, in this case, is doubled (relative to a bare rotor of same diameter as the ducted rotor).

Considering the COE split for the conventional turbine system as in Table 9.3, and the rough comparisons developed here regarding structure loads, drive-train torque and power performance, a rudimentary COE model can be created. There is as yet no final conclusion from these outline calculations except that they should be undertaken from day one in any design development of a ducted turbine and embodied in COE estimates that gradually become more refined and solidly based as work progresses. They will, at an early stage, inform where the key competitive strengths and weaknesses of the design lie and, if appropriate, develop the case for investment.

With many innovations, all too often, some quite sophisticated and sometimes extensive analyses are presented relating to one area of design (say, computational fluid dynamics (CFD) on aerodynamics) and the concept is promoted to investors without a rounded top-level view of all important aspects of design and often in consequence with unrealistic expectations of competitive potential.

References

1 The Nordex N60/1300 kW Wind Turbine and the 'Gearless' Enercon E-58, http://crestdl.lboro.ac.uk/outside/studyWithCrest/caseStudy/wp03cs01.pdf (accessed April 2011).

2 Considerably Higher Yields – Revolutionary Rotor Blade Design. WindBlatt, Enercon (3, 2004).

3 Johansen, J. and Sørensen, N.N. (2006) Aerodynamic Investigation of Winglets on Wind Turbine Blades Using CFD. Report No. Risø-R-1543(EN), Risø National Laboratory. ISSN: 0106–2840, ISBN: 87-550-3497-7.

4 Johansen, J., Madsen, H.A., Sørensen, N.N. and Bak, C. (2006) Numerical Investigation of a Wind Turbine Rotor with an Aerodynamically Redesigned Hub-region.

5 Nheolis 3D Wind Turbines, http://www.nheolis.com/en (accessed April 2011).

6 Zhu, W.J. (2009) Aero-acoustic computations of wind turbines. PhD dissertation, Technical University of Denmark.

7 Whelan, J., Thomson, M., Graham, J.M.R. and Peiro, J. (2007) *Modelling of Free Surface Proximity and Wave Induced Velocities Around a Horizontal Axis Tidal Stream Turbine.* Proceedings of the 7th European Wave and Tidal Energy Conference, Porto, Portugal, 2007.

8 Di Maria, F., Mariani, F. and Scarpa, P. (1997) *Chiralic Bladed Wind Rotor Performance.* Proceedings of the 2nd European & African Conference on Wind Engineering, Palazzo Ducale, Genova, Italy, June 1997, pp. 663–670.

9 Scerbina, A. (Inventor) (2009) Vertical axis rotor-type wind turbine. US Patent 0, 202,356 A1, Published August 2009.

10 Garvey, S.D. (2010) Structural capacity of the 20 MW wind turbine. *Proceedings of the IMechE, Part A: Journal of Power and Energy*, **224** (A8), 1083–1115. doi: 10.1243/09576509JPE973

11 Jack, C.H.B., inventor and proprietor (1992) Free Rotor. Patent WO1992020917 A1, published November 1992

12 Watson, W.K. (inventor) (1988) Space frame wind turbine. US Patent 4,735,552, Published 5 April 1988.

13 Jonkman, J., Butterfield, S., Musial, W. and Scott, G. (2009) Definition of a 5-MW Reference Wind Turbine for Offshore System Development. Technical Report NREL/TP-500-38060, National Renewable Energy Laboratory.

14 Snel, H. (2003) Review of aerodynamics for wind turbines. *Wind Energy*, **6** (3), 203–211, Special Review Issue on Advances in Wind Energy,.

15 Lynette, R. (1988) Comments in public discussion. BWEA 10, London, March 1988.

16 Carter Wind Energy, http://www.carterwindenergy.com (accessed April 2011).

17 Quarton, D.C., Schwartz, H.U. and Wei, J. (1996) *Monitoring and Analysis of a Carter 300 Wind Turbine.* Proceedings of 18th BWEA Conference, Exeter, 1996.

18 Armstrong, J.R.C. (1996) *Progress with the WEG MS4-600 Turbine.* Proceedings of 18th BWEA Conference, Exeter, 1996.

19 Takahashi, S., Hata, Y., Ohya, Y. *et al.* (2012) Behavior of the blade tip vortices of a wind turbine equipped with a brimmed-diffuser shroud. *Energies*, **5**, 5229–5242. doi: 10.3390/en5125229

Part III

Design Themes

11

Optimum Blade Number

11.1 Energy Capture Comparisons

The design speed of a rotor is often limited by acoustic noise and ultimately by aerofoil drag. As is evident from Equation 2.1, choice of rotor speed in conjunction with aerofoil design lift in turn determines rotor solidity. Figure 1.10 demonstrates that the maximum attainable C_p increases with blade number because the tip losses reduce. However, at any given total rotor solidity, more blades imply more slender blades which may become problematic structurally. Thus, multi-bladed rotors tend either to be high solidity and low speed, as for water pumping applications, or have rims (see Chapter 18) that can provide structural stability to the blade system. Although the optimum speed for a multi-bladed rotor system reduces and C_p increases with increasing blade number (Figure 1.10), the requirement for high torque for water pumping applications dictates a suboptimal design speed and limits C_p max to around 0.3.

Large wind turbines for generating electrical power require comparatively high running speeds for efficient operation (design tip speeds most commonly in a range from 60 to 80 m/s) and hence have comparatively low solidities (~5%). Such wind turbines have almost exclusively been one-, two- or three-bladed.

Evaluation of the relative merits of one-, two- and three-bladed rotors is very challenging. Each has quite different implications for the general design of the hub area, the dynamics of the machine and the operational strategies. Consider first energy capture issues in terms of rotor power coefficient characteristics.

Figure 11.1 may be derived from Equation 1.82. Assuming aerofoils with a peak lift-to-drag ratio of 120, as may be typical of current state-of-the-art designs, the overall maximum C_p (based on a uniform axial induction of 1/3) for any given number of blades and the associated optimum tip speed ratio are indicated.

For one-, two- and three-bladed rotors, C_p max is predicted to be 0.46, 0.485 and 0.50, respectively. In round numbers, the associated optimum λ values are, respectively, 14, 11 and 9. Historically, one-, two- and three-bladed rotors were designed to run at quite different maximum tip speeds. In general, the one- and two-bladed rotors ran at much higher tips speeds than their three-bladed counterparts, but the logic of a one-bladed rotor being designed for higher rotational speed than two-bladed designs or any real consistency with the data of Figure 11.1 was rarely observed. For example, the Gamma 60 two-bladed wind turbine [1] had a remarkably high tip speed of 138 m/s.

At present, three-bladed rotors are very suitably designed for near-optimum design λ with typical design tip speed ratios around 9. However, to design a two-bladed

Innovation in Wind Turbine Design, Second Edition. Peter Jamieson.
© 2018 John Wiley & Sons Ltd. Published 2018 by John Wiley & Sons Ltd.

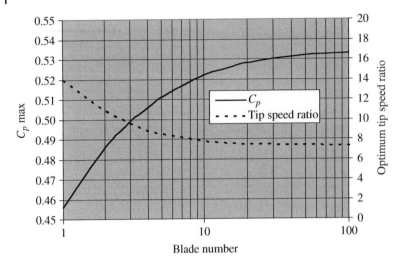

Figure 11.1 Optimum rotor performance at lift-to-drag ratio of 120.

or one-bladed rotor optimally implies higher design tip speeds than are presently acceptable (at least on European land-based sites) in terms of acoustic noise emission. If the one- and two-bladed options are considered for offshore on the basis that acoustic noise will be less restrictive, there is a serious energy penalty with the single-bladed rotor. This is associated with power levels below rated wind speed that may be over 8% less than for a three-bladed rotor of the same diameter. Although ideally the power loss with a two-bladed rotor, as compared to three-bladed, is much more marginal, it should not be overlooked as it may be just as significant economically as some potential benefits of a two-bladed rotor (say, in simpler rotor erection) which often loom larger in general debate.

If design λ is set at 9 so that one-, two- and three-bladed designs may compete on land with similar rated tip speed and similar acoustic noise emission, then the one- and two-bladed rotors are more distinctly suboptimal and the associated maximum C_p values (based on Equation 1.82) are, respectively, 0.44, 0.48 and 0.50.

11.2   Blade Design Issues

Consider now issues of blade design, blade mass and rotor mass. At any given design tip speed ratio, with similar blade aerofoils, blade materials and structural concepts, the running speed, solidity and hence total blade surface area for one-, two- or three-bladed rotors will be similar.

Suppose that the design blade bending moment of the single bladed rotor is an out-of-plane bending moment of magnitude M, and the chord width at some chosen radial station is c_1. Then, approximately (neglecting tip effects), the design moments of a blade of an equivalent n-bladed rotor is M/n. The underlying assumption is that optimum rotors are designed with blades having similar aerofoil choice, geometric and structural similarity and equal overall rotor loading. The same rotor loading is a broadly reasonable assumption if operational loads dominate, as all optimum rotors have much the same thrust coefficient.

Also, if the total rotor solidities and hence total projected planform areas are similar, then at any representative section, the chord width of the n-bladed rotor is $c_n = c_1/n$. The associated section modulus Z_n is then proportional to the square of the section thickness and the width of the cap spar. If the sections are geometrically similar, aerofoil section thickness and cap spar width will in turn be proportional to chord width. Under these assumptions, the stress on a section on an n-bladed rotor is

$$\sigma_n = \frac{M}{nZ_n} = \frac{k_0 M n^2}{c_1^3} \propto n^2 \quad (k_0 \text{ is a constant}) \tag{11.1}$$

Thus, σ_n is n^2 times greater than σ_1, the stress on the blade of a single-bladed rotor. The single-bladed design is then clearly the most structurally efficient, essentially because all the installed blade surface area is in a single beam.

However, consider a more realistic scenario (possibly offshore) where the one-, two- and three-bladed rotors are designed, respectively, for their optimum tip speed ratios. Assume also that the surface geometry of the one-, two- and three-bladed rotor blades is similar but the internal thickness of the cap spars are determined to equalise surface bending stress. Recalling from Equation 1.72 that the optimum chord width is inversely proportional to the square of design tip speed ratio, λ_n, and the number of blades, n, and assuming also that the internal thickness of the spar cap, d_n, is small compared to the aerofoil section depth, it is readily shown that:

$$\sigma_n = \frac{k_0 M}{n d_n c_n^2} = \frac{k_0 M n}{d_n c_1^2} \left\{ \frac{\lambda_n}{\lambda_1} \right\}^4 \tag{11.2}$$

The mass m_n of all n blades of an n-bladed rotor can be expressed as

$$m_n = k_1 n \rho R c_n d_n \propto n \left\{ \frac{\lambda_n}{\lambda_1} \right\}^2 \tag{11.3}$$

where ρ is an average blade density.

The mass ratio of three blades of a three-bladed rotor to two of a two-bladed rotor is predicted from Equation 11.3 to be 1.06 and one-, two- and three-bladed rotors operating at optimum tip speed ratio at the same design blade bending stress are compared in Table 11.1. The optimum tip speed ratios of Table 11.1 are based on Figure 11.1. The assumptions employed in deriving Table 11.1 are that the main structural element of the blade is a beam represented as a simple cap spar (see Figure 3.4). All blades are assumed to be geometrically similar in external geometry. Thus, the depth of the beam (equivalent to aerofoil thickness) and width of the beam are assumed to be proportional to blade section chord width but the internal thickness of the caps is varied as necessary to maintain a constant stress.

Under these circumstances, the chord of an n-bladed rotor, $c_n \propto 1/n\lambda_n^2$ and the thickness of cap, $d_n \propto n\lambda_n^4$. Thus, since rotor diameters and blade lengths are assumed to be the same, the mass of the cap spars of an n-bladed rotor (assumed proportional to the mass of the rotor) is proportional to $nc_n d_n$.

In Table 11.1, for comparison purposes, the chord, spar thickness and mass of one of the blades of the three-bladed rotor are set to unity. Note that the two-bladed

Table 11.1 Rotor mass comparison.

Blade number	1	2	3
Optimum λ	14	11	9
Chord ratio	1.24	1.00	1.00
Spar thickness ratio	1.95	1.49	1.00
Rotor mass ratio	2.42	2.99	3.00

and three-bladed designs appear to have the same chord. This is a consequence of Equation 2.1. For the two-bladed design, $B\lambda^2 = 2 \times 11^2 = 242$, whereas for the three-bladed design, $B\lambda^2 = 3 \times 9^2 = 243$. However, the two-bladed rotor has 50% higher loading per blade and needs additional internal structure. Hence, the rotor masses are almost identical. The rotor (blade) mass of the single-bladed rotor is definitely the least, but this takes no account of the requirement for a counterbalance weight.

It will be obvious that in a real-life comparison of detailed designs of one-, two- and three-bladed rotors, designing loads may be quite specifically related to details of the operational strategies and arise in different load cases for each design. Also, the structural solutions may not be very close to similarity. Moreover, in general, deflections need to be considered as well as stresses, although it can be shown that, for a prescribed deflection limit, the mass is proportional to $(B\lambda^2)^2$ which also leads to similar masses for the two- and three-bladed rotors. Although the derived blade mass estimates should not therefore be taken too seriously, it appears that two- and three-bladed rotors, designed for their respective optimum tip speed ratios, will have similar rotor masses (Table 11.1). This certainly challenges casual assumptions which still have some currency, for example, that two-bladed rotors will universally be lighter than three-bladed rotors by as much as in the ratio 2/3.

Two-bladed rotors have been suggested for offshore application. Perhaps especially on jacket structures with structural capacity strongly affected by water depth, much larger rotors (in relation to a defined rated power) can be supported. The much lower specific rotor loading implied may well be appropriate to increase capacity factor and optimally exploit the expensive electrical infrastructure. In that event, the reduced rated wind speed associated with the relatively larger rotor diameter will increase the proportion of operation at full rated power and minimise the energy deficit of a two-bladed rotor as compared with three. This, in conjunction with possible benefits in transportation, erection and maintenance, may make a case for the two-bladed rotor.

11.3 Operational and System Design Issues

It is normal to shut down wind turbines in very high winds in order to protect them structurally. The rotor, if stall regulated, may be stopped and locked in a fixed position by a mechanical brake. Extreme rotor torques (arising perhaps in the 50-year return gust) can be very high to the extent that the wind turbine designer may decide to accept the possibility of rare events where the rotor will transiently slip on the brakes rather than pay for an excessive brake capacity. However, most commonly the rotor will be allowed

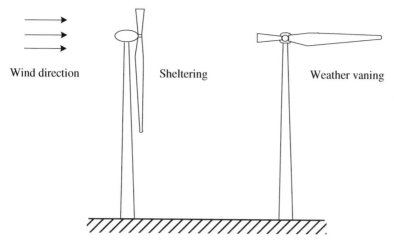

Figure 11.2 Parking strategies of a one-bladed rotor.

to idle, that is, be free to rotate, but with the blades pitched at high angles preventing rotor speeds of more than a few revolutions per minute.

Parking strategies are designed to avoid the much higher blade loads and tower loads that would result if normal operations were attempted under extreme wind conditions. It was established (Section 1.5) that an optimally operating rotor will have a thrust coefficient, $C_t = 8/9$ and heavily loaded operating rotors (see Equation 1.79) may experience thrust coefficients above unity. When the rotor is stopped or at very low speed with blades in the operating position (say, a stall regulated rotor), the thrust coefficient is reduced to a level similar to the rotor solidity, say ~0.05, that is by a factor of 20 for typical large wind turbines. The thrust coefficient is yet further reduced on a pitch-regulated rotor with the blades in idling position.

The one-bladed design allows unique parking strategies (Figure 11.2) – downwind parking with the blade sheltered behind the tower or sidewise parking with a single blade acting as wind vane – which may minimise storm loading impact. These strategies, however, rely on being able to secure a chosen parked blade position.

Disadvantages of a single-bladed rotor design include the following:

- Added mass to provide a counterweight to balance the rotor statically;
- Reduced aerodynamic efficiency (as discussed in Section 11.1) due to single-bladed rotors having the greatest tip loss effect;
- Generally complex dynamics requiring a hinged connection of blade to main shaft in order to relieve loads.

The single-bladed rotor designs of Riva Calzoni, Aeroman, Messerschmidt and others were of too high a tip speed to be acceptable in the modern European market from an acoustic point of view. The high tip speed is not an intrinsic requirement of the single-bladed concept; but, as has been indicated, it is required to optimise the design.

In an overall evaluation of one-, two- and three-bladed rotor technology, it is rather hard to justify single-bladed designs in view of relative energy loss and dynamic complexity.

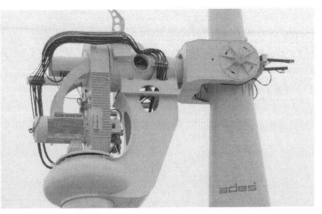

Figure 11.3 ADES single-bladed pendular wind turbine. Reproduced with permission of Manuel Lahuerta.

However, just when the industry seemed to have finally put the single-bladed turbine to bed, a radically innovative single-bladed turbine with a pendular drive train was announced by the Spanish company Aplicaciones de Energias Sustitutivas S.L. (ADES). The generator stator (Figure 11.3) is bolted to the gear box, which is set on a bearing and suspended from the nacelle. The generator rotor is connected to the low-speed shaft via gearing inside the gear box. Thus, the generator, gear box casing and gearing act as a pendulum, smoothing the inherently variable torque input from the single blade.

According to ADES,

> the pendulum not only balances the torque, it also accumulates potential energy by its upward movement and releases it when it drops. Therefore, it acts as a regulator that tends to even out the amount of energy injected into the grid, attenuating peaks and softening power valleys.

The blade has a hinged connection to the hub, as is usual (perhaps essential) with a single-bladed rotor. This relieves the out-of-plane blade root bending moment and decouples it from the yawing moment. This wind turbine is pitch regulated and operates downwind normally in free yaw with a yaw motor to adjust nacelle position for start-up or unwinding power cables (see discussion of free yaw in Chapter 14). According to an ADES data sheet, the 35-m-diameter model is offered for a range of rated output powers from 127 kW (rated wind speed 9 m/s) to 335 kW (rated wind speed 12.5 m/s) with corresponding variable speed ranges up to 50 and 69 rpm, respectively. This implies a range of rated tip speeds from 92 to 126 m/s depending on the chosen generator rating. Thus, some designs will operate at tip speeds much higher than any other present commercial wind turbine. As discussed previously, high tip speeds are necessary for optimised single-bladed wind turbine designs.

The minimal, if unusual, nacelle of the ADES design hints at a striking potential advantage of this concept. According to ADES data for the 2 MW design, the rotor mass is 11 000 kg, the pendulum mass (including generator) is 20 000 kg and the nacelle mass

Figure 11.4 Nordic N 1000 two-bladed wind turbine (Paraje Ojos de Agua, Uruguay). Reproduced with permission of Nordic Windpower Ltd.

is 8000 kg, giving a tower top mass of only 39 000 kg. Industry averages for tower top mass of conventional three-bladed horizontal-axis wind turbines (HAWTs) of 2 MW rating have tower top mass around 100 000 kg. However, a final evaluation of weight saving in the ADES designs must consider, as has been strongly emphasised in Chapter 10 and elsewhere, the basic relationships between tip speed, torque and mass.

Two-bladed rotors (Figure 11.4) are better balanced dynamically than one bladers but still disadvantaged dynamically compared to three-bladed rotors. The rotational inertia of a two-bladed rotor about the yaw axis of the wind turbine is periodic, being maximum when blades are horizontal and minimum when they are vertical. The three-bladed rotor has a constant inertia about the yaw axis independent of the angular position of the rotor blades. In consequence, there are potentially higher cyclic loads on a two-bladed wind turbine.

In some such designs, the hub connection to the blades is rigid and the design implications of higher cyclic loads are accepted. Increasingly, in three-bladed rotor design, individual pitch control of blades is seen as a significant means of employing more advanced control techniques to alleviate differential loading on a rotor, and this option can also be used to relieve the additional cyclic loading on a two-bladed rotor with a rigid hub.

However, in many designs, borrowing a concept well established in helicopter technology [2], a teeter hinge is introduced which preserves a rigid connection of the blades to each other but allows the blade pair to rock relative to the shaft axis, sweeping out a circle of rotation in a plane in general at an oblique angle to the rotor shaft axis.

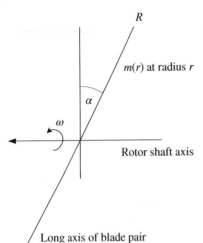

R

m(r) at radius *r*

α

ω

Rotor shaft axis

Long axis of blade pair

Figure 11.5 Teetered rotor dynamics.

The teeter motion is typically constrained within an angular range of perhaps ±7°. In uniform oblique wind flow, the teeter cyclic amplitude is almost independent of blade mass. The centrifugal effect provides the restorative moment within each pendular cycle, but the cyclic amplitude is limited by aerodynamic effects.

Interestingly, the free teeter motion of a teetered rotor is resonant at its rotational frequency.

Referring to Figure 11.5, in which the blade mass per unit length at radius r is m (r), the teeter angle α, and the shaft rotational speed ω, the rotational inertia about the teeter axis is

$$I_{yy} = 2 \int_0^R m(r)r^2 dr$$

and the restoring moment due to centrifugal force is

$$M_c = 2 \int_0^R (m\omega^2 r)r \sin \alpha \, dr \cong 2\alpha\omega^2 2 \int_0^R m(r)r^2 \, dr = \alpha\omega^2 I_{yy}.$$

Hence, the equation of free motion reduces to $\ddot{\alpha} + \omega^2 \alpha = 0$, indicating that the free teeter motion is resonant at the rotor frequency, ω. Under steady conditions with constant teeter angle amplitude, this implies that the rotor rotation is in a fixed plane at an angle $((\pi/2) - \alpha)$ to the shaft axis.

In much of normal operation, teeter action will substantially relieve both cyclic- and turbulence-induced fatigue loading on rotor and tower top. However, teeter stop impacts occur in critical load cases, often in emergency braking and perhaps occasionally in extreme turbulent wind events, and this can negate much of the generally substantial loading advantages of a teetered rotor. The capital cost and reliability issues of a teeter system are also a concern which is avoided with a rigid hub three-bladed rotor.

Some points in favour of a two-bladed rotor are as follows:

- Structural advantage in the blade leading to lower overall rotor blade cost if the tip speed is not much higher than a competing three bladed design;
- Various possibilities of easier rotor erection;

- The possibility of an efficient continuous blade structure passing across the hub centre if a teetered hub design is employed.

According to Lowson [3], the most significant mechanism for broadband noise radiation is due to turbulence passing the trailing edge of the blade. The two-bladed rotor would therefore have lower acoustic noise emission in principle than a three-bladed design because the total length of trailing edges is less. This however can only apply comparing two- and three-bladed rotors at the same design tip speed, which is generally disadvantageous for the two-bladed rotor design and has not occurred in practice.

When there is an energy deficit in a particular (say. Two-bladed) design compared say to a baseline (three-bladed) design, the argument often appears that the energy can be recovered at the comparatively negligible added cost of having slightly longer blades. This type of argument is very common over a wide range of topics where some small energy deficit is at issue but is unsound in principle for the reasons given in Section 10.6. In essence, increasing diameter implies either higher tip speed or higher drive-train torque at fixed tip speed, with implications that are much more significant than any added blade cost. Increasing diameter allowing tip speed to rise will also increase acoustic emission. Similar issues arise where, for example, an improved control system strategy reduces loads allowing a larger rotor to be considered. There may well be an overall cost benefit, but impacts on tip speed or drive-train torque must be accounted for.

The issues of two blades versus three are complex and there may ultimately be little to choose between them on overall technical merit. However, three-bladed wind turbines have been deemed much more acceptable from the standpoint of visual impact. They have dominated the European market for that reason and also the world market because so much of well proven wind technology has emanated from Europe.

11.4 Multi-bladed Rotors

Before the development of the modern wind industry in the late 1970s with its focus on electricity generation, and indeed before widespread electrification in the developed world, wind pumps were well-established technology used mainly for farmland irrigation. In the 1920s, around 6 million were in use in the United States, especially in the Prairies. In order to develop sufficient torque to operate a water pump under light wind conditions, a low-speed high-solidity rotor is required and multi-bladed rotors with blade numbers up to about 40, were developed. An example of an early "Prairie" wind pump is the Crux Easton Wind Engine [4] of John Wallis Titt.

Among modern wind turbines, multi-bladed rotors in a variety of design formats are common at small scales with rated powers up to a few kilowatts. Some multi-bladed rotors for much larger scale systems have been considered. An innovative five-bladed rotor system of NorseTek directed at much lighter rotors in the megawatt range is discussed in Chapter 22.

It is common but incorrect to assume that multi-bladed rotors are essentially inefficient. In fact, for optimum designs, due to the reduction in tip loss with increasing blade number, the opposite is clearly true (see Figure 11.1). It is simply a historical circumstance that, for the original water pumping applications, the high starting torque provided by high-solidity rotors was required and the focus was on simple, durable, inexpensive rotor construction rather than on aerodynamic efficiency.

References

1 Avoloi, S., Arcidiacono, V., Botta, G. *et al.* (1993) *Dynamic Analysis of Gamma-60 Wind Turbine Generator and Control Systems: Design and Experimental Validation.* Proceedings of EWEC, Lubeck-Travemunde, March 1993.
2 Johnson, W. (1994) *Helicopter Theory*, Dover Publications. ISBN: 0486682307
3 Lowson, M.V. (1993) Assessment and Prediction of Wind Turbine Noise. ETSU W/13/00284/REP, 1993.
4 http://windengine.tripod.com/corepage.htm (accessed September 2010).

12

Pitch versus Stall

12.1 Stall Regulation

For many years in the 1980s and 1990s most wind turbine design types were stall regulated, although about half the market by volume was pitch regulated mainly due to the large market share of Vestas.

A stall-regulated wind turbine runs at approximately constant speed in high wind speeds, yet not producing excessive power and achieving this without any change to the rotor geometry. Maintaining nearly constant rotor speed is critical and depends on the connection of an asynchronous electric generator to the grid. In this respect, the grid behaves like a large flywheel, holding the speed of the turbine nearly constant irrespective of changes in wind speed. As wind speed increases, providing the rotor speed is held constant, inflow angles and angles of attack over the blade sections also increase (see Figure 1.8). The blade sections become progressively more stalled and this limits power to acceptable levels without any additional active control.

Stall is an essentially three-dimensional phenomenon involving unsteady aerodynamic effects, and it has remained a challenge for rotor designers to predict the performance and loads of stall-regulated rotors. Blade element momentum (BEM) theory may be used, but then empirically based adjustments to the aerofoil data especially for the inboard regions of the blade are necessary in order to get realistic predictions of power and loads. There has been much discussion of stall delay effects on the inboard rotor [1–3]. A variety of aerodynamic devices, vortex generators, stall strips, and so on [4] have been used to improve the power curve characteristics of stall-regulated rotors.

Power performance in the operating region where the stall process is used to limit maximum power is sensitive to air density (that varies with temperature and air pressure which are related to site location and altitude) and aerofoil surface roughness. Thus, power curves can be significantly changed by rain or contaminants such as dirt and insect adhesion (see Figure 13.7). The blade-to-hub connections are often engineered to allow small changes of pitch setting which may be used to tune the peak power. Analysis shows that at any given tip speed ratio, power is proportional to the cube of rotor speed and the fifth power of diameter. Thus, in addition to the complexity of stall as a process, details of the rotor geometry and the precise running speed in relation to the slip of the generator can also have a marked impact on the maximum power.

Innovation in Wind Turbine Design, Second Edition. Peter Jamieson.
© 2018 John Wiley & Sons Ltd. Published 2018 by John Wiley & Sons Ltd.

The power coefficient, C_p, of a wind turbine is conventionally defined from the relation:

$$P = 0.5\rho U^3 \pi R^2 C_p \tag{12.1}$$

where P is power and ρ is air density. Substituting for the tip speed ratio $\lambda = \omega R/U$ in Equation 12.1:

$$P = 0.5\rho\omega^3 \pi R^5 \left\{\frac{C_p}{\lambda^3}\right\} \tag{12.2}$$

In Equation 12.2, the term $\{C_p/\lambda^3\}$ is a unique function of tip speed ratio, λ. Hence, under conditions of aerodynamic similarity, that is, at any given value of λ including that corresponding to maximum power, power is proportional to the cube of rotor speed and the fifth power of the rotor diameter. These results are commonly termed the 'fan laws' in connection with industrial motored impellers used to pump air or other gases.

A point (P_1, V_1) on a power curve at any given rotor speed ω_1 can then be mapped to a corresponding point (P_2, V_2) on a power curve at a different rotor speed ω_2, according to the equations:

$$P_2 = \left(\frac{\omega_2}{\omega_1}\right)^3 P_1 \tag{12.3}$$

$$U_2 = \left(\frac{\omega_2}{\omega_1}\right) U_1 \tag{12.4}$$

Hence, with only the source power curve available, say, corresponding to operation at 22 rpm, this mapping allows a set of curves to be generated for a range of speeds, as for example in Figure 12.1, where power curves corresponding to speeds ranging from 22 to 25 rpm in intervals of 0.1 rpm are displayed. The curves are essentially similar and corresponding points, for example, the locus of the maxima shown in Figure 12.1,

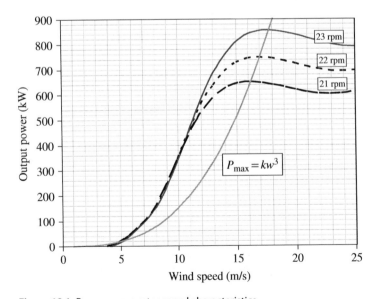

Figure 12.1 Power versus rotor speed characteristics.

are associated with the same tip speed ratio (but different wind speeds on account of Equation 12.4) and have values varying as the cube of the rotor speed. The curves all cross each other in pairs in a region around 11 m/s mean wind speed, but no three or more curves intersect at a single common point.

Stall has a significant advantage over pitch control or active aerodynamic control in that it takes effect automatically. This sets an automatic cap on loads, whereas, with active pitch control, there is the possibility for the control system to be caught out in turbulent wind events with the blades at some inappropriate position incurring an exceptionally high load. There are however many downsides to stall regulation. It can only realise an approximately satisfactory power regulation and not control to a precise set point in the way that pitch regulation can, especially when used in combination with variable speed. Avoiding excessive power in a stall-regulated rotor design usually involves compromising the power curve just below the rated power level and implies some associated loss in energy capture. There is also some increase in acoustic noise emission around stall; and a stall-regulated rotor design may be too noisy at the aerodynamically optimum pitch setting angle, leading to further compromise and reduction of energy output. The generator must provide reaction torque in gust events to maintain essentially constant speed and this means that the peak drive-train torque rating may need to be substantially greater than for a pitch-regulated design.

Stall regulation has also been considered in combination with variable speed [5, 6]. This regulation concept removes the difficulties associated with many of the sensitivities of peak power to rotor speed, contamination, air density, and so on, but still leaves a demanding duty in providing the necessary reaction torque with safe margins.

12.2 Pitch Regulation

The main alternative to stall-regulated operation is pitch regulation. This involves turning the wind turbine blades about their long axis (pitching the blades) to regulate the power extracted by the rotor. In contrast to stall regulation, pitch regulation changes the intrinsic rotor geometry. This involves an active control system sensing blade position, measuring output power and rotor speed and instructing appropriate changes of blade pitch.

The objective of pitch regulation is similar to stall regulation, namely, to regulate output power in high operational wind speeds. Pitch control offers the possibility of regulating quite accurately to a chosen power set point. Stall regulation is always approximate.

Fixed speed operation in combination with variable pitch control was used in early designs, of the 1980s, especially by Vestas, for power limiting. The control proved to be quite problematic in high wind speed operation. In order to protect against excessive loading, manufacturers of fixed-speed, pitch-controlled wind turbines such as Windmaster and Wind Energy Group (both no longer trading) would reduce power set point or shut down wind turbines in wind speeds well below the 25 m/s mean wind speed, which is now virtually an industry standard for shut-down wind speed.

In response to such problems, Vestas developed OptiSlip, a system allowing up to 10% slip above rated power, equivalent to introducing a limited variable speed capability so that the rotor could store energy in gusts reducing demand on pitch activity. In consequence, much better power control could be achieved on full load, but the variable slip system did not confer the advantages of variable speed (potential to regulate acoustic noise or augment energy capture) in below rated wind speed operation which remained as essentially fixed speed operation. Suzlon still adheres to this technology (the system, FlexiSlip, provides up to 17% slip), but otherwise it is fading from mainstream use. The wind turbine market is currently dominated by the doubly fed induction generator (DFIG) concept, with various types of synchronous generators (permanent magnet generator (PMG) especially) about to play an increasing role.

The problems of combining fixed-speed operation with variable pitch regulation have been bypassed with the almost universal adoption of variable speed control in combination with pitch regulation on large modern wind turbine systems. In broad terms, the generator reaction torque can be controlled to respond rapidly to immediate changes in aerodynamic loading, the large rotational inertia of the rotor ensures that rotor speed changes are gradual and the pitch system can act comparatively slowly to the trends in aerodynamic input torque and limit speed changes.

12.3 Fatigue Loading Issues

In the early 1980s James Howden and Co Ltd in Glasgow, Scotland, manufactured wind turbines with rotating tips (Figure 12.2). It was common at that time in Danish stall-regulated designs to employ rotating tips as air brakes to prevent overspeed in fault conditions; but the Howden concept was less usual, employing active control of the blade tips for power regulation. The question arose in which direction to move the blade tips – nose back from wind into stall or leading edge into wind maintaining

Figure 12.2 Howden 26 MW wind farm in the Altamont Pass, California.

attached flow until stall in negative incidence occurred. Field tests on an instrumented 330 kW wind turbine in both modes of operation were conducted and data collected.

Much later in the 1990s, the Wind Turbine Company of Seattle [7], in the context of a particular design, asked GL Garrad Hassan to conduct a study evaluating the comparative merits of pitching (whole blades) into stall or into attached flow as is the norm with modern large, pitch-controlled wind turbines.

Interestingly, the results from both studies were unclear, showing that in overall terms similar levels of fatigue damage accumulated whether pitching into stall or attached flow. This can be understood qualitatively comparing steady-state, out-of-plane blade bending moments or rotor thrust characteristics as a function of mean wind speed. Figure 12.3 compares steady-state rotor thrust characteristics of typical stall- and pitch-regulated designs. The thrust values are normalised[1] to unity at the rated thrust of the pitch-regulated design.

In the case of the stall-regulated design (Figure 12.3), the thrust loading curve, like the power curve, is rising but with moderate gradient above rated wind speed. Thus, instantaneous changes in wind speed, regarded only in terms of the steady-state characteristic, do not produce large changes in load level (the source of fatigue). Nevertheless, in reality, the intrinsically unsteady nature of stall implies that there will constantly be much larger changes in load level than can be inferred from the quasi-steady-state characteristic.

With a pitch-regulated design involving pitching into attached flow, the mean thrust decreases rapidly (Figure 12.3) with increase of wind speed above rated wind speed. Pitching into attached flow reduces lift and thrust, which can even become negative in large rapid pitch excursions as may occur in emergency braking of the rotor. The steep slope of this characteristic implies large changes in load level associated with wind speed changes and hence is a source of system fatigue loading.

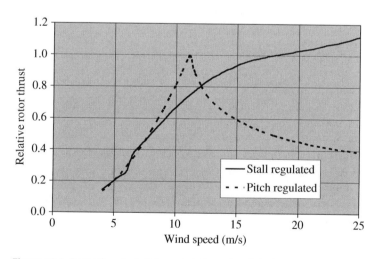

Figure 12.3 Rotor thrust of pitch- and stall-regulated designs.

---

1 The normalisation of the data here to unity at the rated wind speed of the pitch-regulated design is not on account of any confidentiality issue but because the source data referred to designs of different diameters and rated powers.

Thus, in a stall-regulated operation, the gradient of the steady-state thrust (or out-of-plane blade bending) characteristic is moderate but unsteady effects contribute significant fatigue variations, whereas in attached flow, unsteady effects are avoided but a steep gradient of the steady-state load characteristic contributes significantly to fatigue damage leaving no clear picture as to which mode of operation may be best.

12.4   Power Quality and Network Demands

12.4.1   Grid Code Requirements and Implications for Wind Turbine Design

It was noted in Chapter 8 that the general preference for variable speed wind turbines and the comparative decline, especially in new wind turbine designs, of systems based on induction generators without fully rated power converters has been much influenced by electrical network demands in the form of grid codes [8, 9]. 'Grid code' is however a loose term. Strictly, it applies to the technical requirements for all users of the transmission system and therefore to high-voltage systems, generally above 100–150 kV. In the United Kingdom, for example, there is a separate 'Distribution Code' with less onerous technical requirements for users of distributions systems (lower voltage). In some countries there are also relevant technical requirements contained in connection agreements, standards and even power purchase agreements. It may be preferable to describe these items as technical requirements in order to avoid confusion, but the common term in the wind industry for any requirement of an electricity system operator is grid code.

There is a strong tendency for electricity system operators, faced with wind generation for the first time, to copy technical requirements from others. Copying from E.ON and National Grid seems particularly popular. This is sometimes done without considering the particular technical requirements of the particular electricity system and may result in the following:

- Requirements being more onerous than necessary (e.g. a requirement for fault ride-through (FRT) capability, even though the wind penetration is going to be very small for the foreseeable future, and the loss of all wind generation due to a disturbance would have negligible effect on the electricity system);
- Requirements that are completely inappropriate or meaningless for wind generation;
- Conflicts and incompatibilities between documents;
- The operators misunderstanding their own requirements.

Despite this copying, there is a poor level of standardisation between countries; and the European Wind Energy Association (EWEA) is attempting to draw up a standard grid code, concentrating initially on definitions, terminology, presentation of requirements and proving compliance.

Defining requirements is one thing, proving compliance is another. Efforts are being made to make the compliance process easier, including type testing of turbines or components, to remove the need for testing on installed wind turbines. Electricity system operators can be reluctant to accept computer simulation results and so, for example, in Spain there is a recent requirement for all wind farms (even existing ones) to have FRT capability, and to demonstrate this by site tests which typically require substantial test equipment contained in a large vehicle. Some operators will accept simulation results, but have been disappointed by the reluctance of the wind turbine suppliers to provide any detailed description of controller algorithms.

The main grid code requirements relevant for wind turbines [8–10] are as follows:

1) *Voltage and Frequency Tolerance*
   Wind turbines must operate normally within specified voltage and frequency limits (e.g. ±10% voltage, and 47.5–51.5 Hz, with short-term operation down to 47 Hz). The frequency range may be unachievable for existing stall-regulated turbine designs. The voltage range is less of a problem, because, excepting for small projects, the grid connection involves a single large transformer which can be fitted with a tap changer to keep the voltage at the turbine terminals within tight limits.

2) *Fault Ride-through (FRT)*
   The turbine has to be able to remain connected during a major disturbance caused by a fault on the electricity system. This can cause the voltage at the point of fault to drop to zero for periods of around 100 ms, and then take a further period of a second or so to recover to within 10% of the nominal value. The voltage at the turbine terminals will not drop to zero, but nevertheless it is very difficult for simple induction machines to remain connected. DFIGs can cope, but need additional items and control complexity to do so. Wind turbines with fully rated converters can manage the disturbance with little difficulty. Stall-regulated turbines will inevitably overspeed during the fault and then shut down and are therefore not compliant.
   There are several power-electronic products that can be installed in the wind turbine or centrally in the wind farm to make even simple induction generators comply with the requirements. In some cases (as in the United Kingdom), there are also requirements [11] on the behaviour immediately after the fault, for example, that active power output must be restored to 90% of its final value within 1 s, which may be quite demanding for pitch systems.

3) *Reactive Power and Voltage Control* Indicative requirements in order of increasing difficulty are as follows:
   a) To keep the power factor anywhere within a given range which is achievable by all turbine design options, with additional switched capacitors if necessary.
   b) To adjust the power factor (or reactive power export/import) anywhere within a given range, as required by electricity system operator. This is feasible with switched capacitors and reactors, but easier with power electronics. The power electronics can be in the turbine as part of the variable speed drive, or an additional item in the turbine, or a central item within the wind farm.
   c) To adjust the power factor as before, but with dynamic requirements in the speed of achieving response. This probably rules out using switched capacitors and reactors, so systems with fully rated converters and DFIGs have an advantage. Also, the Supervisory Control and Data Acquisition (SCADA) system may not have sufficient communication speed to achieve the demanded dynamic response and this has required some manufacturers to modify their SCADA systems.
   d) To additionally control reactive power in order to contribute to keeping voltage close to a voltage set point, instead of achieving a specific power factor. This needs a measurement of voltage at the point of connection or equivalent.

4) *Frequency Regulation*
   This involves adjusting output power in response to a measurement of system frequency, in order to contribute to network frequency control and is only really

possible for pitch-regulated wind turbines. At present, there are no requirements on speed of response or accuracy. So, in principle, any pitch regulated turbine can comply.

5) *Power Control*
   a) *Cap*: limit the output power to some level when instructed, probably due to some constraint on the transmission system.
   b) *Ramp*: limit the rate of change of output (megawatt/minute). This is easy to do for a positive ramp (increasing wind speed). For a negative ramp (decreasing wind speed), it can be done by forecasting the decrease in wind speed and reducing the wind farm output slowly in advance.

Both caps and ramps are readily achievable, in principle, by a pitch-controlled turbine. It is more efficient, however, if the control is exercised at the wind farm level rather than at turbine level. For example, limiting the ramp rate of each turbine will lose more production than limiting the ramp rate of the whole wind farm. This is even more the case for the summated output of multiple wind farms, but it is not done at present perhaps due to organisational difficulties.

Thus, in summary:

- A *fixed-speed stall-regulated* wind turbine cannot achieve FRT, provide frequency regulation or power control and has difficulty coping with a wide frequency range. It is likely to need extra equipment for reactive power and voltage control.
- A *fixed-speed wind turbine employing active stall* cannot achieve FRT, is likely to have trouble with frequency regulation and voltage control and may have trouble with frequency and voltage limits. It is likely to need extra equipment for reactive power and voltage control.
- A *wind turbine employing DFIG and pitch regulation* meets all requirements; but FRT is not straightforward, requiring careful design and some extra hardware.
- *Wind turbines with fully rated converter systems and pitch regulation* can adequately perform all five functions.

This overview of grid codes related to wind turbine types should clarify how grid code demands have favoured the development of wind turbine systems with fully rated converters.

References

1 Breton, S.-P. (2008) Study of the stall delay phenomenon and of wind turbine blade dynamics using numerical approaches and NREL's wind tunnel tests. Doctoral theses at NTNU, p. 171. ISBN: 978-82-471-1019-5.
2 Himmelskamp, H. (1947) Profile Investigations on a Rotating Airscrew. MAP Volkenrode Report and Translation No. 832.
3 Snel, H., Houwink, R., van Bussel, G.J.W. and Bruining, A. (1993) *Sectional Prediction of 3D Effects for Stalled Flow on Rotating Blades and Comparison with Measurements*. 1993 European Community Wind Energy Conference Proceedings, Lubeck-Travemunde, Germany, pp. 395–399.
4 Grabau, P. (inventor) (2000) LM Glasfiber A/S (applicant). Wind turbine blade with vortex generator. Patent WO/2000/015961, Published March 23, 2000.

5 Leithead, W.E. and Connor, B. (2000) Control of variable speed wind turbines. *International Journal of Control*, **73** (13), 1366–5820, 1189–1212.

6 Muljadi, E., Pierce, K. and Migliore, P. (1998) *Control Strategy for Variable Speed Stall Regulated Wind Turbines.* NREL/CP-500-24311-UC Category 1211. Presented at American Controls Conference, Philadelphia, PA, June 1998.

7 http://www.windturbinecompany.com/technology/index.html (accessed April 2011).

8 European Wind Energy Association (2010) Powering EUROPE: WIND ENERGY and the Electricity Grid, November 2010.

9 Ibsch, M. and Nohme, K. (2009) *Role of Regulations and Standards for the Grid Connection of Wind Turbines–Integrating More Wind Energy.* Proceedings 8th International Workshop on Large-Scale Integration of Wind Power into Power Systems as Well as on Transmission Networks for Offshore Wind Farms, Bremen, October 14–15, 2009.

10 Ciupuliga, A.R., Gibescu, M., Fulli, G. *et al.* (2009) *Grid Connection of Large Wind Power Plants: A European Overview.* Proceedings 8th International Workshop on Large-Scale Integration of Wind Power into Power Systems as Well as on Transmission Networks for Offshore Wind Farms, Bremen, October 14–15, 2009.

11 National Grid Electricity Transmission plc (2010) *The Grid Code*, Issue 4 Revision 4, 18 October 2010, http://www.nationalgrid.com/uk/Electricity/Codes/gridcode (accessed 26 September 2017).

13

HAWT or VAWT?

13.1 Introduction

Figure 13.1 shows a vertical-axis wind turbine (VAWT) of the Darrieus type often referred to as '*egg-beater*' alongside the other main type of vertical-axis design (also originally patented by Darrieus), the H-type. The VAWT may possibly claim priority over the horizontal-axis wind turbine (HAWT) as the oldest type of windmill.[1] This concept was the basis of windmills operating in tenth century Persia. These mills were multi-bladed with blades made from bundles of reeds, and mills of this type continued to operate in areas of the Middle East into the twentieth century.

In the context of modern wind technology for large-scale electricity generation, some prototype large VAWTs of the Darrieus type were tested. In 1988, Eole [2], a 4 MW Darrieus-type VAWT, was developed in a collaborative project with industry, the Canadian government and the utility Hydro-Quebec and this turbine operated for 6 years with 94% availability [3]. During the 1980s, prototype H-type turbines [4–6] based on the ideas of Musgrove, were developed by VAWT Ltd in the United Kingdom. However, the only significant commercial exploitation of VAWT technology was by FloWind Inc. At its peak in 1987, FloWind had installed over 500 turbines in California, representing a total rated capacity of over 100 MW. The designs were apparently not competitive with HAWT technology and FloWind went bankrupt in 1997.

The world's first instance of electricity production from wind involved a HAWT design with the Scottish design of James Blyth constructed in July 1887 beating the better known US design (1888) of Charles F. Brush for priority. However, in 1891, Blyth produced a different, more enduring design based on the VAWT concept. These early designs charged batteries and successfully electrified the designers' homes.

13.2 VAWT Aerodynamics

VAWT designs are sometimes loosely categorised as lift- or drag-based designs. Whilst this is often reasonable, there are pitfalls in that devices such as the Savonius rotor

---

1 Although Murray (see Chapter 0) documents the first written mention of a windmill as early as the fifth century BC, D.G. Shepherd, one of the authors in Spera [1], after considering a number of earlier references as of uncertain provenance, concludes that the first accepted establishment of the use of windmills was in the tenth century in Persia.

Innovation in Wind Turbine Design, Second Edition. Peter Jamieson.
© 2018 John Wiley & Sons Ltd. Published 2018 by John Wiley & Sons Ltd.

(a)   (b)

Figure 13.1 Modern VAWT designs. (a) Reproduced with permission of RES Ltd. (b) Reproduced with permission of Sandia Corporation.

[7], often categorised as drag devices, in fact employ significant lift. A simple analysis (Manwell *et al.* [8], p. 104), suggests that the maximum power coefficient of a purely drag-based turbine is given as

$$C_p = 0.5C_D\lambda(1 - \lambda)^2 \tag{13.1}$$

In Equation 13.1, C_D is the drag coefficient of the turbine blade surface. Maximum C_p occurs at a tip speed ratio, $\lambda = 1/3$ and is of a magnitude $(2/27)C_D = 0.15$ for an assumed $C_D = 2$. Note that this is half the value $(4/27)C_D$ derived by Wilson [1] for a translating drag device (as, e.g. the Spinnaker sail of a sailing ship running before the wind). 'D' sections, such as are the blades of a Savonius rotor, when flow impinges on the convex side, produce very strong lift forces (Figure 2.2). Early workers Wilson *et al.* [9] recognised the contribution of lifting forces to rotor performance; and it is no surprise that the Savonius rotor, although among the lowest efficiency wind devices, can achieve a rather better maximum C_p than would be predicted purely based on theoretical drag performance.

A critical aspect of VAWT operation, as compared to HAWT, emerges comparing the nature of the shaft torque produced when operating in an ideal uniform wind field. In a uniform wind field, the torque produced by each blade of a HAWT is constant independent of the blade angular position. This is true because the local velocity triangle at any blade element (see Figure 1.16) has no dependence on blade angular position nor

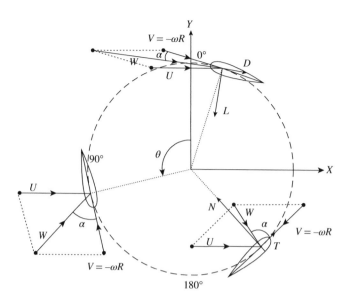

Figure 13.2 VAWT aerodynamics.

does blade position appear in the defining system of blade element momentum (BEM) equations for a HAWT (Equations 1.35–1.39). This means that each blade element of a HAWT can operate optimally at all angular positions in a cycle of rotation.

Operation of a VAWT is fundamentally different. Considering either the H-type or Darrieus-type VAWT, the local flow triangle comprising the tangential velocity of the blade element and the wind vector varies strongly with blade angular position.

This is illustrated in the flow diagram of Figure 13.2, where U represents the steady wind flow, V the tangential velocity of the blade section, W the resultant inflow and α the angle of attack. This diagram is much simplified and ignores induced velocities, but it shows clearly that, even in perfectly uniform wind, the flow geometry at a blade section varies strongly with position in the rotor circle and specifically that the angle of attack varies cyclically. This variation of aerodynamic characteristic over the cycle of rotation implies that the section cannot operate optimally at every angular position and suggests that the aerodynamic performance of a VAWT is fundamentally inferior to a HAWT. The cyclic variation of angle of attack also implies that dynamic stall is of intrinsic importance for VAWT design, whereas many aspects of HAWT aerodynamics can be addressed without much consideration of dynamic stall effects.

In real turbulent wind flow, the inflow and hence the performance of a HAWT blade will naturally vary with its angular position. However, the variations are stochastic rather than strongly periodic with each rotor cycle. A consistent wind shear or upflow will produce quasi-deterministic cyclic variations, but all such effects on the HAWT produce minor torque variations compared to those of the intrinsic torque cycle of the VAWT. Consequently, a substantially better average power performance of a HAWT rotor can be realised compared to a VAWT.

A second distinctive aspect of VAWT aerodynamics is revealed in Figure 13.3 adapted from Freris [10]. The VAWT sweeps out a cylinder in the flow stream rather than a disc as with a HAWT. Considering the flow through that cylinder as a set of streamtubes,

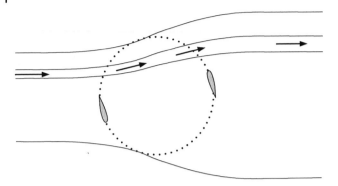

Figure 13.3 Actuator cylinder model for a VAWT.

then each blade crosses a given streamtube twice. Extracting energy from the same streamtube in two passes is very loosely equivalent to a single pass with a blade of higher solidity. Also, because the VAWT blade cannot operate at maximum lift to drag in all positions of the rotor cycle, it has in effect a lower average lift-to-drag ratio. By a loose analogy with Figure 1.18 for a HAWT, this suggests that the optimum tip speed ratio of a VAWT is lower than that for a HAWT.

Comparing the performance of H-type and Darrieus VAWTs, although the performance of the blade elements of the Darrieus turbine reduces where they are near the axis of rotation, so also does the swept area; and the blade elements near the axis may still retain some degree of aerodynamic function. In contrast, the cross arms of an H-type design, which are essential to support the blades, are entirely parasitic in aerodynamic function contributing drag (especially around the point of interconnection to the blades) which reduces rotor power.

Comparing a HAWT with H-type and Darrieus VAWTs, the relevant swept areas and typical associated C_p max are illustrated in Figure 13.4. The suggested impact of the blade support system in reducing C_p max is also illustrated in Figure 13.4, and such structures (Figure 13.1) are quite substantial.

The C_p values in Figure 13.4 are suggested as realistic ones as measured on real systems rather than on ideal theoretical ones. Based on Paraschivoiu [11] for a VAWT

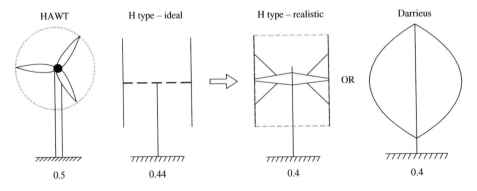

Figure 13.4 Swept area and maximum C_p.

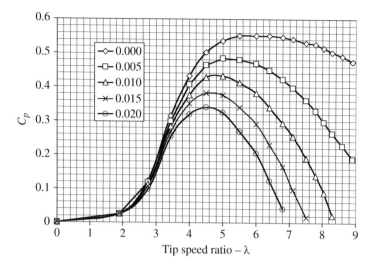

Figure 13.5 Effect of drag on VAWT performance.

of optimum solidity,[2] $\sigma = Nc/R = 0.2$ where N is the number of blades, c average chord width and R rotor radius at equator, the optimum tip speed ratio at zero drag (Figure 13.5) is 6 when a C_p max of 0.56 is achieved. The values in the legend of Figure 4.3 represent the drag at zero incidence of the symmetric NACA 0012 aerofoil section artificially using values ranging from zero to more realistic values approaching 0.02. Thus, with realistic finite aerofoil drag levels (considering only aerodynamic rotor drag and ignoring any parasitic drag from blade support structures) a C_p max limit of around 0.45 is indicated. This makes it entirely credible that a practical limit (in the presence of parasitic drag on blade supports and vortex shedding associated with tip effects) not much above about 0.4 should apply. It also confirms that aerodynamically optimum tip speed ratios will be around 4 and not exceed 5 in practical, efficient VAWT designs. VAWT aerofoils in themselves, being most usually symmetrical and not cambered, may have somewhat lower maximum lift to drag than HAWT aerofoils but this is a minor consideration. The key point is that the VAWT aerofoil cannot operate at maximum lift to drag in most of the cycle of rotation so that the VAWT has a very low average lift-to-drag ratio per rotor cycle and hence a low optimum tip speed ratio.

According to Wilson (analysis of a giromill [12]), an H-type VAWT with cyclically pitching blades has the same limit as a HAWT, that is, the Betz limit. There is even a suggestion that the fundamental limit could exceed the Betz limit a little [13] if, as has been adopted in some methods of VAWT aerodynamic analysis, the blades in the upwind semi-cylinder and in the downwind semi-cylinder are treated as acting on two separate actuator discs in series when the associated C_p limit is found to be 16/25 as opposed to the Betz value of 16/27.

---

2 Paraschivoiu [11] notes that there is no advantage in higher solidities and also comments 'Although lower values of solidity may widen the useful operating range of the turbine … they also reduce maximum available power.' This confirms that there is a price to be paid in power performance and energy capture if the VAWT is run at relatively higher speeds in order to reduce drive train torque and cost.

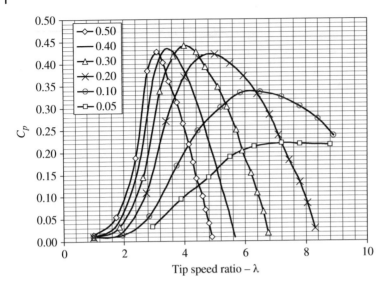

Figure 13.6 Optimum solidity and tip speed ratio of a VAWT.

Figure 13.6 illustrates the effect of varying rotor solidity, $\sigma = Nc/R$ over a range from 0.05 to 0.5. There is clearly a severe penalty on C_p max and by implication on energy capture if the design tip speed ratio of a VAWT is pushed much above optimum. In general, the result (not shown here) of increasing blade number (at constant solidity) will increase C_p max but drive optimum tip speed ratio to lower values as is the case in HAWT design.

The intrinsically inferior aerodynamic performance of a VAWT, coupled with an intrinsically lower optimum tip speed ratio, implies energy loss and higher drive-train torque rating for a VAWT. This will be characterised as the *energy-torque issue*[3] and be examined in more detail in Sections 13.3 and 13.4.

For the reasons discussed to do with the cyclic variation of angle of attack involving dynamic stall and the double crossing of each streamtube, VAWT aerodynamics, in respect of basic steady-state performance, is more complex than HAWT. Although VAWT technology has not survived in mainstream large-scale electricity generation, the aerodynamic theory and general design of VAWTs has been very much researched. A study [14] of 1992 commissioned by VAWT Ltd to review VAWT technology (Darrieus especially) cited over 1200 published references, and many other publications on VAWT technology and aerodynamic theory have appeared since.

13.3 Power Performance and Energy Capture

All the FloWind designs were stall regulated, and power curves taken from a Sandia report [15] of the most widely used FloWind 19-m VAWT are now compared with a conventional HAWT. A feature of stall regulation, equally applicable to HAWTs and

---

3 This phrase is not in general usage in the industry and is coined here as summarising the single most important aspect of VAWT design at least in the context of potential for large-scale electricity generation.

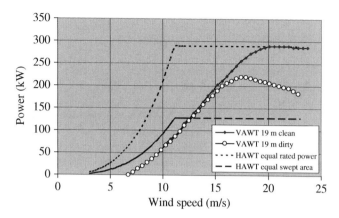

Figure 13.7 Power curve comparisons of HAWT designs with the FloWind 19 m.

VAWTs, is the sensitivity of the stall point to roughness. Significantly different power curves can result when blades are dirty or wet. Figure 13.7 reveals a substantial difference in the steady-state maximum power of the FloWind 19 m between ~220 kW (blades dirty) and 290 kW (blades clean). This is a generic issue with stall regulation which is avoided in designs with pitching bladed operating in attached flow. Thus, the problem is not specific to VAWT design except to the extent that stall regulation is often advocated as the preferred operational mode. Impact of dirt and rain on performance has triggered the development of aerofoils with reduced roughness sensitivity, as was discussed in Section 2.4.

In Figure 13.7, the FloWind 19-m design is compared with state-of-the-art HAWTs[4] with typical power density (ratio of rotor swept area to rated power) of 0.4 kW/m² and rated wind speed around 11 m/s. The very high cut-in wind speed of about 7 m/s and the very high rated wind speed around 17–20 m/s imply a poor level of performance compared to state-of-the-art HAWTs. FloWind, at their height, were a major supplier to the Californian market of the mid-1980s to mid-1990s and had every incentive to improve the designs substantially should it have proved readily possible.

Based on the power curves of Figure 13.7, at a good offshore site with annual mean wind speed of 9 m/s, a well-designed HAWT will produce about twice the annual energy. An actively controlled VAWT could, in principle, have a much better power curve than the FloWind 19 m. This, however, would be added cost and complexity, whereas most VAWT designs claim the simplicity of stall regulation as a benefit. Also, there has been no consensus on effective aerodynamic mechanisms for active control of VAWTs. The need for a rotor safety system of a stall-regulated VAWT implies a mechanical shaft brake. Considering that speed of VAWTs is lower and torques higher and that stall regulation requires extra load margins compared to pitch/active regulation, the brake of a VAWT can be a substantial item.

The stall-regulation characteristics of the VAWT designs of VAWT Ltd in terms of power curve shape are much better [4–6] than the FloWind 19 design. This may reflect some generic factor about H-type versus Darrieus, but it is primarily related to the high

---

4 The HAWT designs are conceptual as no commercial designs exist of exactly the stated ratings, but they are realistic representations of the power curves that can be achieved in practice with such technology.

design speed chosen by Flowind which compromises power performance specifically to mitigate problems with high torque rating and system weight.

13.4 Drive-Train Torque

The cost of a wind turbine drive train is primarily related to its input torque rating. In this context, 'rated' strictly means the maximum torque rating for safe design and not the torque associated with nominal rated power, although that may still be a very useful measure for comparative purposes.

The data of Table 13.1 is available from [4–6]. The developments of this period can reasonably be regarded as a serious effort to make effective VAWT designs of the potentially most efficient H type.

The data of Table 13.2 shows tip speed ratios for optimum operation of around 3 and 4 for the VAWT designs of VAWT Ltd. It may be rather difficult to characterise with much precision the optimum tip speed ratio of a VAWT or establish the ratio of design speed of a HAWT to VAWT for similar power ratings and swept areas. Nevertheless, large HAWT blades for variable speed wind turbines are typically designed for a tip speed ratio of ~9. Thus, the design speed of a HAWT may exceed that of an equivalently rated VAWT by a factor of 2 or more, and torque, weight and drive train cost will vary inversely as speed.

The torque rating of a VAWT of a given swept area may be reduced by having a high aspect ratio of height to diameter. This enables a relatively smaller diameter and faster rotation at any given 'tip speed' (speed of blade element furthest from the shaft axis in the case of a VAWT). This was recognised in the development of Flowind designs. According to the comprehensive study of Darrieus wind technology [14] commissioned by VAWT Ltd (later Renewable Energy Systems):

> The aspect ratio of Darrieus rotors has gradually increased. Early machines had aspect ratios of close to 1.0, perhaps because this shape can minimise the length of blade and column for a certain swept area. Later designs have acknowledged, among other things, that the cost of the drive train (approximately 50% of the total machine cost …) is very dependent on the value of the low speed torque; if the aspect ratio is increased then the rotor speed increases (to maintain the same relative wind speed and tip speed ratio) and the torque decreases if the power is constant.

Obviously this can only be exploited to a certain degree without the design becoming unwieldy structurally, and practical issues of dynamics of a whirling shaft becoming

Table 13.1 Performance of VAWT UK designs.

Design	C_p max	Associated tip speed ratio
VAWT 260	0.4	4
VAWT 450	0.3	3
VAWT 850	0.31	3.2

Table 13.2 Torque rating of FloWind 19 m compared to HAWTs.

VAWT 19 m	Dirty	Clean	
Power	219	289	kW
Tip speed	51.8	51.84	m/s
Diameter	19.2	19.2	m
Shaft speed	5.4	5.4	rad/s
Torque	40.6	53.5	kNm
HAWT of equal maximum power			
Power	219	289	kW
Tip speed	75.0	75.0	m/s
Diameter	26.4	30.3	m
Shaft speed	5.7	4.9	rad/s
Torque	38.5	58.4	kNm

problematic. High aspect ratio as a solution to reducing rated torque of a VAWT is discussed further in Chapter 15.

Nominating a maximum steady tip speed for the HAWT designs of 75 m/s as typical of a large efficient HAWT operating in variable speed, and with rotor diameters determined to correspond to the industry average rating to swept area ratio of $0.4 \, kW/m^2$, the torque rating of the FloWind 19 (clean and dirty blades) may be compared to HAWTs of the same maximum power. Table 4.3 shows that there is little difference in torque rating. However, the VAWT designs (Figure 13.7) have a very high rated wind speed. Thus, the FloWind design may be seen as a VAWT design at one end of the torque–energy spectrum, with torque similar to that of a HAWT and simple cheap manufacture, being stall regulated, but with poor power performance.

The opposite end of the spectrum is illustrated with the VAWT 450 design of VAWT Ltd. Rated wind speed was 11 m/s and variable speed operation was adopted ensuring that, for the chosen rated power (130 kW), a power curve and energy capture comparable[5] to an efficient HAWT would be obtained. Thus, FloWind, in running an essentially fixed-speed VAWT at higher than optimal speed, avoids the usual torque penalty of the VAWT but, as was evident from Figure 13.6, pays an unaffordable price in energy loss associated with high cut-in wind speed and the general inadequacy[6] of stall regulation.

Considering a HAWT of 130 kW rating, and preserving as before the industry average rating to swept area ratio of $0.4 \, kW/m^2$, implies a rotor diameter of 20.2 m. In variable speed mode with a maximum tip speed of 75 m/s, the associated maximum rotor speed is 70.4 rpm. The VAWT 450 operated in variable speed mode up to 27 rpm. This implies

---

5 This is a generous statement as far as VAWT technology is concerned. The power regulation achieved by the reefing system of the VAWT 450 was imperfect, with the power still rising in high wind speeds.
6 VAWT Ltd argued that avoiding the blade reefing system and being able to rely on stall regulation was a useful step forward for VAWT design. However, considering the power curve of the FloWind 19 m (Figure 4.7) in the clean blade state, it is apparent that the stalling characteristics of the VAWT are very delayed, leading to very high rated powers and very poor power curves. Power curve data of the VAWT 450 indicated no maximum on peak power before shut-down speed (25 m/s).

a factor of $(70.4/27) = 2.6$ on torque rating. Thus, a premium is necessarily applied to weight and cost of the VAWT drive train. Although the gearbox and generator were at ground level in this design, a further consideration was the long torque tube required to transmit the torque to ground level. In the progression to the VAWT 850 design on account of the mass and cost of such a torque tube, the concept of a ground-level drive train to facilitate maintenance was abandoned and a gearbox was mounted at the level of the rotor cross arms.

13.5   Niche Applications for VAWTs

A variety of VAWT designs survive at small scale. There is a possible niche market for VAWTs emerging in urban environments. VAWTs can sometimes be better integrated in building designs and, with a relatively lower rotational speed than HAWTs, can make a case for greater safety and less noise and vibration. Another significant technical issue is that in addition to the axisymmetry of a VAWT providing inherent insensitivity to wind direction in a horizontal plane, the VAWT is well suited to coping with up-flows which commonly arise at the edge of buildings (see Figure 18.3).

With the development of the Turby VAWT [16] at Delft, it was discovered that the performance of the turbine seemed to increase when the flow into the rotor was not perpendicular to the axis of rotation. This phenomenon was investigated further by van Kuik, van Bussel *et al.* [17–19]. It was found that, with increasing skew angle, a greater part of the downwind blade emerges from the wake shadow of the upwind blade, allowing it to see an incoming flow with higher energy. The model used was an H-rotor with two blades of 0.7 m span and 0.06 m chord. In an up-flow of about 25°, the power coefficient was found to increase to a maximum of about 1.3 times its value in horizontal flow. Remarkably, the power coefficient was equal to or greater than its value in horizontal flow over a range of up-flow angles within ±40°.

This property of VAWTs could prove advantageous for a floating wind turbine system, where motion of the support structure is unavoidable. However, quite large angles ~20–30° are required to gain much benefit in rotor power and with the rotor tilted, fatigue is introduced from blade self-weight effects, something which the VAWT design is often credited with avoiding. Thus, while HAWTs lose efficiency in vertically skewed flow, VAWTs gain efficiency. However, any such energy gains of the VAWT associated with enhanced up-flow performance are still swamped by the inherent deficit in power performance compared to an efficient HAWT.

13.6   Status of VAWT Design

13.6.1   Problems

The discussion of Sections 13.3 and 13.4 is in no sense a final evaluation, but it should clarify the core torque–energy issue that is critical for the economics of VAWT technology and provides a clear indication why previous VAWT designs have not been successful commercially. This is not to rule out future VAWT technology but to show the key, major challenge confirmed by commercial history in realising viable VAWT technology.

The overall comparison of wind turbine concepts is always very complex. VAWTs usually employ simpler blade designs which may allow cheap manufacturing solutions. However, this is offset by a need for greater installed blade area, both on account of lower design speed and higher solidity and reduced aerodynamic efficiency compared to HAWTs. The H-type VAWT moreover requires additional blade support structures. It has been suggested that VAWTs would come into their own again at very large scale as the self-weight blade loading is constant over each revolution, whereas there is a fatigue cycle in radial tension and in-plane bending of the blades of a HAWT which becomes increasingly significant as HAWTs are upscaled. However, VAWTs also have other challenging fatigue loading issues.[7] Thus, it is very doubtful if this advantage would really be of much significance in the overall loads comparison of large HAWTs and VAWTs and, for example, allow a 20 MW VAWT of the traditional H type to compete advantageously with an equivalent HAWT design. Moreover, there is no comfort or logic that VAWTs will upscale differently from HAWTs with essentially cubic upscaling having been observed (see Section 4.6).

VAWT enthusiasts may possibly feel that the evaluation of VAWT design presented here is rather negative. Hopefully, it will be understood that the author has absolutely no axe to grind on this issue. It is matter of fact (both historically, in commercial design and in terms of state-of-the-art modelling) that the optimum design tip speed ratio of a HAWT is around twice that of a VAWT and that the maximum C_p is 15–20% greater. This observation does not acknowledge or evaluate any system benefits of the VAWT concept, but it reveals immediately that such benefits must overcome substantial implied disadvantages.

13.6.2   Advances in VAWT Understanding and Technology

Apart from niche markets for small VAWTs, innovations are needed if a viable large-scale VAWT design is to be realised. The 'V'-type VAWT [20], for example, with angled blades converging near base level realises the original VAWT design intention of having generating equipment and maintenance sites at base level and a low mass centre benefiting support structure design.

Recent work of Ferreira [21] substantially challenges previous aerodynamic understanding of VAWTs at a fundamental level. This suggests that VAWT design can be better optimised and perhaps the performance gap compared to HAWTs may be reduced. Text adapted from Ferreira provides the following explanations regarding improved aerofoil design for VAWTs comparing the 26.2% thick DU12W262 aerofoil with a NACA 0018. Historically, the symmetrical aerofoils of the NACA 00xx series have been the mainstream choice for VAWT designs.

Unlike the HAWT, the optimisation of a VAWT rotor is not for an actuator disc in uniform flow but for an actuator cylinder (as in Figure 13.3) where the flow is not only non-uniform but, most importantly, each streamtube crosses the actuator cylinder twice. The load optimisation in each streamtube is therefore a balance between the

---

7 For example, wind shear and general turbulence will create bending moments at the cross arm to blade connections of an H- type VAWT. In a rotational displacement of the blade, centrifugal force will increase on the blade part that is most displaced radially outward but decrease on the corresponding part that is displaced radially inward. Thus, a destabilising moment is produced that tends to augment the original rotation. Design of VAWT joints is, in general, very demanding.

loading in the upwind section and the downwind section. An aerofoil must therefore be optimised for two points of operation for each streamtube. Considering a continuum of streamtubes that cross the actuator, the requirement is that the aerofoil be optimised over a continuous range of angles of attack (e.g. at a tip speed ratio of 4, this range is ~18°). An analysis of the typical distribution of angles of attack over the rotation of a VAWT shows that the aerofoil operates most of the time at the extremes of the distribution; this is consistent with the concept of balancing the loading between an upwind and downwind actuator. The aerodynamic specification of the aerofoil of a VAWT can thus be interpreted as a continuous optimisation of glide ratio over the range of operation but weighted at the extremes. This can be more simply reduced to the optimisation of Cl slope over Cd [22, 23]. This glide ratio view is important for the structural optimisation of the aerofoil. For a HAWT aerofoil, in a single lift point operation, increased aerofoil thickness does not generate higher lift, but it incurs additional drag and consequently thicker aerofoils exhibit lower lift-to-drag ratios. However, increased aerofoil thickness does result in higher lift slope (as can be demonstrated by potential flow theory), up to a certain limit when viscous effects become dominant. Consequently in a VAWT rotor design, the aerofoil drag penalty due to thickness can be partially compensated for by the increase in lift slope due to aerofoil thickness.

Figure 13.8 shows how a 26.2% aerofoil (DU12W262) designed for a VAWT matches the performance or outperforms typical NACA aerofoils. The comparison is based on use of the aerofoil design code XFOIL with selected cases also validated experimentally. The NACA 0015 (not shown in Figure 13.8 to avoid clutter) outperforms the thicker NACA 18 or 25. This is because for the NACA 4-series, when viscous effects are accounted, the maximum lift slope is found at the NACA0015, and for thicker aerofoils the lift slope starts to decrease while drag keeps on increasing. However, the DU12W262 has significant benefits beyond a slight edge in peak performance. Firstly, the peak occurs at a tip speed ratio between 4 and 5. Most VAWTs (excepting the Flowind design which was severely penalised in peak C_p through having a high design tip speed ratio) had been designed for tip speed ratios in the range 3–4. A higher

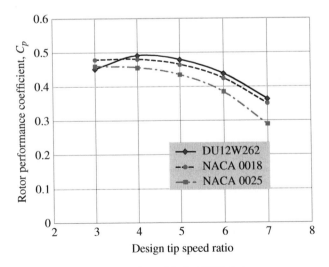

Figure 13.8 Comparison of aerofoils for VAWT design.

design tip speed ratio will always reduce drive-train torque, weight and cost, but, with conventional NACA aerofoils, incur performance penalties increasing strongly with increase in design tip speed ratio. Secondly, the 26% thick DU aerofoil is much more comfortable structurally than a 15% NACA and much superior in performance to a 25% NACA. Overall, with more advanced aerofoil design as developed by Ferreira, peak rotor C_p is now much closer to best HAWT values, although it should not be forgotten that drag on whatever structure is created to support the VAWT rotor blades will still reduce rotor C_p ahead of any drive-train inefficiencies affecting overall performance which are, in principle, common to HAWTs and VAWTs.

VAWT advocates make much of the simplicity of the design in its stall-regulated form without pitch or yaw systems. Yet, given that stall-regulation characteristics of VAWTs are not particularly good (partly on account of dynamic stall effects maintaining lift) and that overspeed protection (non-existent in the simple concept without blade pitching) is a significant demand requiring special air brakes or heavy-duty shaft brakes, VAWT designs with pitching blades (full or partial span) may have merit. Cyclic pitching will not only improve C_p max but, in increasing the average lift-to-drag performance of the aerofoils per cycle of rotation, will increase optimum design tip speed ratio and thereby reducing drive-train torque.

An alternative solution to the VAWT torque issue exploiting the potential economic benefits of multi-rotor systems has been mentioned and is reviewed in Section 15.7.

Since the intrinsically low speed of a VAWT implies relatively high shaft torque, assuming conventional power take-off on the rotor axis, this invites the question if there is a better concept for power generation. Options such as discussed in Section 6.13 may be worth consideration.

References

1 Spera, D.A. (ed.) (2009) *Wind Turbine Technology: Fundamental Concepts in Wind Turbine Engineering*, 2nd edn, ASME Press.

2 Dery, J. (1985) *Eole, Aerogenerateur a Axe Vertical de 4 MW a Capchat, Quebec, Canada*. 9th Biennial Congress of the International Solar Energy Society, June 1985.

3 http://www.mti-energy.com/mkby/vawt-history.pdf (accessed April 2011).

4 Powles, S.J.R., Anderson, M.B. and Tan, C.C. (1989) *Two Years of Operation of the 25 m VAWT at Carmarthen Bay*. Proceedings EWEC 89, July 1989.

5 VAWT Ltd (1990) The Demonstration of A Vertical Axis Wind Turbine on a Remote Island, Draft, October 1990.

6 Mays, D., Morgan, C.A., Anderson, M.B. and Powles, S.J.R. (1990) *Experience with the VAWT 850 Demonstration Project*. Proceedings, EWEC 90, September 1990.

7 Savonius, S.J. (1931) The S-rotor and its application. *Mechanical Engineering*, **53**, 333–338.

8 Manwell, J.F., McGowan, J.G. and Rogers, A.L. (2008) *Wind Energy Explained*, John Wiley & Sons, Ltd. ISBN: 13: 978-0-471-49972-5 (H/B).

9 Wilson, R.E., Lissaman, P.B.S. and Walker, S.N. (1976) *Applied Aerodynamics of Wind Power Machines*, University of Oregon.

10 Freris, L.L. (1990) *Wind Energy Conversion Systems*, Prentice Hall International (UK) Ltd, p. 97. ISBN: 0-13-960527-4.

11 Paraschivoiu, I. (2002) *Wind Turbine Design: With Emphasis on Darrieus Concept*, Polytechnic International Press, Canada. ISBN: 978–2553009310.

12 Wilson, R.E. (1978) Vortex sheet analysis of a Giromill. *Transactions, ASME Journal of Fluids Engineering*, **100**, 340–342.

13 Newman, B.G. (1986) Multiple actuator-disc theory for wind turbines. *Journal of Wind Engineering and Industrial Aerodynamics*, **24** (3), 215–225.

14 Schienbein, L. (1992) *Darrieus Wind Turbine Technology Assessment Study*, Renewable Energy Systems, Ltd, Report for December 1992.

15 FloWind Corporation (1996) Final Project Report: High-energy Rotor Development, Test and Evaluation. Sandia Report No. SAND96-2205, Sandia National Laboratories, Published September 1996.

16 (2006) Turby Information and Specifications Brochure, January 2006, http://www.wind-power-program.com/Library/Turby-EN-Application-V3.0.pdf as viewed December 2017.

17 Mertens, S., van Kuik, G. and van Bussel, G. (2003) Performance of an H-Darrieus in the skewed flow on a roof. *Journal of Solar Energy Engineering*, **125**, 433–440.

18 Ferreira, C.S., Dixon, K., Hofemann, C. *et al.* (2009) The VAWT in Skew: Stereo-piv and Vortex Modelling. AIAA 2009-1219, January 2009.

19 Ferreira, C.S., van Kuik, G. and van Bussel, G. (2006) Wind Tunnel Hotwire Measurements, Flow Visualization and Thrust Measurement of a VAWT in Skew. AIAA 2006-1368, January 2006.

20 http://www.enviro-news.com/news/funding_for_new_wind_turbine_design.html (accessed April 2011).

21 Ferreira, C.S. (2009) The near wake of the VAWT, 2D and 3D views of the VAWT aerodynamics. PhD thesis. TU Delft.

22 Ragni, D., Ferreira, C.S. and Correale, G. (2015) Experimental investigations of an optimized airfoil for vertical-axis wind turbines. *Wind Energy*, **18** (9), 1629–1643.

23 Ferreira, C.S. and Geurts, B. (2015) Aerofoil optimization for vertical-axis wind turbines. *Wind Energy*, **18** (8), 1371–1385.

14

Free Yaw

14.1 Yaw System COE Value

The cost of energy (COE) split of Table 9.1 suggests that the yaw system may have a value of 3% or 4% of the capital cost of the machine and correspondingly less in terms of the lifetime COE. Noting that typically a third or more of the cost of the yaw system may lie in the yaw bearing, and that a yaw bearing is not dispensable, it will be apparent that the capital value saved in a horizontal-axis wind turbine (HAWT) with a free yaw system cannot be great. This argument is somewhat different considering a vertical-axis wind turbine (VAWT), which intrinsically does not require to yaw. A VAWT avoids the complete capital cost of a yaw system, inefficiencies associated with imperfect tracking (of a HAWT), reliability impacts of yaw system faults and associated general lifetime maintenance costs. The avoidance of a yaw system in a VAWT has therefore a more substantial net COE benefit than in any free yaw solutions of a HAWT, but of course there are other very significant counterbalancing design issues with a VAWT (Chapter 13).

14.2 Yaw Dynamics

Basic yaw dynamics are now briefly discussed.[1] The source of yaw system loading is asymmetry in loads over the rotor plane. This may arise in the incident wind field or in geometric and structural aspects of blade and rotor. Aerodynamic yaw moments arise principally from differential rotor loading producing imbalance in the out-of-plane blade bending moments of the (three) blades. Differential in-plane shear forces, acting on a moment arm determined by the distance from tower centreline to rotor plane, also contribute. Most other load components can also contribute to yaw torque to some extent, for example, the rotor torque if there is a tilt in the rotor shaft axis. Because the most significant loads are differential loads, they are intrinsically harder to measure and predict than disc-averaged quantities like rotor torque or thrust. The following discussion is not specific to free yaw but it will enlighten issues around free yaw, especially the question of yaw stability.

It is easy to show that a harmonic in the blade out-of-plane bending moment of order n, in general, produces harmonics of order $n + 1$ and $n - 1$ in the nacelle and yaw system.

---

1 BEM theory can be adapted to encompass yaw behaviour, but it is most challenged in yawed flow and specifically in regard to assumptions about the wake and effects of unsteady aerodynamics.

Innovation in Wind Turbine Design, Second Edition. Peter Jamieson.
© 2018 John Wiley & Sons Ltd. Published 2018 by John Wiley & Sons Ltd.

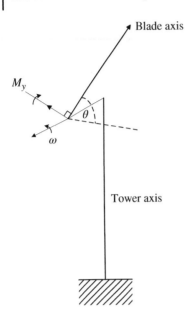

Blade axis   **Figure 14.1** Transformation of blade harmonics.

M_y

θ

ω

Tower axis

In particular, this means that the first (once per revolution) blade harmonic is a determinant of the zeroth harmonic of the yaw system, that is to say, the steady yaw moment.

Consider a two-bladed turbine in a steady wind where the blades are pitching cyclically once per rev with a maximum flap moment (which is transitorily a yawing moment) occurring at 0° and a minimum at 180°, these angles being measured with respect to a horizontal blade position. This will obviously result in a net torque about the yaw axis. Without actually applying any cyclic pitch change, it can be understood that a once per rev aerodynamic flap harmonic in the blade will tend to produce the same effect. This is readily generalised in mathematical terms.

In Figure 14.1, the rotor is rotating at frequency, ω and the position of one blade is at angle, $\theta = \omega t$, with the associated out-of-plane bending moment represented as the vector M_y.

Let the nth harmonic of M_y be $M_n \sin(n\omega t)$. Then, the component of this vector in the direction of the yaw (tower centreline) axis is

$$M_n \sin(n\omega t) \cos(\omega t) = \frac{1}{2}\{M_n \sin([n+1]\omega t) + M_n \sin([n-1]\omega t)\} \qquad (14.1)$$

Thus, the nth harmonic of a blade loading is, in general, shifted to become $[n+1]$th and $[n-1]$th harmonics in the reference frame of the nacelle. Unfortunately, the once per rev ($n=1$) aerodynamic moment of the blade that is the source of the aerodynamic steady yaw moment ($n=0$) is particularly complex to predict. It is this aerodynamic load that is particularly affected by stall hysteresis, which may be associated with operation in yaw error and with the general three-dimensional unsteady aerodynamic behaviour of the blade. Thus, the yaw tracking behaviour is particularly difficult to predict. This is not in itself an argument against free yaw, but it is a practical issue that the designer must confront. It is generally necessary to observe free yaw behaviour of a prototype and tune the system. Also, the yaw stability characteristics are likely to change between low and high wind speeds according to differences in the development of stall on the wind turbine blades.

Further, with flexible blades, especially if free yaw is used in the downwind lightweight flexible blade configuration, there is a phase shift between the elastic response of the blades and the aerodynamic loading. This may enhance yaw stability but may also cause the rotor to track the wind direction with a specific angular offset related to wind speed and direction of rotor rotation. Teetered rotors (see also Section 11.3) are somewhat different from rotors with rigid or near-rigid attachment of blades to hub. They often exhibit better yaw tracking and lower yaw rates because the blade out-of-plane moments are decoupled from the yaw torque. Standard blade element momentum (BEM) theory will not predict restoring yaw moments that are due to purely aerodynamic effects and a skewed wake model is required (see Section 1.10.5).

14.3   Yaw Damping

A wind turbine in free yaw is at risk of excessively rapid yaw movements which may introduce undesirable gyroscopic loads into the blades. Thus, yaw damping is often required. This is an unsurprising outcome considering that actively yawed wind turbines, although operating at yaw rates that are unproblematic as far as gyroscopic loads are concerned, still require strong yaw damping to limit vibrations in the yaw bearing.

Friction damping may be considered and may seem simpler than, for example, a geared motor, to provide the damping. However, in several specific studies [1] it was found that the geared motor solution was as cost-effective as various friction-based alternatives. Thus, given that a yaw bearing is always required and that the components are now very similar to an active yaw system, the cost saving in a system with free yaw may be very little. There can still be operational benefits associated with free yaw. Some upwind wind turbines that are generally stable in yaw attitude may have active yaw control capability but can be allowed to operate in comparatively free yaw with some damping applied and with intervention of the control system when yaw rates or yaw errors exceed set limits.

14.4   Main Power Transmission

It is customary in a wind turbine with a powered yaw system to allow a few complete revolutions of the rotor in yaw in which the main power cables (connecting generator in the nacelle to switchgear, transformer and grid) may simply be allowed to twist. The full revolution(s) in yaw may take place over months or even years. A counter will measure rotations in yaw; and when a limit is reached, the yaw system will, under suitably moderate wind conditions, turn the rotor nacelle as required in the opposite direction and hence untwist the main power cables. If a powered yaw system is dispensed with, then there is either the further complication of having yaw position measurement and sensors and a light-duty motor to perform this function or there is a requirement to transmit the primary power output of the wind turbine through slip rings which, at the high power levels of megawatt-scale wind turbine systems, may be troublesome and expensive items.

14.5   Operational Experience of Free Yaw Wind Turbines

In the mid-1990s, GL Garrad Hassan and Partners conducted a review of free yaw for the UK Department of Industry [2]. This included a parametric study which concluded as follows:

- Using standard BEM theory, fixed-hub rigid-blade downwind machines would be predicted to be unstable in yaw at high wind speeds.
- Teetered downwind machines would be predicted to be stable in yaw due to the phasing of teeter response introducing a stabilising thrust component.
- Blade flexibility can introduce sufficient phase shift into the contribution of the blade flapwise moments to the yaw moment to improve the yaw stability.
- The introduction of skewed wake aerodynamics can reinforce the yaw stability at low wind speeds but has a negligible effect at high wind speeds.
- The introduction of stall hysteresis (in a manner similar to blade flexibility) introduces a phase lag into the flapwise moment which produces a stabilising yaw moment.

The review highlighted the sensitivity of yaw tracking, in general, to both aerodynamic and structural dynamic factors in the rotor design.

The review of operational experience of turbines in free yaw noted that many free yaw machines have been well behaved in yaw tracking accuracy – often, rather better than theoretical modelling will predict. This was true of the Carter 300 kW machines, which operated in the mid-1990s at Great Orton Wind Farm in the United Kingdom. Often, the yaw tracking behaviour is more variable in light winds when factors such as friction in the yaw system may inhibit response. Two-bladed wind turbines with teetered hub are, in general, problematic in free yaw operation. An interesting account was included of the free yaw behaviour of the 3 MW, 78-m-diameter, flexible, two-bladed, teetered, downwind, pitch-regulated WTS-3 Maglarp turbine.

Ganander [3] discussed this. The wind turbine was unable to remain pointing into the wind and underwent a saw tooth yaw motion with a range of about 5° and a period of 30 s (the machine yawing out of the wind, at a rate which depends on wind speed, and then being brought back by the yaw drive). This behaviour was related to the teeter degree of freedom; when the teeter hinge was locked, the machine was able to follow the wind.

Thus, the Maglarp machine tried to yaw out of the wind when teetered and that rather surprisingly locking the teeter hinge improved matters. A possible explanation for this could be the location of the mass centre relative to the teeter hinge, which will give rise to a 1P teeter motion resulting in the thrust vector driving the machine off wind. However, the in-plane shear force component would be expected to eventually stabilise the motion. Ganander noted that a test had been performed where the rotor was allowed to move freely but that the yaw error became so great that the operators lost their nerve before an equilibrium position was reached.

Rather different experience of free yaw behaviour was noted in trials at Risø. The history behind the two-bladed, teetering rotor project at Risø [4, 5], is that the original Windane 12-m-diameter machine (which operated in free yaw) was installed in California on a highly turbulent (complex terrain) site and suffered several failures. This was not expected because the same machine installed at Risø had survived all wind conditions including the autumn storms when turbulence intensity levels are as high as 15%.

An investigation of why the machines failed suggested that the very limited (2°) teeter motion of the original machine was at fault.

The project then explored different levels of teeter stiffness and with and without a tilt degree of freedom in order to 'optimise' the concept. The measurement programme was backed up by the development of an aeroelastic model specifically aimed at the downwind, two-bladed, teetered, free yaw concept. Apparently, the teeter stiffness was the most significant parameter regarding yaw stability with small levels of stiffness and damping producing the best yaw behaviour. Too small a level of stiffness could cause an oscillatory transient yaw behaviour.

General discussion of the behaviour of the machine indicated that it was independent of wind speed (suggesting the dominance of the teeter motion which causes the thrust to realign the machine in determining the yaw behaviour).

14.6 Summary View

In spite of the limitations in benefit from free yaw, there can be value in relief of certain system loads and in sensitive tracking of the wind if the system is able to perform well.

Many upwind turbines, at least in certain operational wind speed ranges, are intrinsically stable in yaw; and partial free yaw has been an option possibly of practical benefit. In this circumstance, all the mechanisms and control systems for powered yaw exist; but the turbine is essentially allowed to free yaw until yaw error or yaw rates limits are exceeded, when the active yaw system will intervene. Related to this are systems with 'soft' yaw characteristics [6] which may be realised with compliant hardware and/or actuators with intelligent control algorithms. Overall, there are many subtle issues to consider in yaw system design, but it is very clear that there is no case for advancing free yaw capability in a HAWT design as offering, in its own right, a significant cost reduction. In other words, there is no such thing as a free yaw!

References

1 Jamieson, P. and Jaffrey, A. (1997) Advanced wind turbine design. *Journal of Solar Energy Engineering*, **119**, 315–320.
2 Rawlinson-Smith, R.I. (1996) Free Yaw Behaviour of Wind Turbines. GH Report 517/BR/001, ETSU, Agreement No: W/13/00484/11, March 1996.
3 Ganander, H. (1989) *Importance of Yaw System of Two Bladed HAWT, Practical and Theoretical Results*. European Wind Energy Conference and Exhibition, Glasgow.
4 Kretz, A. and Rasmussen, F.R. (1990) An Aeroelastic Model with Applications to a Two Bladed Wind Turbine. Risø-M-2884.
5 Rasmussen, F.R. and Kretz, A. (1992) Dynamics of a Two Bladed Teetering Rotor. Risø-R-617(EN).
6 Engström, S. (2001) Soft Yaw Drives for Wind Turbines. DEWI Magazine, No. 18 (Feb. 2001).

15

Multi-rotor Systems (MRS)

15.1 Introduction

Multi-rotor technology has a long history and the multi-rotor concept persists in a variety of modern innovative systems, but the concept had generally fallen out of consideration in mainstream design from a perception that it is complex and unnecessary as very large single wind turbine units are now technically feasible. Systems such as in Figure 15.1 conceived by Honnef[1] around 1930 arose out of a vision of how wind power may be deployed as large-scale units at a time when steel was the only practical material for rotor blades and when, especially due to the enormous weight of large-scale steel blades, very large capacity single turbines were not feasible.

The largest modern wind turbines depend on high-strength composite materials for their rotor systems. These composites have a much higher strength-to-weight ratio than steel and allow much larger rotors to be realised.

15.2 Standardisation Benefit and Concept Developments

In the 1970s, Heronemus,[2] advanced many visionary concepts for multi-rotor systems (MRS) including offshore floating multi-rotor arrays for hydrogen production. According to Woody Stoddard, referring to Heronemus' writings of 1968:

> Bill Heronemus not only predicted the worldwide energy difficulties which were to come …, but saw the grand scale of future of renewable energy development. This included solar thermal, land-based and offshore wind, and ocean thermal energy. He had encyclopaedic knowledge of power plants, oceanography and engineering, and he was able to present his arguments in a practical and convincing way.

A significant advantage of MRS was identified in standardisation of rotor and drive-train components and in 'scale-ability': the fact that very large system capacities could be realised without overstretching the capability to upscale individual rotors.

---

1 Honnef's vision encompassed direct-drive rim generators and urban and offshore applications.
2 Jon McGowan of the University of Massachusetts remembers Bill Heronemus as a father figure of wind power in the United States, a visionary of the hydrogen economy.

Innovation in Wind Turbine Design, Second Edition. Peter Jamieson.
© 2018 John Wiley & Sons Ltd. Published 2018 by John Wiley & Sons Ltd.

Figure 15.1 Multi rotor systems. Reproduced from Honnef (1930).

For many years, blade manufacturers and other component suppliers have been drawn into the challenge of producing ever larger units. Yet, if asked what would most enhance reliability and lower production cost, the most common answer is if the size of unit demanded by the market stabilised at a single or a few preferred sizes. This potentially huge benefit of standardisation could be realised with the MRS where larger capacities may be achieved with more rather than larger rotors. The cost benefit of standardisation is well documented in experience curves [1] describing cost reduction of many industrial components.

15.3   Operational Systems

In the 1980s and 1990s, Henk Lagerweij of Lagerwey Wind built a variety of MRS and gained experience of key engineering issues such as yawing and rotor interaction. According to Lagerwey, regarding a particular six-rotor system, the system was satisfactory but customers preferred the conventional solution. The Lagerwey design was developed as a 'tree' structure of cantilevered elements with rotors supported on branches off a main support structure. This is potentially subject to vibrational problems, especially if extended to a large array for which a closed lattice support structure seems much preferable.

15.4   Scaling Economics

Around 1995, Jamieson [2] was contracted to conduct a study of advanced wind turbine concepts for the UK Department of Industry. Fundamental means of improving cost of energy from wind turbine systems were sought. One such fundamental issue related to increasing the ratio of rotor swept area, equivalent to the value energy, to system mass which is essentially a cost indicator.

The report [2] suggested that one way to realise this benefit was through MRS. This work introduced another major advantage of MRS, not previously clearly recognised:

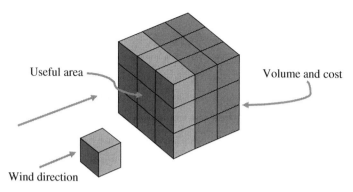

Useful area

Volume and cost

Wind direction

Figure 15.2 Basic scaling characteristics.

that a multiplicity of small units is fundamentally more cost-effective in the rotor and drive-train systems than a single large unit. The reason for this is essentially about elementary relationships between surface area and volume (Figure 15.2).

In Figure 15.2, the darker sides represent the area (proportional to energy capture and value) exposed to wind and the lighter sides the associated volume (proportional to mass and cost) associated with the useful area. A system of nine individual units (light grey) has one-third the volume (and, by inference, mass and cost) of a single large unit (equivalent to all 27 units of the larger cube). In general, when a system of n rotors is compared with a single large rotor of equivalent capacity, the ratio of total mass and cost of rotors and drive trains of the MRS to mass and cost of the conventional single rotor is as $1/\sqrt{n}$. This implies that the comparison is made at the same stage of technology development and not comparing small old components with new large ones as is the norm with wind technology but instead considering small components made optimally with the latest technology.

Of necessity, in the way that the market has developed, old, small-scale technology is invariably compared with new large-scale advanced technology and real commercial data on system and component masses, at first examination (see more extensive discussion on scaling in Section 4.5), appeared to contradict this simplistic scaling law. Thus, the second key ingredient in demonstrating the scaling advantage was careful analysis of commercial turbine data which then, far from contradicting the scaling potential, supported it. The key issue is that whilst historically blade mass, for example, appears to follow an almost quadratic curve, the underlying relationship (Figure 4.5) is cubic. Therefore, if latest technology advances are employed to revisit the design and manufacture of smaller blades, it is possible to descend a new cubic curve exploiting latest technology advances and realise the full economic advantage of MRS.

There are evidently new challenges in the support structure and a yawing mechanism for the multi-rotor array. However, excepting for some additional drag on structural elements, aerodynamic loading is generally no more than for an equivalent large turbine and better distributed with some extremes much reduced. Thus, shedding $(1 - (1/\sqrt{n}))$ of the cost of rotor blades and drive-train systems compared to a single large rotor provides a massive margin to offset any added cost in structure and yaw mechanism. Considering, for example, a 20 MW MRS comprising 100 rotors each rated at 200 kW, provided the designs are fully optimised taking account of latest technology advances, there is a realistic prospect of saving up to 9/10 of rotor and drive-train mass and cost as compared to a single 20 MW unit.

15.5   History Overview

In summary, multi-rotor history comprises the following:

- Conceptual solution to large unit capacity when rotor size was limited by steel as the only blade material considered practicable (Honnef 1930 and others).
- Recognition of the potential of MRS for large unit capacity, the benefit of standardisation avoiding upscaling of turbine units and the development of many concepts for such systems (Heronemus 1970–2002 and Stoddard 1973–2007).
- Some practical experience of design and operation of MRS (Lagerweij 1970–1995) and also research projects in the Netherlands.
- Recognition of the scaling principles as implying a fundamental economic advantage of MRS and development of a case based on critical analysis of commercial data to justify this potential benefit [2].

Commencing 2012, in the collaborative European project of the 7th Framework Programme, Innwind.EU, a number of innovations including the MRS [3] were evaluated in terms of potential to reduce the cost of energy from offshore wind. Innwind results are further discussed in Sections 15.6 and 15.7. In April 2016, the leading Danish wind turbine manufacturer Vestas announced the installation of a four-rotor MRS prototype for testing and concept development (see Figure 15.5 and Section 15.7).

15.6   Aerodynamic Performance of Multi-rotor Arrays

In the system developed by Lagerwey, there was no power loss from adverse rotor interactions, but the rotors were not closely spaced. The aerodynamic performance of a large array of closely spaced horizontal-axis wind turbines (HAWTs) has not been much researched, although, in an elegant wind tunnel experiment, Smulders *et al.* [4] showed that a pair of rotors as closely spaced laterally as 2.5% of diameter had no net loss in power performance and even a slight gain. With counter-rotating rotors, the adjacent tip vortices were observed to cancel, but this did not differentiate net power produced from the co-rotating case.

In October 2007, Ocean Wind Energy Systems (OWES) commissioned the Southwest Research Institute (SwRI) to conduct testing of a seven-rotor array in the NASA Langley Full Scale Tunnel in Virginia. SwRI developed a support structure design specifically for the wind tunnel testing, which allowed the six rotors surrounding the central rotor to move out on radial tracks so that a wide range of rotor spacing (2–16% diameter) could be evaluated. No significant effect on system power performance [5] was detected over this range of spacing. A few tests at ±10° yaw error showed the expected degradation of performance in yaw but apparently no more per rotor than for a single isolated rotor.

Within Innwind.EU, major analytical work on rotor interaction was conducted by the National Technical University of Athens (NTUA). Two different aerodynamic tools were employed: a RANS flow solver (flowNS) [6] in which a rotor is simulated as an actuator disc and a Vortex solver (GENUVP) [7] in which the blades are represented individually. Results for power and thrust loadings from both methods were generally found to be in good agreement. The aerodynamic evaluation, including predictions obtained using the computational fluid dynamics (CFD) model, CRES-flowNS, was reported in Refs [3] and [8] and some results are shown in Figure 15.3.

Comparing the MRS array of 45 rotors with 45 × (power of a single rotor operating in isolation) a net power gain ~8% was indicated, while similar modelling of a seven-rotor array had suggested a net power gain of ~2.6%. This relates to a so-called blockage effect observed in tidal current arrays. Flow cannot pass as easily around the turbines in the heart of the array, as they are in close proximity to other turbines and a greater pressure difference across the turbines may be maintained allowing additional power extraction. Also, flow in the free spaces between the rotors may induce additional flow over the outer parts of the rotor blades and enable some increase in power performance. Initial modelling was at 5% diameter rotor minimum spacing, but results were later confirmed as similar at a near-minimum spacing of 2.5% diameter. The increases in total power were logically accompanied by proportionally similar increases in total thrust. Extra power can be harvested at all wind speeds below rated wind speed where additional thrust has negligible design impact, but, if necessary, peak thrust at or near rated wind speed can be regulated by blade pitching. In relation to what may be called the envelop area covering all the rotor disc swept areas and the spaces between them, there is no gain in power. However, the spaces are useful in that they contribute some power augmentation without incurring any cost in installed blade area. Overall, the most important factor is that there are no substantially detrimental aerodynamic interactions in a planar array of very closely spaced rotors.

15.7   Recent Multi-rotor Concepts

Companies promoting MRS [9–11] have reappeared from time to time. There is little sign that the arguments about scaling economics have been explicitly recognised as a main motive for the multi-rotor concept. However, the implicit benefit of scaling characteristics to system cost may perhaps have explained the persistence of multi-rotor ideas in the background of wind technology development. Recently, however, there has been more significant interest. Brose had tested a small 24-rotor system at Kaiser Wilhelm Koog with excellent performance results verified by DNV GL Energy (Garrad Hassan). Laboratory experimental work [12] and some field testing on small arrays of ducted rotors are under way at Kyushu University (Figure 15.4, right).

From an industry perspective, a most significant development is in the interest of Vestas in MRS. Their four-rotor system (Figure 15.5), according to Erik Miranda who led the design team, is for testing and learning and does not in itself represent a final direction for Vestas with MRS. This particular design has potential at sites where access for large components is problematic. The yawing of the rotors in pairs avoids any major design challenge in providing yaw capability.

In a presentation to the Deepwater Conference hosted by the US DoE in Washington (2004), Jamieson [13] presented a study commissioned by OWES showing a comparison of extreme loads on a closely spaced multi-rotor set supported by a lattice structure

CFD – actuator disk, x = 0

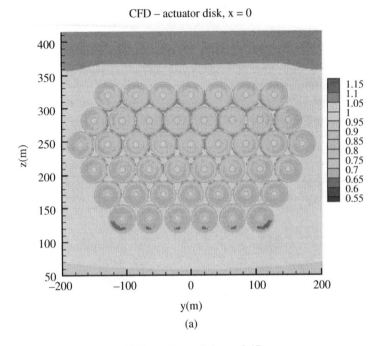

(a)

CFD – actuator disk, x = 0.1D

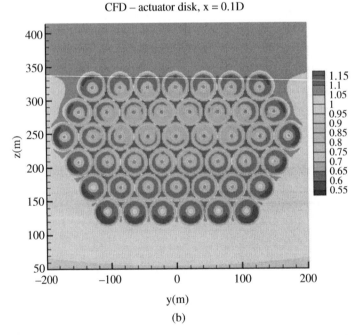

(b)

Figure 15.3 Normalised axial velocity contours (TSR = 9) for a 45-rotor MRS. (a) Rotor plane $x = 0$ and (b) downstream position $x = 0.1D$.

Figure 15.4 MRS at Kyushu University.

Figure 15.5 The Vestas four-rotor MRS.

Figure 15.6 OWES 16-rotor multi-rotor array. Reproduced with permission of Marcia Heronemus Pate.

developed from finite element analyses. This suggested, in general, that the MRS structure loading was more benign than that for a single equivalent large wind turbine.

OWES commissioned further work from GL Garrad Hassan and Aerotrope [14] who developed in 2008 a design of support structure and yaw system for a 16-rotor MRS. Figure 15.6 is a visualisation of that design but one based on conservative load estimates and giving a realistic impression of the solidity of rotor blades and support structure. The reduction in the mass of rotors and drive trains of the MRS as compared to a single large wind turbine was enough that all the rotors, drive trains and a frame to support them could be carried on a tubular tower and yaw bearing rather similar to the support structure of an equivalent single large rotor. This gave a strong indication that the need to yaw a multi-rotor array should not greatly erode its potential economic advantage. These findings about loads and yawing feasibility were consolidated in the more recent Innwind.EU project discussed in Section 15.9.

15.8   MRS Design Based on VAWT Units

In the discussion of vertical-axis wind turbine (VAWT) design in Chapter 13, the VAWT concept was seen to be seriously challenged in having lower aerodynamic efficiency and higher torque (weight and cost) compared to a HAWT. However, in a multi-rotor array, some new arguments appear in favour of a VAWT. Specifically, they are as follows:

- The VAWT may integrate better into a structure with rectangular stacking being possible and also in allowing mechanical interconnection of rotor outputs.
- The VAWT is much less sensitive to a yaw error in a multi-rotor array which, considering also interference from the support structure, may be more of a problem for an array of HAWTs than for a single HAWT unit.
- In certain regions where the wind rose has predominant directions, a multi-rotor array based on VAWTs may not need to yaw.

An MRS design based on VAWT units as in Figure 15.7 was developed by the Israeli company, Coriolis Wind, and a test system was erected in May 2009 incorporating three full-scale VAWT units. The Coriolis design reveals another unique feature of VAWT design which was briefly discussed in Chapter 13.

A VAWT can avoid a torque penalty relative to a HAWT if its blade length to diameter ratio is large enough. Suppose an H-type VAWT is designed to have the same rated power and rated torque (and hence the same rated rotor speed) and also (through having a larger swept area) to achieve the same rated wind speed and similar energy capture. Elementary analysis will show that this is achieved when

$$D_V = \left(\frac{V_V}{V_H}\right) D_H \tag{15.1}$$

and

$$\frac{L_V}{D_V} = \frac{\pi}{4}\left(\frac{V_H}{V_V}\right)^2 \left(\frac{CP_H}{CP_V}\right) \tag{15.2}$$

Figure 15.7 Coriolis multi-rotor concept. Reproduced with permission of Dr Rafi Gidron.

In Equations 15.1 and 15.2, D is diameter, V, the maximum steady-state tip speed, L, blade length (applicable only to an H-type VAWT) and CP, the power coefficient at rated wind speed. The suffixes V and H refer to VAWT and HAWT, respectively. In Equation 15.2, the VAWT and HAWT are assumed to have the same drive-train efficiency. Considering that the M5000 Multibrid wind turbine for the Alpha Ventus offshore project [15] has tip speed, $V_H = 90$ m/s, whereas a recent VAWT design, VertAx [16], has tip speed, $V_V = 44$ m/s, and assuming that at rated wind speed $Cp_H \sim 0.46$ and $Cp_V \sim 0.4$, Equation 15.2 indicates that the blade length to diameter ratio of a VAWT with parity in torque and energy is ~ 3.8. If the HAWT considered had been designed for land-based projects with a tip speed of, say, 75 m/s, the ratio reduces ~ 2.6.

Moreover, the ratio of total blade length installed in the VAWT compared to the blade length total of an equivalent HAWT (assuming equal blade numbers on each) is

$$\frac{2L_V}{D_H} = 2\left(\frac{L_V}{D_V}\right)\left(\frac{V_V}{V_H}\right) \tag{15.3}$$

Equation 15.3 predicts a ratio of 3.6, for $V_H = 90$ m/s (as for the Multibrid M 5000 offshore wind turbine) or 3.1 for a typical land-based HAWT with $V_H = 75$ m/s. To make a single VAWT unit in this way with a very high blade length to diameter ratio would be challenging structurally and is evidently also expensive in terms of blades and blade support structure.

In the Coriolis concept (Figure 15.7), sets of vertically stacked VAWT modules, each well supported by the surrounding structure, are connected mechanically to a single (direct-drive) generator. Thus, making use of the interconnecting support structure for the whole array, the Coriolis design effectively achieves the favourable aspect ratio and low torque that would be problematic for a single-unit VAWT. Unfortunately investment has not been forthcoming to continue development of the Coriolis design, but it illustrates very well the different issues and also some promise in employing VAWTs in MRS design.

15.9 MRS Design within the Innwind.EU Project

15.9.1 Loads, Structure and Yaw System Design

Issues of loads and yaw system design were revisited in the Innwind.EU project. The proposed design of MRS comprised 45 rotors of 444 kW rating providing a rated system output of 20 MW. In order to develop such a design, it was obviously essential to have a means of load prediction. DNV GL Energy had extended their long-established Bladed software to deal with a small array of tidal turbines. They further extended this software with a development enabling the 45 rotors to be modelled in an all-encompassing turbulent wind field. Each rotor operates independently in the usual variable speed, variable pitch mode characteristic of most recent large-scale designs. An interesting outcome of this is that under turbulent wind conditions, the rotors, being independently controlled, as would be the case in a conventional wind farm, are generally neither in phase in terms of blade angular position nor running at exactly the same speed. These random frequency and phase variations have the effect of cancelling many periodic load variations fed into the structure through destructive interference. This is shown very clearly

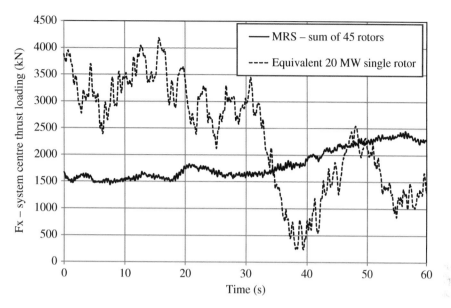

Figure 15.8 Comparison of system centre thrust forces.

in Figure 15.8, where the total thrust at the centre of the MRS is compared with the thrust at hub centre of a 20 MW wind turbine of the type designed in the UPWIND EU project [17]. The huge difference in load range can be understood considering that the comparison is of the effect of many small cantilevered blades as compared with three giant ones.

Thus, a rather paradoxical result arises. Increased turbulence will, as is normally the case, imply greater fatigue loading on any given wind turbine in the array. However, due to the averaging effect in summing loads from all the turbines, the greater the turbulence intensity, the less total variable loading is passed into the structure. Increased turbulence will, as always, increase fatigue of the support structure from wind acting directly on the structure members but paradoxically reduce fatigue from rotor loadings!

The Centre for Renewable Energy Studies (CRES) in Athens undertook to develop a support structure for the 20 MW MRS. A triangular lattice rotor support structure was designed based on 15 standard tube sizes, making iterative selections to meet design criteria and with an objective of minimising total mass. In general, to meet a prescribed allowable bending stress, the largest diameter, thinnest tubes that are not subject to buckling will minimise mass. However, with much reduced impact of rotor loading being intrinsic in the MRS concept, self-induced aerodynamic loading of the structure related to the project areas of structural members remains a major source of loading and optimisation of tube selection requires account of this. The design criteria included sufficient robustness that the structure could survive failure of the most critically loaded member. This design did not provide for yawing in any direct way, but could be used rigidly mounted on a floating platform capable of yawing in the sea. In order to have more direct comparability with the Innwind reference design [18], which was mounted on a jacket foundation, a rotor support structure with mechanical yawing capability was subsequently developed.

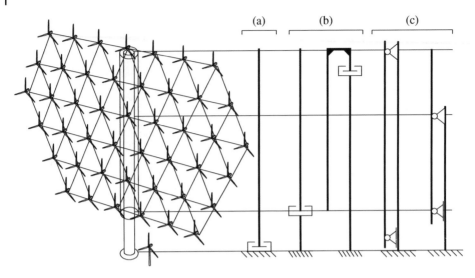

Figure 15.9 Yaw bearing concepts for a 20 MW, 45-rotor MRS.

The Hamburg University of Applied Sciences (HAW Hamburg) contributed a study on yaw bearing concepts for MRS. In Figure 15.9, the arrangements considered included (i) a single bearing at base level, (ii) a single bearing at low and high level of the rotor frame structure and (iii) twin bearings at the top and bottom of a full height tower and at intermediate levels.

HAW Hamburg confirmed as a validation exercise that the chosen software, RSTAB, a commercial analysis program for 3D beam structures, predicted results similar to those of CRES for the fixed rotor support structure. After consideration of all five options, the type (c) solution on the extreme right of Figure 15.9, designated the semi-tower option, was preferred.

Comparing the semi-tower design of Table 15.1 with the reference design fixed at base level, the reduction in mass of the space frame carrying the rotors is striking and is due to much more of the structure being in tension when it is hung on bearings. Thus, even with the significant added structure of the bearing support tower, mechanical yawing capability has been added to the MRS without much overall mass increase. Although the work was quite independently conducted in each case, the earlier study of Aerotrope of an MRS structure at 5 MW scale (Figure 15.6) had also suggested the semi-tower option as the most effective for yawing capability.

15.9.2 Operations and Maintenance

The concept for turbine maintenance is to have a top-level crane rail with a travelling crane that can handle complete rotor modules in a highly automated way. It is then feasible to avoid any major vessels (floating cranes or jack-up barges) being required for maintenance operations on turbine units. A major consideration in operations and maintenance (O&M) is to ensure that a local turbine fault in an MRS will not disable a large proportion of output capacity and wiring arrangements, for example, have been optimised to minimise the number of turbines affected by a cable fault within

Table 15.1 Comparison of designs for yawing and fixed base.

System component	Semi-tower design Mass (t)	Fixed base design Mass (t)
Yaw bearing connection top	390	—
Yaw bearing connection bottom	17	—
Yaw bearings	78	—
Tower	1520	—
Space frame with rotor-nacelle assemblies	1850	3760
Overall support structure	3855	3760

the interconnections of the 45-rotor array. The need to access site, possibly with very large-scale equipment (jack-up barges or floating cranes) and only under favourable weather conditions can lead to substantial loss of availability and be very expensive for single large turbines. Individual turbine faults in an MRS can be ignored for longer periods, some spares may be held on site and there would be no requirement to bring large-scale equipment to site. Therefore, unscheduled maintenance may largely be avoided. It is however important that minor procedures/repairs can be conducted very efficiently as there are many more rotors and greater probability of faults.

There is a clear case that the smaller turbines, with a faster manufacture and development cycle in greater quantity production and somewhat simpler in concept without individual yaw systems, can be more reliable per unit than a large turbine. A further factor, tending to reduce O&M cost in comparison to a single turbine wind farm, is that no single turbine will have as great a capacity as an MRS. Thus, in the Innwind comparison of a 500 MW wind farm of 25×20 MW MRS with one of 50×10 MW reference wind turbines, the number (25) of maintenance sites associated with the MRS is half of the number (50) associated with the DTU 10 MW.

In the Innwind study, an established modelling tool for prediction of O&M costs and optimisation of O&M planning of conventional wind farms [19] was adapted to capture the most significant differences in O&M that are specific to the MRS concept. This was supported by a reliability analysis conducted by DTU [20]. The outcome was a prediction of 13% reduction in O&M costs for the MRS substantially due to avoiding the use of jack-up vessels.

15.9.3 Cost of Energy Evaluation

An overall cost of energy evaluation not only of the MRS but also of a number of innovations considered in the Innwind project led to results as in Table 15.2. The cost of energy evaluation tool was developed as an independent task [21] within Innwind. It is largely based on the NREL model [22] with adaptations to deal with specific innovations.

Table 15.2 shows a 16% reduction in levelised cost of energy (LCOE) of the MRS wind farm (80.2 €/MWh) as compared to one based on RWT technology (95.6 €/MWh). The 16% reduction in cost of energy attributed to the MRS is a base level. It does not

Table 15.2 Levelised cost of energy comparisons.

LCOE impact	%
MRS	−16.0
Low induction rotor	−6.0
Advanced two-bladed rotor	−7.6
Smart rotor with flaps	−0.5
Carbon truss blade structure	−0.6
Bend-twist coupled rotor	−0.8
Superconducting generator	−0.4
PDD (Magnomatics) generator	−3.2

include any credit for the power gains (~10% considering both rotor interactions and turbulent wind response) or for a predicted 14% reduction in O&M cost. The leading three concepts are complete system concepts and naturally less benefit is to be expected from innovations in components and subsystems. However, the 3.2% reduction predicted for the Magnomatics PDD (pseudo direct drive; discussed in Chapter 25) is notable and much due to reduction of part load losses as compared with conventional drive trains.

To assess robustness of the cost of energy predictions, sensitivity studies were conducted (Figure 15.10). Sensitivities to energy capture and O&M are expected to be the greatest for almost any wind generating system. Comparing the MRS with a single turbine, cost sensitivity to the support structure is greater but to the rotor-nacelle systems is much less. The very weak sensitivity to yaw bearing cost gives assurance that the yaw bearing system is unlikely to be very problematic for economic viability of the MRS concept.

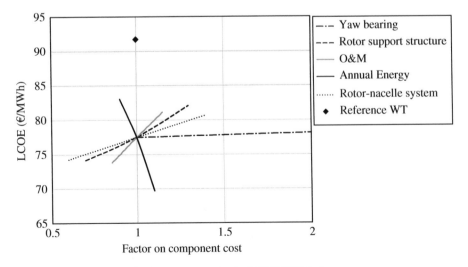

Figure 15.10 Sensitivity of cost components in the 20 MW MRS design.

15.10   Multi-rotor Conclusions

Based on scaling law impacts, advantages of standardisation and the huge commercial benefit of de-risking turbine technology involving faster development and reliability improvement of turbines at a proven manageable scale, there remains a good case for more detailed investigation of MRS. Challenges specific to the MRS are mounting rotors on a spatially extended structure, realising yaw capability of the whole system, dealing with the aerodynamic interaction of closely spaced rotors and having high overall system reliability with a high total part count. Based on the results of Innwind, none of these challenges appear as show-stoppers. Moreover, aerodynamic performance and maintenance appear as potentially advantageous rather than problematic.

The MRS may thus provide economically a single system of large net capacity, perhaps beyond 20 MW, and therefore of greater unit capacity than is ever likely to be realised on the basis of the single rotor concept. The next critical stage of design development is to develop a detailed design especially in respect of maintenance. It is much more likely that the optimum unit turbine size will relate to handling assembly and maintenance issues rather than to intrinsic turbine characteristics like aerodynamics or drive-train efficiency.

The emphasis in a multi-rotor array is on multi. As the scaling relationships indicate, the economic advantage increases with rotor numbers and there is no point in a system with a small number of rotors, at least not directly as a solution to economic upscaling in the context of reducing cost of energy from offshore wind. Some systems have been proposed using only a few rotors [23]. They have merits, but the case for them is different and relies less on the fundamental area to volume benefit.

The essence of multi-rotor economics has been shown to be area–volume relationships. This is a recurrent issue in engineering. Heat transfer rate is area related, whilst heat capacity is volumetric. Thus, fuel is pulverised or radiators have fins to increase total surface area associated with a given volume. The problem in seeking to maximise the area-to-volume ratio of wind turbine rotors by extending the multi-rotor concept to a multitude of microscale rotors is the huge loss in aerodynamic performance that would result in very low Reynolds number.

However, ideas such as a 'wind tree' where vibrating 'leaves' can extract energy through, for example, the piezoelectric effect have been considered. A recent system exploits 'rotor vibrations' of a system with multiple units of around 10 cm diameter. It has a reported claim [24] of around 18% efficiency, whereas a conventional wind turbine at that scale (it is suggested) would be about 1% efficient. However fanciful such ideas seem, it should be noted that there is a profound logic to them. It is a matter of finding an effective way of integrating power from multiple sources and also to have good aerodynamic function at very low Reynolds number, something thoroughly mastered by insects if not by humans.

References

1 Stewart, R.D., Wyskida, R.M. and Johannes, J.D. (eds) (1995) *Cost Estimator's Reference Manual*, 2nd edn, John Wiley & Sons, Inc. ISBN: 978-0-471-30510-1

2 Jamieson, P. (1995) The Prospects and Cost Benefits of Advanced Horizontal Axis Wind Turbine Design. Garrad Hassan Report No. 317/GR/02, February 1995 (internal company report).

3 Deliverable 1.33 Innovative Turbine Concepts – Multi-rotor System, http://www .innwind.eu/publications/deliverable-reports (accessed 26 September 2017).

4 Smulders, P.T., Orbons, S. and Moes, C. (1984) *Aerodynamic Interaction Between Two Wind Rotors Set Next to Each other in One Plane*, EWEC, Hamburg.

5 Ransom, D., Moore, J.J. and Heronemus-Pate, M. (2010) Performance of wind turbines in a closely spaced array. *Renewable Energy World*, **2** (3), 32–36.

6 Chaviaropoulos, P.K. and Douvikas, D.I. (1998) *Mean-flow-field Simulations Over Complex Terrain Using a 3D Reynolds Averaged Navier–Stokes Solver*. Proceedings of ECCOMAS '98, Vol. I, Part II, pp. 842–848.

7 Voutsinas, S.G. (2006) Vortex methods in aeronautics: how to make things work. *International Journal of Computational Fluid Dynamics*, **20** (1), 3–18.

8 Chasapogiannis, P., Prospathopoulos, J.M., Voutsinas, S.G. and Chaviaropoulos, P.K. (2014) Analysis of the aerodynamic performance of the multi-rotor concept. *Journal of Physics Conference Series*, **524** (1), 012084.

9 http://www.selsam.com/ (accessed April 2011).

10 www.windharvest.com (accessed April 2011).

11 Rainey, D.L. and Weisbrich, A.L. (1998) *WARP Technology for Low Cost and Environmentally Friendly Marine Based Wind Power Plants*. Proceedings of BWEA 20, Cardiff, September 1998.

12 Goeltenbott, U., Ohya, Y., Karasudani, T., Yoshida, S. and Jamieson, P. (2015) *Aerodynamic Analysis of Clustered Wind Lens Turbines*. Proceedings of the International Conference on Power Engineering-15 (ICOPE-15), Yokohama, Japan, December, 2015.

13 Jamieson, P. (2004) *Future Vision of the Technology*. Presentation to US, DOE Deepwater Wind Energy Workshop, Washington, DC, October 2004.

14 http://www.aerotrope.com/ (accessed April 2011).

15 http://www.alpha-ventus.de/ (accessed April 2011).

16 http://vertaxwind.com/ (accessed April 2011).

17 UPWIND (2011) *Design Limits and Solutions for Very Large Wind Turbines*, EWEA.

18 Deliverable 1.21 Reference Wind Turbine Report, http://www.innwind.eu/ publications/deliverable-reports (accessed 26 September 2017).

19 Dalgic, Y., Lazakis, I., Dinwoodie, I. *et al.* (2015) Advanced logistics planning for offshore wind farm operation. *Ocean Engineering*, **101** (1), 211–226.

20 Gintautas, T. and Sørensen, J.D. (2015) Integrated system reliability analysis. Task 1.34 Report for Innwind.EU, September 2015, Deliverable 1.34, http://www.innwind .eu/publications/deliverable-reports (accessed 26 September 2017).

21 1.22 Definition of Performance Indicators (PIs) and Target Values. Deliverable 1.22

22 Fingersh, L., Hand, M. and Laxson, A. (2006) Wind Turbine Design Cost and Scaling Model. Technical Report NREL/TP-500-40566, National Technical Information Service.

23 de Vries, E. (1997) Multi rotor wind turbines. *WindStats Newsletter Autumn* 1997 **10** (4), 8.

24 http://www.worldchanging.com/archives/003739.html#piezo (accessed April 2011).

16

Design Themes Summary

In evaluating any wind turbine system, innovative or otherwise, after basic feasibility hurdles have been cleared, the next step is a long hard look at energy capture performance and the basics of torque and speed.

Of the design themes listed, the merits of pitch versus stall and two blades versus three have been enduring hot topics in the industry. For all the historical investment of industrial and academic effort, in the end these options may be little differentiated in overall technical merit and cost of energy (COE). Thus, aspects that are non-technical such as visual appearance or, indirectly, technical aspects like the specifics of grid compliance have made the choice. Variable speed has been key in enabling particularly effective pitch control. There are added costs and reliability issues with electronics and only small energy gains compared with fixed-speed, high-slip or two-speed systems. Thus, it is doubtful if a variable speed, pitch-regulated wind turbine has a significant COE advantage over a well-designed stall-regulated wind turbine. However, variable speed offers important operational flexibility and, with modern power converters, excellent grid compliance.

There seems to be a clearer picture that the vertical-axis wind turbine (VAWT) as a single unit is seriously disadvantaged compared to a horizontal-axis wind turbine (HAWT) being either massive and expensive or much inferior in energy capture; or, to a lesser degree in each, having some net disadvantage in both. VAWT technology has not been successful in the markets for large-scale electricity generation. This is not to dismiss the VAWT but suggests that a conceptual shake-up is needed if VAWTs are to be successful in large-scale commercial applications.

It also seems that little cost reduction per se can be achieved by having free yaw in a HAWT. All sorts of free yaw, limited free yaw, soft yaw or intelligent yaw may nevertheless be of net benefit.

Upwind versus downwind has not been included in the topics reviewed and it is not much of a debate in the abstract. There is usually little point in considering downwind operation of conventional HAWTs with blades sufficiently stiff to suit upwind operation. The main case for downwind operation is to harmonise with rotor designs with flexible blades such as the high-speed rotor discussed in Section 7.2. There is however some interest in downwind turbines in Japan from a different angle regarding suiting up-flows on hilly sites. Also, while the tower shadow effect may be more severe for blades in the wake of the tower, the nacelle blockage can slightly enhance performance in displacing central flow to the more aerodynamically active parts of the blades. So the subject is not entirely closed!

Innovation in Wind Turbine Design, Second Edition. Peter Jamieson.
© 2018 John Wiley & Sons Ltd. Published 2018 by John Wiley & Sons Ltd.

Whilst the multi-rotor system concept has persisted in a small way, the mainstream industry has paid little attention to it. It began as an expedient at a time when steel was the only practical material for rotor blades. The potential as an economic solution to upscaling with huge savings on quantities of composites and highly engineered drive-train materials has scarcely been recognised. A major step forward took place within the Innwind.EU project. Verification of aerodynamic performance of a large array of closely spaced turbine, evaluation of structure and yaw system feasibility and cost, operations and maintenance (O&M) issues (especially avoiding use of jack-up vessels) and associated costs were all encouraging rather than hinting at likely showstoppers. Perhaps the next most critical step is a detailed design for operational logistics and maintenance to ensure that the most frequent very minor faults are dealt with effectively. Considering the economic promise which derives both from the fundamentals of the square-cube law and from the benefits of standardised mass production, multi-rotors surely deserve further investigation.

The preceding comments are clearly opinions – albeit based on some reasoning and analysis as has been presented in previous chapters. There are no final answers – new ideas may breathe new life into apparently inferior options. The primary intention of the design theme reviews is to define the main issues around key design options, suggest possible approaches to evaluating the concept options and ventilate logical arguments that may reinforce or conflict with popular views or received wisdom.

Part IV

Innovative Technology Examples

Having considered design background and having reviewed some of the main themes in design choices of wind turbines, a few examples of innovative technologies are now presented. These are as described – just examples. There is no attempt to single out the most significant wind technology innovations in either technical or commercial terms, still less to be comprehensive in an industry where new patents appear almost daily. Also, many innovations, even after they are evident in the market place, cannot be discussed effectively for reasons of commercial confidentiality. The following considers a mix including some quite unusual concepts and also a few more mainstream developments.

17

Adaptable Rotor Concepts

17.1 Rotor Operational Demands

In exploring innovative rotor systems, the basics for safe and effective regulation of a wind turbine rotor must firstly be considered. Rotor regulation systems have two prime functions:

1) Prevention of dangerous overspeed of the rotor,
2) Regulation of output power above rated wind speed within satisfactory limits preventing damaging overpower levels in the drive train or overloads in other system components.

Beyond these prime safety functions, the aim is then to find an optimum balance between maximisation of energy output and minimisation of design driving loads.

Preventing overspeed of a rotor in a fault case where there is no reaction torque provided by the generator is a much more onerous duty than regulating power above rated wind speed. Figure 17.1 shows a typical power performance characteristic, C_p, as a function of tip speed to wind speed ratio, λ, for three fixed-pitch angle settings, 1° (the optimum of a particular rotor design for maximum C_p), 10° and 20°. The blade pitch action is assumed to be towards negative incidence, maintaining attached flow until stall occurs at negative angles of attack. Each characteristic passes through zero at a certain maximum value of tip speed ratio, λ_{max}, which corresponds to the maximum possible value of rotor speed attainable in the absence of any generator or braking system reaction torque or frictional losses. The tip speed ratio corresponding to λ_{max}, the highest tip speed ratio and the maximum rotor speed that the rotor can attain with any given set pitch angle, is shown in Figure 17.2. The peak limiting tip speed ratio of approximately 20 at a pitch angle of 1° corresponds to the curve with highest maximum C_p. Pitch angles greater or less than 1° result in reduced maximum C_p and reduced λ_{max}. In reality, the limiting tip speed ratio may be reduced in circumstances where the tip speed is very high and associated with a Mach number in the transonic region. This is somewhat academic as such tip speeds would usually be associated with failure of the rotor or generator under excessive centrifugal loading. Although large rotors will generally be designed for an extreme overspeed of about 30% in excess of maximum steady-state operating speed and operated within much more limited overspeed levels, understanding extreme runaway conditions can be important for small rotors [1] where rather high overspeeds may be tolerated to avoid complex rotor regulation systems.

Innovation in Wind Turbine Design, Second Edition. Peter Jamieson.
© 2018 John Wiley & Sons Ltd. Published 2018 by John Wiley & Sons Ltd.

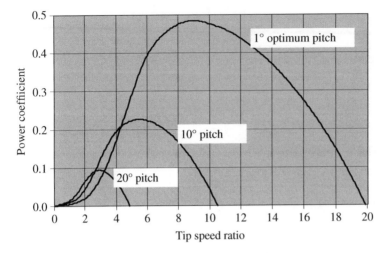

Figure 17.1 Power coefficient characteristic for selected pitch angles.

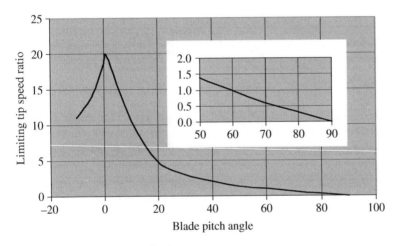

Figure 17.2 Limiting tip speed ratio.

The limiting blade tip speed of the rotor is given by

$$\omega R = \lambda_{max} V \qquad (17.1)$$

Suppose a normal maximum steady-state operating tip speed is V_t and that an ultimate tip speed limit in an extreme fault condition is a factor, f, greater than this. Then adequate safety requires

$$\lambda_{max} \le \frac{f V_t}{V}$$

Considering typical design values of V_t as 75 m/s, f as 1.3 and an extreme gust of 70 m/s, then:

$$\lambda_{max} \le 1.4$$

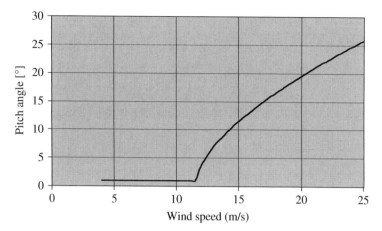

Figure 17.3 Typical steady-state pitch schedule.

It is clear from the inset diagram of Figure 17.2 that restraining λ_{max} to values in a range below 1.4 typically requires very high pitch angles, at least $>50°$. Idling rotor pitch angles would generally be greater than this ~80–90° in order to keep the rotor at very low idling speeds under all wind conditions. On the other hand, power limiting regulation (Figure 17.3) requires steady-state pitch angles typically in a range from 0° to 30° and perhaps up to about 40° in dynamic control action.

Thus, the aerodynamic demand for overspeed protection is much more onerous than the requirements for power regulation above rated wind speed.

17.2 Management of Wind Turbine Loads

All adaptable rotor concepts, including smart rotors, concern managing loads (output power may be included in the broadest definition of loads) so as to meet an overall objective which is generally to reduce cost of energy. First, it is useful to identify the loads that are essential to the prime function of the wind turbine of extracting energy from the wind.

Torque: to produce power in a rotary system implies accepting a level of steady-state torque. Usually, power take-off is from a central shaft. The rated (maximum steady state) power is then the product of rated torque and associated rotor speed. At any chosen system scale, tip speed is always limited either by acoustic noise emission or by the variety of factors discussed in Section 7.2 relating to the high-speed rotor concept. The limit on rotor speed determines an essential level of steady-state torque that must be accepted to generate the required rated power level as

$$Q_{rated} = P_{rated}/\omega_{limit} \tag{17.2}$$

Thrust: to extract wind power by any means whatever also implies accepting a certain level of thrust. For a wind turbine with active pitch control which sheds loading by fining pitch, using pitch control to reduce the peak steady-state thrust that would occur at rated wind speed and thus accepting a small loss in power production may be profitable. However, there will always be a level of steady-state thrust that must be accepted to generate adequate power. For the system to extract energy in doing work on the air passing through the rotor plane, power extracted at the chosen thrust limit is some fraction of the source power in the air given by

$$P = U_0(1 - a)\, T_{\text{limit}} \tag{17.3}$$

In Equation 17.3, U_0 is the far upstream wind speed that is associated with the chosen thrust limit. Thrust (a consequence of resistance to air flow through the rotor) is the basis of doing work on the air and torque is the basis of extracting such work with a rotor. In conclusion, the wind turbine designer cannot avoid minimum levels of steady-state torque and thrust in any design that extracts energy. However, all other loads, transient, extremes in operation and in idling and fatigue loads can often be regulated without necessarily involving compromises in energy capture. This can have significant benefits and is very much what adaptable and smart rotors aim to do.

17.3   Control of Wind Turbines

Rotor aerodynamics, structural dynamics, drive-train dynamics and control system dynamics are all critically interactive in the design of a wind turbine. Control [2] is largely left as a specialist subject beyond the scope of this book. The discussion here serves specifically to relate control to adaptable rotor concepts and is in no way proportionate to its overall importance in wind turbine design.

In the hierarchy of operation of a wind turbine, safety is the primary concern and it is an aim of the engineering design to have systems that are mechanically and electrically failsafe to protect the system in the event of any foreseeable faults developing. In normal operation, all sizeable wind turbine systems in use for the production of electricity have active control systems which rely on computers. The overall management of operation is described as supervisory control and this level of control addresses decisions about start-up, shut-down, yawing braking procedures and also responses to faults that are not at a level to trigger safety system actions.

Within the supervisory control system are the so-called closed-loop control systems which, in the case of the prevalent variable speed, pitch-regulated wind turbine, will regulate the pitch of the blades and, typically, through control of the power converter, will also regulate the reaction torque of the generator. This control system will operate on the basis of feedback from pitch angle, rotor speed and generator torque/power measurement. The control objective may be primarily to regulate the electrical output power to a desired set point or to maximise power by tracking a prescribed torque-speed characteristic.

However, the output power is only one wind turbine 'load' and it is now well recognised that control systems through their intrinsic interaction with the complete machine dynamics are, in general, capable of doing much more. Appropriate pitch action can de-tune tower vibrations [3–5], and may substantially reduce tower fatigue damage. Such control of loads can be rendered yet more effective if additional sensors (typically accelerometers) are provided to measure critical vibrations. There is usually some compromise in extending control objectives beyond managing output power – power performance may be degraded a little, pitch system duty may be made more onerous.

Nevertheless, there is often potential net benefit to the wind turbine system in managing loads as well as power and this is now a very active area of interest in wind turbine control system development. This potential benefit can be augmented if more adaptable rotors are developed with added actuator capability – topics discussed in Sections 17.5–17.7.

17.4 LiDAR

17.4.1 Introduction

LiDAR (light detection and ranging) is a modern laser-based technology for wind field measurement that may be also considered in control of large wind turbines as discussed in a seminal paper [6] of 2006. It is of proven usefulness in general wind field measurements for research (e.g. wake measurements [7]) and for wind farm development, especially offshore where anemometer masts are extremely expensive and cannot readily be moved to new locations. LiDAR may be nacelle, hub or blade mounted. A short-range beam can provide inflow information close to each blade [8]. LiDAR certainly provides an opportunity to have good spatial and temporal definition of the wind field ahead of a rotor. However, in the application to a single wind turbine unit, two main issues arise:

- LiDAR is relatively expensive and thus only justifiable on very large turbines.
- Conventional control involving reaction torque is very fast. In effect, it avoids problematic wind speed measurements and, in responding to measured power as a rotor disc averaged quantity, is highly effective.

17.4.2 The LiDAR Operational Principle

A LiDAR works by sending out a laser beam and measuring the light reflected back from particles or aerosol droplets in the air which are assumed to move with the wind velocity. The Doppler shift in frequency between the original and reflected light provides an accurate measure of the component of the wind velocity in the direction of the laser beam. The measurement represents an average velocity along a section of the laser beam – it is not possible to obtain a measurement focused precisely at a single point.

Two main types of LiDARs are currently available: pulsed- and continuous-wave (CW) systems. The pulsed system is capable of measuring wind speeds simultaneously at a number of distances (ranges), by timing the interval between the emission of the pulse and the arrival of the reflection. The CW system focuses the beam at a particular range, so that the strongest reflection is obtained from a focal point a particular distance along the beam. For both systems it is possible, in principle, to steer the laser beam to allow it to scan the approaching wind field over the swept area of the rotor, or to make use of multiple beams to sample the wind at different points simultaneously.

There are several commercial LiDAR systems suitable for turbine-mounted applications. The main contenders at present would be a CW LiDAR with a single beam scanning in a circle ahead of the turbine, and a pulsed LiDAR with two or more fixed beams pointing ahead of the turbine at an angle to the centreline and capable of measuring at several points simultaneously along each beam. LiDAR may be blade mounted and a short-range beam can provide inflow information close to each blade, as demonstrated in an R&D context in [9].

At one stage, a LiDAR systems supplier claimed 20% improved energy capture due to improved yaw control. Loss of power in yaw may be roughly estimated proportional to the cube of the cosine of the yaw angle. Thus, if a turbine spends all of its life in 10° yaw error and obtains two-thirds of its energy in wind speeds below rated power, the associated energy loss compared to perfect yaw tracking is less than

$$\Delta E = \{1 - \cos^3(10)\} \times (2/3) \times 100 = 3\%.$$

Inverting this calculation, to lose 20% of annual energy due to yaw error alone, a permanent yaw error of about 28° would be required. State-of-the-art turbines with normally good yaw control generally operate inside a band of $\pm 10°$ yaw error without any great sophistication in yaw control. Most commonly, the turbine is locked in a fixed yaw position and when the $\pm 10°$ limits are approached (as measured by a crudely calibrated nacelle-mounted wind vane), the yaw motors are activated and restore the nacelle position to an estimated zero yaw. The industry has matured in the past 5 years in experience of LiDAR, and largely such extreme claims have disappeared.

It has also been argued that improved (software) control using LiDAR can enable energy gains without any change in system hardware. This latter possibility can almost be dismissed. Simulation studies at Strathclyde University considered the performance in overall energy capture for various controllers (optimal and very much suboptimal) operating without any direct knowledge of the wind field. This is compared with ideal control where complete knowledge of the wind field is provided. Results [10, 11] indicate that, even with total predictive knowledge of the incoming 3D wind field, there is barely a percent of extra energy to be gained compared to present state-of-the-art control and this shows very clearly that claims of a few percent energy gain and more, as may have been suggested as possible benefits of using LiDAR, are quite unrealistic.

17.4.3   Evaluation of LiDAR for Control of Wind Turbines

The main identifiable benefit in using LiDAR for wind turbine control is in reduction of blade and tower base fatigue loads.

Simulation studies of DNV GL [12] suggest that reductions in the lifetime damage equivalent blade rotor bending moment in a range from 6% to 12% may be achievable and associated reduction in the tower base overturning moment in a range 16–20%. While encouraging, such benefits from LiDAR control must be considered in the context that similar benefits may be realised less expensively using feedback from sensors measuring blade loads. For frequencies lower than blade passing (most control action and fatigue damage is associated with such lower frequencies), feed-forward control using LiDAR [13, 14] has been considered, although feed-forward of loading information from one blade to the next derived from conventional sensors may be as useful for predictive benefit as LiDAR.

At present, wind turbine design certification standards (see Section 9.8) include some rather artificial load cases as in gusts of preset shape and coherent extreme wind direction changes. Such design load cases may not be very realistic, but they represent an understandable concern that a turbine design should have margins of safety to deal with unpredictable extreme events. LiDAR may well be very useful in characterising structures in real wind [15] and informing further development of design load standards.

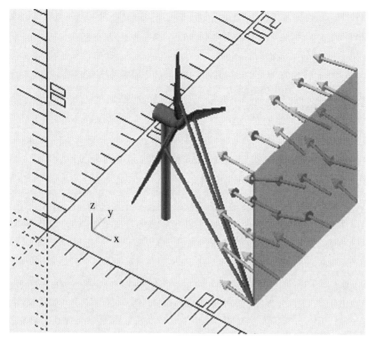

Figure 17.4 Wind vectors measured with converging beams.

17.4.4 An Example of Future Innovation in LiDAR

Theodore Holtom of Windfarm Analytics has been developing an innovative converging beam LiDAR (Figure 17.4) which has the potential to give much more accurate definition of wind vectors and turbulence than diverging conical scans, and so on, especially in non-uniform wind conditions. This work has been funded by Innovate UK and conducted in partnership with Sgurr Energy, Fraunhofer UK and Thales.

As was discussed, the level of benefit to control of advance information about the wind field and the issues of whether LiDAR can do better than blade sensors have not been resolved. However, high-quality advance information about large structures in the wind or extreme events and major wind field asymmetry may benefit supervisory as well as closed-loop control action, for example, in dealing with extreme events or in better management of start-up or shut-down sequences which are usually related both to short-term averages of wind speed and transient gust levels.

17.5 Adaptable Rotors

A thematic area of current research is in developing more adaptable rotors. The broad aim is to increase energy capture whilst containing design driving loads. Although significant progress has been made, there is limited capability to increase energy directly by improved aerodynamics of blade sections and improved aerodynamic optimisation of rotor design. However, systems that can restrict loads can allow larger diameter rotors with more substantial consequent increase in energy output.

Rotor adaptability to serve these ends may arise in a wide variety of ways. Some examples are as follows:

- Differential pitch control of the rotor blades.
- Aeroelastically engineered blades, in general, including tailored structural characteristics to introduce pitch-flap coupling [16].
- Smart rotors with additional controllable surfaces (micro tabs, trailing edge flaps, active vortex generators) [17–19].

In evaluating smart rotor concepts, control dynamics needs careful consideration. Some systems such as micro tabs have suitable actuator capability and may be relatively easy to engineer, but a phase lag in response can be a significant disadvantage. Design studies suggest that trailing edge flaps over a sufficient extent of span can enable impressive rotor fatigue load reductions, perhaps ~60%. Flaps generally have good control characteristics and a key interest is to develop systems with flexible surfaces (unlike typical aircraft designs with rigid flaps) avoiding spanwise breaks in the surface of the trailing edge.

Differential pitch control of blades has received much attention in recent years. It relies on each blade having independently controlled pitch actuators. Early pitch systems of the 1980s were usually collective, typically involving mechanisms that mechanically interlinked them. Among these, an interesting innovation was the ball screw collective mechanism concept (Figure 17.5) of John Armstrong.

The blade pitch is driven by a ball screw moving axially through a hole in the main gearbox shaft. A pitch drive motor drives a ball nut synchronously with the rotor to keep constant pitch. If the motor fails or is stopped, the ball screw drives off its own pitch to stop the rotor. This linkage has been highly effective, being used on WEG

Figure 17.5 Pitch linkage system of the MS2 wind turbine. Reproduced with permission of John Armstrong.

MS2 wind turbines which are still operating in the Altamont Pass wind farm after 24 years.

Initially, it may have been considered a desirable safety feature to have the blades interconnected in a collective system which would prevent any large differences in individual pitch angles. As rotors increased in size, collective pitch mechanisms would have become ever more demanding to engineer. Moreover, when GL rules were introduced in 1998 [20] allowing that a rotor with independently actuated pitch systems could be considered to have two aerodynamic brakes, independent pitch systems became predominant. At first, there was no thought other than to operate the blades in synchronism but the possibility of individual blade pitch control was then readily available. A specific study of such individual pitch control [9] suggests substantial potential benefits, for example, 15–30% reduction of blade root fatigue loads, 20–40% reduction of hub and shaft loads and 10% reduction of tower top moments. Other recent studies [21, 22] confirm numbers of that order and suggest that, without compromise on power performance or speed regulation, whilst the frequency of pitch activity is necessarily increased, both extreme and lifetime fatigue loads of the pitch system can be reduced.

The focus of much current research is to use minimal actuator capability to modify aerodynamic performance targeting the sensitive areas of the aerofoil section (such as the trailing edge) where small geometric changes can significantly modify aerodynamic characteristics.

The aim is to produce a 'smart' rotor where comparatively inexpensive systems with modest actuator capability (see Figure 17.6) can significantly improve regulation of blade and system loading and better manage the trade-off between fatigue damage and maximum energy capture. Such systems are however necessarily supplementary and cannot in themselves change rotor aerodynamic properties enough to address the prime regulation functions of overspeed protection and power limiting. The data of Figure 17.6 is

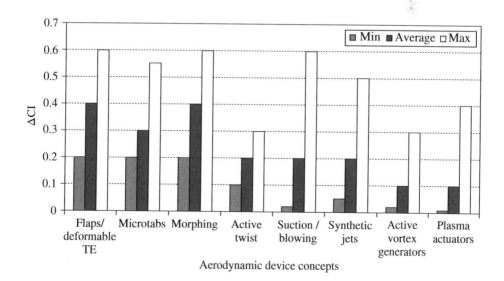

Figure 17.6 Capability of smart blade devices to change lift coefficient.

naturally only indicative as the specific actuator capability depends on the physical size of devices, the extent of blade span over which devices are functional and the capacity of power provided.

More radical concepts are possible. The basis of these is to have a variable diameter rotor that can maximise its size and exposure to wind loading in the light and moderate wind speeds that are most important for energy capture but defend the system against increasing loads in higher wind speeds by reducing swept area. Such concepts are discussed in Sections 17.6 and 17.7.

All the adaptations discussed move in a direction away from rigid rotor geometry, perhaps eventually towards the flexible and highly adaptable technology of the wing of a bird.

17.6 The Coning Rotor

17.6.1 Concept

Hinged blades are the norm in helicopter design where strong restoring moments due to centrifugal forces will balance aerodynamic bending moments and rotor designs with hinged blades were quite common in the early history of wind technology. Until glass-reinforced plastic material systems became established technology for rotor blades in the 1980s, steel was often used in the blade structure (see Figure 4.1 and the surrounding discussion) and blades were hinged to relieve the bending moments associated with the great self-weight of steel.

Thus, early wind turbine designs with hinged blades evolved in much the same way as the earliest multi-rotor systems, essentially for defensive motives, due to the problems of making very large blades in steel. In 1995, GL Garrad Hassan was invited by the UK Department of Trade and Industry (DTI) to conduct a study evaluating the prospects for advanced wind turbine design. In the course of this project [23], it emerged that there could be a quite different positive motivation for having hinged blades (and also for multi-rotor systems as discussed in Chapter 15).

Consideration of a wide variety of innovative designs and a search for basic principles to advance wind turbine design led to the conclusion that there were two, and perhaps only two, main principles for significant improvement of the technology. They are as follows:

1) Increasing the ratio of energy capture (value) to design driving loads (cost),
2) Increasing the ratio of swept area (energy capture value) to system volume (cost).

The first principle has led to variable diameter rotor concepts, to be described, and the second to the multi-rotor concepts discussed in Chapter 15.

The primary motivation for a coning rotor with hinged blades is to remedy the situation (Figure 9.7) where there is a clear mismatch between the wind speed range (3–15 m/s, say) associated with most of the energy production and the higher ranges (12–60 or 70 m/s) where most design driving loads and associated costs are incurred. The idea is to have a rotor (necessarily in a downwind attitude) that could be about

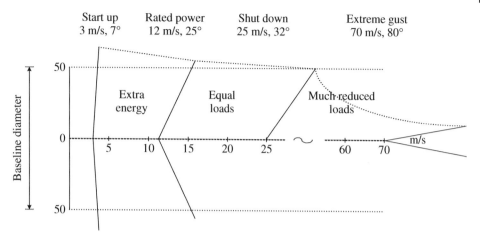

Figure 17.7 Operation of a coning rotor.

25% larger in diameter than a conventional baseline, rigid rotor, design. The aim of the design is to extract much more energy than an equivalent rated conventional system whilst avoiding critical increased loading. The rotor is operated with a coning schedule, typically as indicated in Figure 17.7. The baseline wind turbine is now a state-of-the-art pitch-regulated, variable speed 3 MW wind turbine of 100 m diameter. This is an update relative to an original evaluation (circa 1995) based on a coned rotor design and Danish stall-regulated design as baseline, both rated at 450 kW.

Operation starts at a cone angle of 6° with the rotor at maximum expansion in light winds (say, 3–9 m/s) so as to maximise energy capture. The coning is actively controlled with actuators operating in a load control mode aiming to maximise energy capture in wind speed below rated without exceeding a prescribed blade root bending moment limit. As wind speed increases, the rotor cones freely in a natural equilibrium between aerodynamic and centrifugal forces with additional flap damping provided by the coning actuators operating within the prescribed load limit. This equilibrium angle is determined by the Lock number, γ, expressed here in terms of the chord width, c, the lift slope, $\partial Cl/\partial \alpha$, rotor radius, R and I_b blade flapping inertia:

$$\gamma = \rho c \left(\frac{\partial Cl}{\partial \alpha} \right) \frac{R^4}{I_b} \tag{17.4}$$

The Lock number is scale invariant so that, for example, equilibrium coning angles do not change if scaling adheres to similarity as discussed in Section 4.4. Free coning of a downwind rotor running at constant speed can assist in power regulation, but it can be shown that it is not in itself adequate to regulate power to a set point in wind speeds above rated. There is also a consideration that a free coning rotor is dynamically unstable. Contraction of the swept area reduces rotational inertia of the rotor and, through conservation of angular momentum, the rotor tends to speed up.

The initial study of 1995 considered operation in stall with generator reaction torque being used to reduce rotor speed, which in turn reduces centrifugal forces causing the cone angle to increase. Control dynamics are complex but this was found to be a viable [24], if rather challenging, design option for power regulation. The original study considered only coning and variable speed using control of generator reaction torque to reduce speed and promote coning in high operational wind speeds.

However, more recent consideration of the operational issues suggests that the best option may be pitch control with the blades pitching towards stall. When the concept first developed, it was considered desirable to avoid a pitch system in conjunction with the coning system. This technology is however commonplace in helicopter technology and was successfully implemented in a wind turbine design by WTC, the Wind Turbine Company of Seattle [25]. The choice of pitch to stall in the coning rotor concept is dictated by the consideration that the usual pitching in attached flow would produce a thrust characteristic decreasing with increasing wind speed (see Figure 9.7). This would conflict with a natural coning of the blades being assisted by increasing wind speed and rising rotor thrust.

A further update in design thinking is that, opposed to collective coning systems that were modelled in the original study, individual actuators which would be standard hydraulic cylinders would control coning of the blades. The coning rotor concept is considered to be well suited to a direct-drive permanent magnet generator (PMG) design using the PMG rotor structure as a hub providing mounting points for blade rotor hinges and for connecting the coning actuators to the blades. Another consideration is that conventional hinged or teetered rotor blades, having a very limited angular range, are substantially at risk of impacts on their end stops. Experience has shown that much of the dynamic loading benefit realised by a teetered rotor is negated by extreme loads arising from occasional impacts on the teeter stops. This is avoided in the coned rotor system as it is at a cone angle far from its ends of travel in strong winds; and in failsafe braking action, the coning actuators will still provide damping.

Since the initial work of the mid-1990s, Crawford [26] has developed improved modelling of coning aerodynamics. Crawford and Platts [27] and others have considered further aspects of the operation and engineering of a coned rotor design.

In the original study Jamieson [28] of the coned rotor concept, modelling of loads and performance in turbulent wind was supplemented with other independent evaluations of a number of key issues: acoustic emission [29], visual impact [30], yaw system [31] and PMG design [32].

Key concerns with the design were evaluated. For example, when the blades were parked in their fully coned position and the system was acting as a weather vane in extreme storm conditions, could severe lateral loads arise due to changes of wind direction and general wind turbulence? Simulations with a set of 3D turbulent wind fields, in which the maximum 3-s gust of 70 m/s (extreme gust, class I site) was represented, suggested that lateral loads, although quite significant, were not driving the design of any major components.

17.6.2 Coning Rotor: Outline Evaluation – Energy Capture

Assuming the coning schedule of Figure 17.7, rotor power curves of the coned rotor (fully expanded diameter 125 m and 3 MW rated electrical output) and baseline (diameter 100 m rated at 3 MW) are presented in Figure 17.8. The same drive train with efficiencies representing a direct-drive PMG and power converter is assumed for both designs.

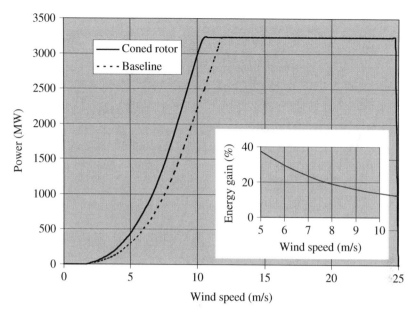

Figure 17.8 Rotor power and energy comparison of coned rotor with baseline.

The percentage energy gain of the coned rotor over the baseline as a function of annual mean wind speed is presented in the inset chart of Figure 17.8.

17.6.3 Coning Rotor: Outline Evaluation – Loads

In general, the aim to limit loads of the 25% larger coned rotor to the levels of the baseline design was successful (Table 17.1). This data is from the original study where comparisons were made at 450 kW with a Bonus 450 stall-regulated wind turbine, then among state-of-the-art designs. Neither turbine employed pitch regulation; and it is suggested that in an updated comparison both would, with pitching to stall, be employed in the case of the coned rotor.

Table 17.1 Comparison of critical bending moments (kNm) Cone 450 and Baseline 450 kW.

Critical bending moments	Lifetime equivalent fatigue loads			Extreme loads		
	Cone	Base	Ratio	Cone	Base	Ratio
Blade	180	187	0.96	521	667	0.78
Hub	198	219	0.90	600	611	0.98
Shaft	197	202	0.98	600	556	1.08
Overturning	113	157	0.72	455	666	0.68
Yawing	81	219	0.37	492	558	0.88

17.6.4 Concept Overview

The major effort in the quite extensive studies of the mid-1990s within the UK DTI project was on modelling of loads and system dynamics including necessarily the complex control system design. Cost evaluations were very encouraging but comparatively superficial. Finite element analyses of hub designs were carried out and outline blade designs were developed, but in general insufficient detailed engineering design was possible to consolidate cost estimates. Even with 25% greater length, the rotor blades of the Cone 450 being much more lightly loaded were estimated to be of less total mass and cost than the conventional baseline design. Allowing for a cost premium on the hub system with the additional linkages and (hydraulic) coning actuators, the complete tower top systems of the Cone 450 were estimated to be 10% more expensive than the baseline design. However, with large energy gains, even allowing for the possibility of greater capital costs, it is easy to see that there was a strong case for further investigation of the concept.

Some of the significant loading and control issues of a pitch-regulated wind turbine with hinged blades have been validated in the field by the WTC design, and the German manufacturer Sudwind also had a design with hinged blades. These companies have found it possibly commercially viable and certainly technically feasible to realise hinged blade designs, yet without obtaining the basic energy benefit associated with a larger diameter rotor protected by a greater cone angle range.

Finally, the greatest potential for a coning rotor concept could be in offshore applications. Offshore, the turbine cost is less consequential than on land and, except for the impact of higher rotor loads on foundation cost, it would be very logical to increase rotor diameter in order to achieve higher capacity factors and greater utilisation of the expensive electrical infrastructure. With a coning rotor design, there is no significant increase in foundation loading, yet increases of up to 25% in diameter depending on wind speed will much increase energy output per megawatt of rated wind farm capacity. Rotor assembly and transportation may also be facilitated by the possibility of having a nacelle assembly including the blades folded together on their hinges.

17.7 Variable Diameter Rotor

The coning rotor actively changes swept area and therefore in effect changes rotor diameter. Another method of achieving a variable diameter rotor was conceived, by Jamieson *et al.* [33], on the basis of the idea of a sliding bladed section contained within the main blade section as in Figure 17.9.

The constant lift distribution required by an optimum blade can be maintained on the smaller chord width of the extendable section by using aerofoils with higher design lift coefficient. Little variation in chord width or twist is required over the outermost 30% of span of an optimum rotor design for a typical tip speed ratio ~9 and so the extendable blade part may conveniently be of constant section. The transition between the end of the main blade and the extendable section is of course of concern aerodynamically and will need careful design to minimise losses.

The strategy of extending to a maximum diameter greater than a conventional rotor of the same power rating in light winds and to a smaller diameter than the same conventional rotor in strong winds is similar to the coning rotor concept. The overall aims are

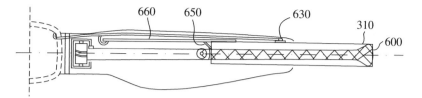

Figure 17.9 Blade of a variable diameter rotor. Extracted from Patent Document US 6,972,498 B2.

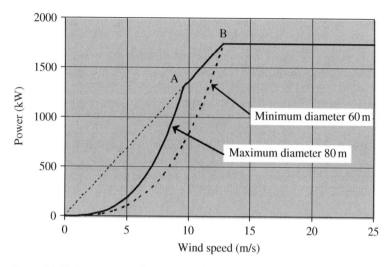

Figure 17.10 Power curve of a variable diameter rotor.

the same to achieve additional energy capture whilst managing designing loads more effectively. There are some interesting differences, however, best illustrated considering the power curve characteristic of Figure 17.9.

The rotor operates as a variable speed rotor of fixed geometry optimised in the fully extended state at ideal tip speed ratio and maximum C_p up to a wind speed corresponding to the point A in Figure 17.10. As wind speed increases, the rotor diameter is contracted. Although the rotor is now necessarily of suboptimal geometry for strictly maximum C_p, perhaps surprisingly, it performs better than a conventional variable speed rotor in the constant speed range below rated power. Increasing wind speed is associated with reducing tip speed ratio. Contracting the rotor increases rotor solidity which suits the reducing tip speed ratio, thereby maintaining a high C_p.

The rotor reaches the rated power of an equivalent conventional rotor (70 m diameter, 1500 kW rated power) having reduced its diameter to 70 m. As diameter reduces further to 60 m and the power reaches the point B, without exceeding any given tip speed limit, the rotor shaft speed may increase. Thus, at the same level of fully rated torque as the conventional wind turbine and same tip speed limit, the variable diameter rotor can operate at a higher shaft speed and produce more power (~1750 kW). There is of course some added cost in uprating the electrical systems to the higher power rating, but trade-off studies have suggested an overall economic benefit in operating in this way. With options for the minimum diameter to be less than conventional and maximum greater, the optimisation of this type of system with respect to choice of these diameters in relation to rated power is very subtle and, naturally, also dependent on the site wind conditions.

Although the blade sliding mechanism is clearly challenging, the energy benefit of variable diameter operation is major, and studies involving blade structural designs including the sliding mechanisms suggested that the engineering problems were surmountable promising net cost benefit for the concept.

References

1 Ebert, P.R. and Wood, D.H. (2002) The near wake of a model horizontal-axis wind turbine at runaway. *Renewable Energy*, **25**, 41–54.

2 Leithead, W.E. and Dominguez, S. (2007) Active regulation of multi-MW wind turbines: an overview. *Power System Technology*, **31** (20), 24–34. doi: 1000-3673(2007)20-0024-11

3 Bossanyi, E.A. (2000) The design of closed loop controllers for wind turbines. *Wind Energy*, **3** (3), 149–163. doi: 10.1002/we.34

4 Leithead, W.E. and Dominguez, S. (2005) *Controller Design for the Cancellation of the Tower Fore-Aft Mode in a Wind Turbine*. CDC/ECC, Seville, Spain, December 2005.

5 Chatzopoulos, A. and Leithead, W.E. (2010) *Reducing Tower Fatigue Loads by Coordinated Control of the Supergen 2MW Exemplar Wind Turbine*. Torque2010–The Science of Making Torque from the Wind, Heraklion, Crete, June 2010.

6 Harris, M., Hand, M. and Wright, A. (2006) Lidar for Turbine Control. Technical Report, NREL/TP-500-39154.

7 Kocer, G., Chokani, N. and Abhari, R.S. (2012) *Detailed Measurements in the Wake of a 2MW Wind Turbine*. EWEA 2012, Copenhagen, April 2012.

8 Bossanyi, E.A., Wright, A.D. and Fleming, P.A. (2012) *Field Test Results with Individual Pitch Control on the NREL CART3 Wind Turbine*. 50th AIAA Aerospace Sciences Meeting and Exhibit, Nashville, Tennessee, 2012.

9 Bossanyi, E.A. (2003) Individual blade pitch control for load reduction. *Wind Energy*, **6** (2), 119–128.

10 Chatzopoulos, A.P. (2011) Full envelope wind turbine controller design for power regulation and load reduction. PhD thesis. University of Strathclyde.

11 Chatzopoulos, A. (2010) *Assessing Energy Capture Capability of a MW Scale Wind Turbine During and Below Rated Operation*. EWEC2010, Scientific Proceedings, Warsaw, April 2010, pp. 251–254.

12 Bossanyi, E.A. (2004) *Developments in Individual Blade Pitch Control*. Proceedings of the EWEA Special Topic Conference 'The Science of making Torque from Wind', DUWIND, Delft University of Technology, The Netherlands, April 19–21, 2004.

13 Schlipf, D., Fischer, T., Carcangiu, C.E., Rossetti, M. and Bossanyi, E.A. (2010) *Load Analysis of Look-Ahead Collective Pitch Control Using LiDAR*. Proceedings of the DEWEK, Bremen, Germany, November 2010.

14 Wang, N. *et al.* (2012) FX-RLS-based feedforward control for LiDAR-enabled wind turbine load mitigation. *IEEE Transactions On Control Systems Technology*, **20**, 1212–1222.

15 Harris, M., Bryce, D.J., Coffey, A.S. *et al.* (2007) Advanced measurement of gusts by laser anemometry. *J. Wind Eng. Ind. Aerodyn.*, **95**, 1637–1647, first submitted 2004.

16 Lyons, J.P., Robinson, M.C., Veers, P. and Thresher, R.W. (2008) *Wind Turbine Technology – The Path to 20% US Electrical Energy*. Power and Energy Society General Meeting, Pittsburgh, July 2008.

17 van Dam, C.P., Chow, R., Zayas, J.R. and Berg, D.E. (2007) Computational investigations of small deploying tabs and flaps for aerodynamic load control. The science of making Torque from wind. *J. Phys. Conf. Ser.*, **75**, 012027. doi: 10.1088/1742-6596/75/1/012027

18 Marrant, B.A.H. and van Holten, T. (2006) *Comparison of Smart Rotor Blade Concepts for Large Offshore Wind Turbines*. Proceedings of the Offshore Wind Energy and Other Renewable Energies in Mediterranean and European Seas, Civitavecchia, 2006.

19 Barlas, T.K. and van Kuik, G.A.M. (2010) Review of state of the art in smart rotor control research for wind turbines. *Prog. Aerosp. Sci.*, **46**, 1–27.

20 Germanischer Lloyd Regulations for the certification of wind energy conversions systems. Edition 1993 with Supplement 1, March 1994 and Supplement 2, March 1998 Chapter 2, Section 1.C.

21 Leithead, W.E., Neilson, V. and Dominguez, S. (2009) *Alleviation of Unbalanced Rotor Loads by Single Blade Controllers*. European Wind Energy Conference, Maseilles, France, March 2009.

22 Leithead, W.E., Neilson, V., Dominguez, S. and Dutka, A. (2009) *A Novel Approach to Structural Load Control Using Intelligent Actuators*. 17th Mediterranean Conference on Control and Automation, Thessaloniki, Greece, June 2009.

23 Jamieson, P. (1995) The Prospects and Cost Benefits of Advanced Horizontal Axis Wind Turbine Design. Garrad Hassan Report No. 317/GR/02, February 1995 (internal company report).

24 Connor, B., Leithead, W.E. and Jamieson, P. (1996) *Control Design of the CONE 450 Wind Turbine*. BWEA 18 Conference, University of Exeter, September 25–27, 1996.

25 http://www.windturbinecompany.com/technology/index.html (accessed April 2011).

26 Crawford, C. (2006) Re-examining the precepts of the blade element momentum theory for coning rotors. *Wind Energy*, **9** (5), 457–478.

27 Crawford, C. and Platts, J. (2006) *Updating and Optimisation of a Coning Rotor Concept*. 44th AIAA Aerospace Science Meeting and Exhibit, Reno 2006.

28 Jamieson, P. and Jaffrey, A. (1997) Advanced wind turbine design. *J. Sol. Energy Eng.*, **119**, 315–320.

29 Lowson, M.V. and Lowson, J.V. (1994) Noise Evaluation of Coning Rotor. Report No. 94/07, Flow Solutions Ltd.

30 Nicholas Pearson Associates & Garrad Hassan (1995) Visual Impact Study – Coning Rotor HAWT. ETSU NAA/668 E/ETSU2/TI.

31 MG Bennett (1995) *Evaluation of Possible Yaw Systems for the CONE-450 Wind Turbine*, MG Bennett & Associates Ltd, Ref. R2637/1A, August 1995.

32 Spooner, E. (1994) *Direct-drive, Variable-Speed Generator for CONE-450 Advance Wind Turbine*, School of Engineering University of Durham.

33 Jamieson, P., Hornzee-Jones, C., Moroz, E.M. and Blakemore, R.W. (2005) General electric company (assignee) variable diameter wind turbine rotor blades. US Patent 6, 972,498 B2, Dec. 2005.

18

Ducted Rotors

18.1 Introduction

There have been numerous designs of ducted or shrouded rotors for wind energy applications, especially small systems for specific markets, but they have not figured in mainstream electricity generation. Especially among innovative wind turbine systems, there has been continuing interest in the idea that a ducted system with a smaller, lighter, perhaps faster and certainly less expensive rotor can extract as much energy as a large rotor in open flow. Of course, these benefits must be traded against the cost of the system that augments the flow. The usual means of augmentation is by placing the rotor in a duct or diffuser which serves to induce extra mass flow through the rotor as compared to open flow. More exotic concepts have also been considered using vanes [1] or a delta wing [2] to induce vortices creating regions of intensified flow. Coming down to earth, quite literally, the least exotic and most commonplace flow augmenter is a hilltop.

The fallacy discussed in Chapter 1 regarding output power as proportional to cube of wind speed through the duct is in no sense 'academic' and has figured in numerous over-optimistic commercial projections that may meet with eventual failure to attract investment and be competitive.[1] Rotors with rims integral with the blades and rotors in ducts or diffusers have appeared from time to time in the history of wind technology. As with tip vanes, a rotor rim, if appropriately designed, may also act as a diffuser which travels with the rotor. Sometimes with rimmed rotors such as the Swift 1.5 kW, 2-m-diameter wind turbine (Figure 18.1), the function of the rim is as much to provide a safer and stronger rotor assembly and to inhibit blade tip noise as to enhance aerodynamic performance. Although an advantage of the integral rim is that it supports and is supported by the rotor blades, a disadvantage is that, in moving with the rotor blades, it adds to total aerodynamic drag on the rotor.

Whilst rims on rotors may have a variety of functions, the most common objective of designs with ducts or diffusers separated from the rotor is to induce additional mass flow through the swept area of the rotor and hence significantly enhance performance for a given rotor diameter compared to performance in open flow. As was discussed at length in Section 1.7.1, any increased power is then proportional to mass flow increase which is linear and not cubic with velocity increase.

---

1 In one case known to the author, ownership of a concept developed in the late 1970s had been traded among several parties over 30 years with business plans consistently relying on a credible speed-up factor (at a late stage validated by CFD) but assuming output augmented as the cube of the speed-up factor.

Innovation in Wind Turbine Design, Second Edition. Peter Jamieson.
© 2018 John Wiley & Sons Ltd. Published 2018 by John Wiley & Sons Ltd.

Figure 18.1 Rimmed rotor design – Swift 1.5 kW.

A key design trade-off is thus between the value of performance benefits and the cost of extra structure. The associated issue in diffuser design is how to get maximum performance enhancement with minimum additional structure. Towards that end, boundary-layer injection [3] has been considered and other systems designed on mixer-ejector principles [4]. Again, the aim is to add to the mass flow in the ambient air that is contributing to the energy extraction process at the rotor plane, in such cases gaining benefit from energy in some of the mass flow that is adjacent to the diffuser but which does not flow through the rotor.

18.2 The Katru Shrouded Rotor System

In October 2006 Varan Sureshan of Katru Eco-Inventions Pty. Ltd., the Australian inventor of the Implux™ wind turbine, a shrouded rotor system (Figure 18.2), asked GL Garrad Hassan (now DNV GL Energy) to assist particularly in developing a rotor design for his system. The author had some involvement in the Vortec project around 1995 and was aware that no rigorous theory to optimise rotors in diffusers existed then. The challenge to solve this problem, at least at the foundation level where ideal inviscid flow is considered, triggered development of the generalised actuator disc theory of Section 1.7 and its application to blade element momentum (BEM) theory.

The Implux design (Figure 18.2) contains a conventional axial flow rotor but mounted on a central vertical axis. It is placed in the throat of an axis-symmetric shroud

(a)

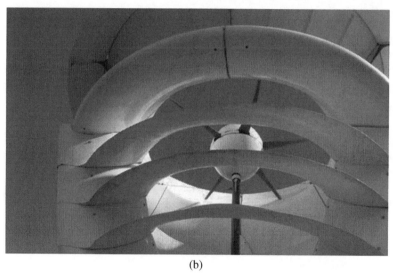

(b)

Figure 18.2 (a,b) Implux shrouded rotor system of Katru Eco-Inventions.

containing internal guide vanes which direct the ambient wind flow upwards through the rotor. The upper structure above the level of the rotor acts as a diffuser and wind flow over the top of the diffuser helps to entrain the wake.

The Implux concept was not targeted primarily at mass flow augmentation. Being designed specifically for deployment on the roof tops of commercial buildings, and noting that strong up-flows associated with regions of flow separation are typical of flows at the top edges of buildings, a key aim was to develop a system that was insensitive to flow direction. With symmetry about a vertical axis, the Implux system, like a vertical-axis wind turbine design, could obviously accommodate any wind direction without a vertical flow component. Flow field simulations using computational fluid dynamics (CFD) (Figure 18.3) showed enhanced performance in up-flows even at quite severe angles around 40°.

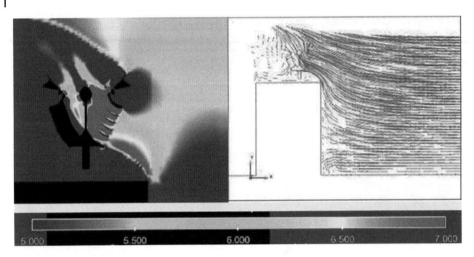

Figure 18.3 Velocity field (m/s) in operation on top of a building.

The CFD simulations also confirmed that, even with the axisymmetric inlet vanes (Figure 18.2(a)) of fixed geometry and open to flow around the structure, there were quite minimal leakage flows bypassing the rotor. This was probably due to suction both from vortices formed at the diffuser edge and from entrainment of the wake.

The vector flow directions (Figure 18.3, on right) show how the design copes well with up-flow considering the inflow angles and the flow angles of the air entraining the wake. For this system, initial analyses suggested that C_t at maximum C_p would not be higher than about 0.7. A preliminary rotor design was then developed using optimum rotor theory based on generalised flow BEM equations (Section 1.8.6).

In the real world of viscous flow operation of a system such as the Katru design, the ideal rotor thrust coefficient value of 8/9 is reduced by inefficiencies in the shroud system but also potentially augmented by external flows entraining the wake. In the development of their design, Katru had first attempted to optimise many geometric parameters of the shroud using CFD modelling, generally modelling rotor resistance as a pressure drop. Later, to validate modelling and to assist specifically in optimisation of rotor design, field tests were conducted. Having a relatively poor wind regime in the region of Australia where the Katru design was being developed, a decision was taken to conduct the testing by mounting the system on a trailer (Figure 18.4) and driving at set constant speeds along straight roads in conditions of still air.

The theoretical validity of this as a means of performance measurement was discussed in Section 1.5.5. However, there are other measurement issues specific to flow augmentation systems. First, note that for a horizontal-axis wind turbine (HAWT) in open flow, the standard for power performance testing, IEC 6400-12 (Wind Turbines – Part 12-1: Power performance measurements of electricity producing wind turbines. IEC 61400-12-1 Edition 1.0, December 2005.) asserts:

> Care shall be taken in locating the meteorological mast ....not...too close to the WTGS (wind turbine generating system) since the wind speed will be slowed in front of the WTGS ... not too far ... since the correlation between wind speed and electric power output will be reduced. The meteorological mast shall be positioned between 2 and 4 times the rotor diameter ... 2.5 ... is recommended.

Figure 18.4 Test vehicle schematic image. *Source*: Varan Sureshan.

Field testing of the Implux design [5] took place over 2013–2014. The anemometers (Figure 18.4) are set on a mast at a distance of 3 rotor diameters (3D) from the rotor plane. In augmented flow systems, however, unlike open flow, the flow is *accelerating* towards the rotor plane. Moreover, the region of flow ahead of the rotor plane and above ambient velocity will probably extend further than the flow expansion region in open flow, although this will depend on effectiveness of the flow augmentation system. No definitive standards exist for testing in such circumstances. If 3D is too close, at least, in overestimating ambient wind speed, results will be conservative.

Considering CFD results for the flow field data in Figure 18.3, it is apparent that in a real operation on top of a building, with the added effect of the building blockage, that 5 or 6D upstream looks to be necessary for valid measurement of ambient wind speed.

Based on test measurements, a_0 was estimated as −0.57 and the optimum induction of the loaded rotor estimated as $a_m = (1 + 2a_0)/3 = -0.047$. Thus, for the specific parameters of the Implux design, when the rotor is operational and producing maximum power at optimum thrust loading, the resistance of the rotor tending to slow the air flow almost cancels the induction of the duct or shroud which is tending to speed it up. This means that the air speed generally over the region immediately upstream of the rotor is close to ambient and hence there is much less sensitivity of measurement accuracy to anemometer location. This suggests that the measurements of maximum power performance will have little error associated with the anemometer location possibly being non-optimum.

After many field test measurements of power performance and of the flow field induction factors with various pre-existing rotor configurations, a new rotor design (2.25 m diameter with six blades) was developed [6] using the generalised BEM described in Chapter 1. Theoretical curves derived from Equation 1.32 are presented in Figure 18.5. Based on a shroud efficiency of 90% and the measured empty duct augmentation, the theoretical C_p limit predicted by the theory for $a_0 = -0.57$ is ~0.94 (Figure 18.5) and would be higher if the shroud efficiency is less than 90%. The measured C_p of 0.85 (Figure 18.5) was excellent in a range two or three times above the performance of small

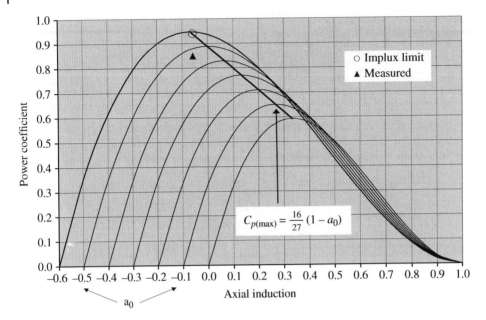

Figure 18.5 Theoretical power coefficient characteristics for various values of a_0. *Source*: Professor Y Ohya.

(rotor $\sim$1 m diameter) horizontal or vertical-axis wind turbines. It was also approximately twice the earlier projections using CFD with less optimum rotor configurations and over-pessimistic estimates of aerofoil characteristics. The 'Implux limit' presented in Figure 18.5 is not in any sense an absolute or rigorous limit. It is based on the measured induction ($a_0 = -0.57$) at a specific rotor loading that produced maximum power in the given configuration. This embodies the assumption that the measured induction is a reasonable estimate of a_0. Thus, this 'limit' serves purely as a guide as to how much better the performance might be if the rotor was fully optimised aerodynamically and had no losses (zero drag).

18.3   The Wind Lens Ducted Rotor

The Wind Lens ducted rotor system has been developed at Kyushu University in Japan with many small systems in operation and more recent (2011) up-scaling to a 100 kW system [7–9] (Figure 18.6 (a)). It has been deployed in an experimental floating offshore system (Figure 18.7(a)) and is also being investigated [10] in multi-rotor configurations (Figure 18.7(b)).

The Wind Lens concept has the typical objective in ducted rotor design of maximising flow augmentation with minimal diffuser structure and makes significant use of viscous flow effects exploiting strong vortex shedding (Figure 18.8) from the rim to create suction in the wake. A programme of experimental testing determined optimum geometric parameters to reach a final design with some performance being traded with minimising duct length and diameter. The system free yaws satisfactorily with the inertia of the diffuser, helping to limit yaw rate.

(a)　　　　　　　　　　　　　　　　　(b)

Figure 18.6 (a) Wind Lens 100 kW turbines at Kyushu University, Ito Campus. (b) An irrigation-greenery plant using wind energy (5 kW Wind Lens turbine farm) in a desert area of northwest China.

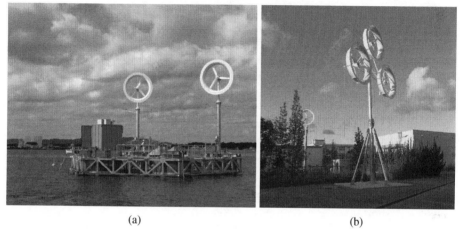

(a)　　　　　　　　　　　　　　　　　(b)

Figure 18.7 (a) Three kilowatt Wind Lens turbines in Hakata bay. (b) Wind Lens MRS at Kyushu University.

Considerable optimisation of the Wind Lens design took place in numerous laboratory experiments and in field testing of 1, 3, 5 and 100 kW units. Following study of systems with a brim, which was in effect a flat hollow disc placed at the duct exit and set normal to the duct axis as in Figure 18.8, a continuous cycloidal shape was substituted as preferable for manufacture in a single piece.

The essence of practical duct design is captured in Figure 18.9, where the maximum power coefficient is determined from measurements for a range of cycloidal duct geometries characterised by the parameters, ratio of duct length to throat diameter L/D and added diameter of brim as fraction of throat diameter. These results are based on wind tunnel measurements using the same rotor of 1 m diameter. The chart in Figure 18.9 shows data for duct types C_0, C_1, C_2 and C_3, which have L/D of 0.1, 0.137, 0.221 and 0.371, respectively. The four lines distinguish results for these ducts with brim height to diameter (h/D) ratios, as in the legend. The picture with Figure 18.9 shows a set of ducts all with brim height $h = 0.05D$.

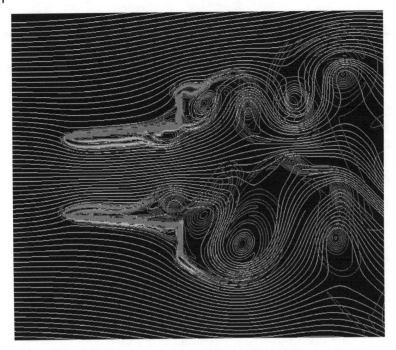

Figure 18.8 CFD dynamic simulation (2D) of Wind Lens flow field. *Source*: Professor Y Ohya.

The rotor diameter is always a little less than throat diameter, some clearance obviously being necessary for practical safe operation, but also because the gap flow can enhance performance. Performance, as measured by maximum rotor C_p, increases with duct length, but with diminishing returns for much extra structure, and is still excellent with short ducts and minimal brim expansion.

The performance of various Wind Lens designs, as evaluated in wind tunnel measurements, is presented in Table 18.1. The parameter 'C_p bare turbine' is the performance coefficient of the rotor in open flow; the area ratio is defined as ratio of area of the duct structure where the diameter is maximum to the rotor swept area (this area being slightly less than the throat area); C_p is the rotor performance coefficient as measured in operation in the duct; C_p^*, the area ratio × the Betz limit = (16/27) is the factor expected by some [11] to be a limit rotor power performance coefficient.

Now C_p^* does always exceed the directly measured rotor C_p, but the test rotor has a C_p in open flow substantially below Betz on account of inevitable friction losses (drag) exacerbated by being a small scale system operating at low Re.

The rotor cannot be optimised for all duct arrangements as they differ in augmentation levels and is not in fact specifically optimised for operation in any of the range of different ducts. C_{pl} is calculated assuming that the rotor is no more efficient in the various ducts than in open flow and factoring its limiting performance by the ratio of Betz limit to open flow C_p value. Interpreting this as an ultimate performance limit also assumes that the duct has no losses. Values of C_{pl} exceed C_p^* significantly. Another interpretation of the C_{pl} data is, considering the maximum value of C_{pl}/C_p^* of 1.388, that the rotor would have to perform about 39% better in the associated duct than in open flow

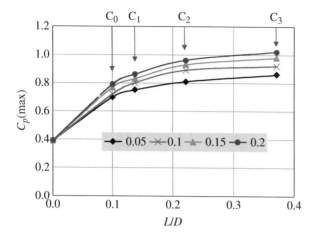

$D_{\text{throat}} = 1.02$ m

$D_{\text{rotor}} = 1.00$ m

$h \quad = 0.05D$

Figure 18.9 Maximum power coefficient for various duct geometries of cycloidal section.

Table 18.1 Key parameters of various duct geometries.

	C_0 B05	C_1 B05	C_2 B05	C_3 B05	C_1 E05 (100 kW)
D_{max}	1.190	1.216	1.262	1.374	1.216
C_p – bare turbine	0.370	0.370	0.370	0.370	0.370
A_r – area ratio	1.416	1.479	1.593	1.888	1.479
C_p – value in duct	0.700	0.750	0.810	0.860	0.760
C_p^* – $A_r \times$ Betz	0.840	0.877	0.945	1.120	0.877
C_{pl} – C_p limit	1.121	1.202	1.298	1.378	1.217
C_{pl}/C_p^*	1.335	1.371	1.374	1.230	1.388

(and more than that if the duct has friction or other losses) for Betz related to the area ratio to be a possible limit on performance.

It may reasonably be argued that this 'limit' of Betz related to the area ratio as derived by inviscid theory [11] should not be expected to apply in complex real flows exhibiting strong viscous interactions but it has already been clearly demonstrated (Section 1.7.1) that neither is the exit-area-ratio-based limit a valid limit for ideal inviscid flow.

The largest Wind Lens system presently installed of 100 kW (Figure 18.6(a)) has a duct length of 2 m, rotor diameter of 12.8 m, maximum diameter of 15.4 m and a measured maximum rotor C_p of 0.76 as determined initially in the small-scale wind tunnel tests and later confirmed in the field. This duct is thus a compact one trading some performance for the advantage of minimal structure.

References

1 Rechenberg, I. (1989) *Development and Operation of a Novel Wind Turbine with Vortex Screw Concentrator*. 2nd Joint Schlesinger Seminar on Energy and Environment, Berlin, pp. 1–11.

2 Sforza, P.M. and Stasi, W.J. (1979) *Field Testing the Vortex Augmentor Concept*. Proceedings of Wind Energy Innovative Systems Conference, Colorado.

3 Oman, R.A., Foreman, K.M. and Gilbert, B.L. (1975) *A Progress Report on the Diffuser Augmented Wind Turbine*. 3rd Biennial Conference and Workshop on Wind Energy Conversion Systems, Washington, DC.

4 http://www.flodesignwindturbine.org/ (accessed April 2011).

5 Sureshan, V. (2014) Implux Wind Turbine Field Testing Progress Report – September 2014.

6 Jamieson, P. (2014) An Evaluation of Testing of the IMPLUX Wind Turbine. A confidential final report on the field testing of the Implux Wind Turbine by Katru Eco Inventions Pty. Ltd. for submission to the Australian Department of Industry, Innovation, Science, Research and Education.

7 Ohya, Y. and Karasudani, T. (2010) A shrouded wind turbine generating high output power with wind-lens technology. *Energies*, **3**, 634–649.

8 Ohya, Y., Uchida, T., Karasudani, T. *et al.* (2012) Numerical studies of flow around a wind turbine equipped with a flanged-diffuser shroud using an actuator-disk model. *Journal of Wind Engineering*, **36** (4), 455–472.

9 Goeltenbott, U., Ohya, Y., Yoshida, S. and Jamieson, P. (2016) The science of making more torque from wind: flow interaction of diffuser augmented wind turbines. *Journal of Physics: Conference Series*, **753**, 022038.

10 Goeltenbott, U., Ohya, Y., Yoshida, S. and Jamieson, P. (2016) *Arrangements of Three Diffuser Augmented Wind Turbines in a Multi-Rotor System*. World Wind Energy Conference 2016, Tokyo, Japan.

11 Sørensen, J.N. (2016) *General Momentum Theory for Horizontal Axis Wind turbines*, Springer International Publishing, Switzerland, p. 23.

19

The Gamesa G10X Drive Train

During the 1980s and 1990s little evolution took place in drive-train design, with most drive trains comprising a two- or three-stage gearbox (according to wind turbine size) and high-speed induction generator (typically 1500 or 1800 rpm) in a clearly modular arrangement. All that has changed (see Section 8.3), with drive-train development now being one of the most active areas of innovation. Drive-train designs, referred to as hybrid, with one or two stages of gearing and permanent magnet generator (PMG) were discussed in Chapter 6 (Section 6.6) and the potential for this to enable a particularly compact and cost-effective drive-train solution was also reviewed (Section 6.12).

In addition to having developed a low-mass two-piece sectional blade and other innovations, Gamesa has adopted hybrid drive-train technology within the G10X, their 4.5 MW, 128-m rotor diameter, wind turbine design.

Gamesa's gearbox system is highly integrated (described by the company as 'COMPACTRAIN') having the rotor main shaft directly connected to the gearbox (Figure 19.1).

The gearbox comprises two planetary stages with an input shaft speed at rated power of 11.84 rpm and with an overall transmission ratio of 37.88. The first stage is a spur gear and the second stage, helical. Tapered roller bearings are used with the first and second stages integrated. The first planetary stage and the second stage are specially designed to absorb deflections caused by main shaft misalignment. There is modularity in the design between the first stage, second stage and high-speed shaft module (which can be replaced in the field if necessary). Forced lubrication is employed using synthetic oil fed through independent channels with oil baths for each bearing and gear. All lubrication pipes are external for ease of maintenance. Gamesa claims a significantly lower parts count for this arrangement compared with competing drive-train layouts. The gearbox mass is 35 tonnes and, including the main shaft assembly, the total mass is around 70 tonnes with overall dimensions of 2.3 m diameter and 3 m length.

The G10X PMG (Figure 19.2(a)) is a radial flux design using high-strength neodymium-iron-boron magnets, as is most usual in PMG designs for large wind turbines. It has 24 poles, a nominal speed of 448 rpm (maximum 627 rpm) and a nominal rating of 4.8 MW. The frame and housings are water cooled and the PMG is designed for a 20-year life that is maintenance free except for lubrication.

Innovation in Wind Turbine Design, Second Edition. Peter Jamieson.
© 2018 John Wiley & Sons Ltd. Published 2018 by John Wiley & Sons Ltd.

Figure 19.1 The G10X gearbox assembly. Reproduced with permission of Gamesa Corporación Tecnologíca, S.A.

(a)

(b)

Figure 19.2 The G10X generator (a) and wind turbine system (b). Reproduced with permission of Gamesa Corporación Tecnologíca, S.A.

The generator mass of 18.5 tonnes combined with 70 tonnes for gearbox and main shaft system is indicative of a low total nacelle system mass compared to the few other very large wind turbine systems of comparable rating and diameter. The drive train shows the merits of the hybrid design type having a nice mixture of integrated design features that can reduce mass and cost with modularity that can facilitate assembly and maintenance.

20

DeepWind Innovative VAWT

20.1 The Concept

Uwe Paulsen of DTU Risø, originator of the DeepWind vertical-axis wind turbine (VAWT) concept [1], wrote

> Statoil's development of their HYWIND concept (see Chapter 7, Section 7.8) and their initiatives for demonstrating the world's first full-scale floating wind turbine was an eye-opener. We identified challenges with the existing offshore wind turbine concepts on design, installation and operation and maintenance … existing offshore wind turbine technology was essentially based on moving onshore wind turbines to the open sea. We concluded that we would need a novel offshore wind turbine concept to overcome current difficulties. We wanted to proof, that our novel concept would open an opportunity for cost improvements in terms of technology and in terms of supply and in this sense filling the gap between offshore and onshore cost in offshore wind energy.

Among many innovations, a key idea in the DeepWind design is the simplification and integration of rotor blades, shaft and support structure into a single element. The design targeted '*potentials for better cost efficiency, installation and O&M options than existing offshore technology … a radical new and simple design with up-scaling potential and reduced O&M costs: a 2-bladed Darrieus rotor with a long spar buoy extending subsea, a bottom based (direct driven) generator, and a sea-bed fixing system*'. Figure 20.1(a,b) compares artistic views of Hywind and DeepWind and Figure 20.1(c) shows a sketch of the DeepWind spar buoy in its final design.

According to Paulsen, the Darrieus rotor was chosen as historically the most successful vertical axis with advantages of simplicity, reasonable efficiency, good economy and a long record of research and development in the past. The aerodynamic work at TU Delft by Carlos Simao Ferreira (see Chapter 13) had indicated that thick aerofoils, such as the DU12W262 [2], could be developed not only of obvious structural benefit but with enhancement of rotor performance.

20.1.1 Blades

Constant section blade profiles can be manufactured by pultrusion technology (Figure 20.2). As with monopiles and many other industrial components, the limiting

Innovation in Wind Turbine Design, Second Edition. Peter Jamieson.
© 2018 John Wiley & Sons Ltd. Published 2018 by John Wiley & Sons Ltd.

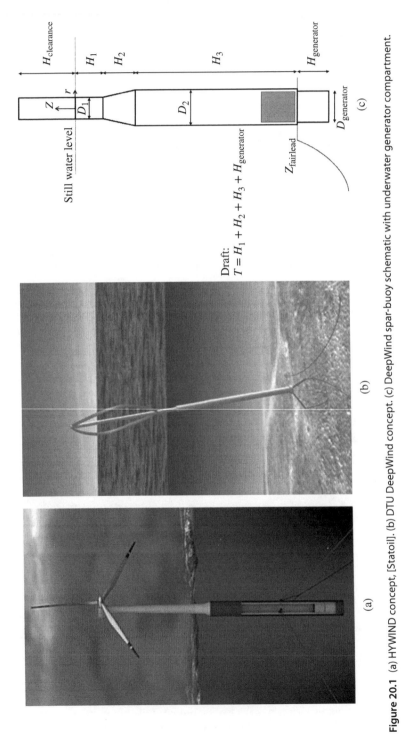

Figure 20.1 (a) HYWIND concept, [Statoil]. (b) DTU DeepWind concept. (c) DeepWind spar-buoy schematic with underwater generator compartment.

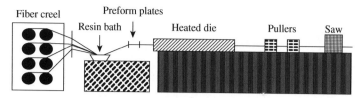

Figure 20.2 Schematic diagram of the pultrusion process.

sizes for manufacture by pultrusion have been pushed up by market demands. A blade with sections up to 3-m chord width can be directly manufactured over a single block with low production costs in production facilities that are relatively small scale. In principle, a production can be set up on a ship, and the blades produced offshore in lengths of kilometres.

This possibility is provided by progression of technology in blade manufacture. Paulsen considers that introducing the continuous glass fibre blade pultrusion technique into full-scale blade production will have a significant impact on mass-production of blades. Not only should the strength of the blades increase but uniformity, quality and the production capacity may also significantly increase and the cost may reduce.

20.1.2 Controls

In radically simplifying the rotor-shaft system, blade pitching would be difficult to engineer and thus conflict with the concept objectives as an undesirable complication. The VAWT concept inherently avoids any requirement for yaw systems and also offers less sensitivity to up-flow. Power control must then necessarily rely on aerodynamic stall; and in the DeepWind concept, this is combined with rotor speed control. The rotor speed control is required to deal with both normal and abnormal operation, aiming to optimise power performance, reduce dynamic stress on the shaft (which replaces the tower of a conventional design) and mooring system, provide safe rotor braking (emergency stop) as well as grid code compliance (reactive power and voltage control, low-voltage fault ride through, and so on [3].

20.1.3 Generator Concepts

Having achieved large cost savings in the rotor-structure system, a major challenge for the DeepWind concept is the design of the generator system with regard to key issues of providing torque reaction, of minimising yet providing for underwater maintenance, and also of meeting the discussed control and safety demands.

As discussed in Section 20.1.2, the turbine operational control requires variable speed generator control. Various direct-driven generator mechanical configurations were considered with placement of the generator at the bottom of the submerged structure. For example:

a) The generator armature is mounted inside the submerged substructure at the bottom and rotates with the rotor. The generator shaft is extended through the foundation bottom and fixed to the torque arms (Figure 20.3(a)).
b) The generator armature is mounted outside the substructure and rotates with the rotor. The shaft is fixed to the torque arms (Figure 20.3(b)).

c) The generator armature is fixed on the seabed and the generator shaft is fixed to the rotating structure (Figure 20.3(c)).

d) Two generators are placed in two turbine gondolas, fixed to the tubular structure. The turbine gondolas each consist of a water turbine and a generator, and because of the rotation of the gondolas in the water they convert the wind turbine rotor power into electricity (Figure 20.3(d)).

There are challenges for the generator concept associated with the submersible generator and power converter design. In the design discussed here, the generator and power converter are enclosed in the same casing, although this need not be so. Seawater under pressure, corresponding to the depth of the generator below the surface, may penetrate the generator and cause damage through corrosion of the metal parts or degradation of the insulation system and water may enter at the shaft entry and the cable bushings.

A further critical challenge is sealing of the shaft, where a concept for submersible magnetic bearings, including mechanical design, magnetic functional concept, and cooling has been explored.

A baseline PI controller was developed and verified under different operating conditions. Controller gains have been tuned to minimise generator torque variations and overspeed. Specifically, the large twice-per-revolution (2p) variations in aerodynamic torque characteristic of two-bladed VAWT need to be isolated from the generator so as to avoid similar pulsations in the mooring line tension, and to enable smoother electric

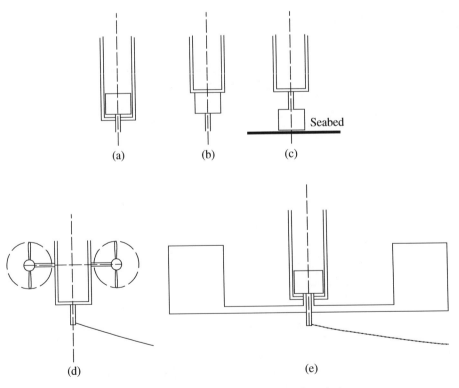

Figure 20.3 (a–e) Generator configurations and torque reaction drag device.

power output. In the baseline controller, this was achieved by means of a notch filter that eliminates 2p variations in the measured generator speed. In effect, the controller trades large 2p torque variations with small 2p speed variations. Aerodynamic power variations can be absorbed by the kinetic energy of the rotor.

20.1.4 Torque Absorption

Concepts for providing torque absorption focus on using a drag device, rotating slowly at the bottom of the structure (Figure 20.3(e)). The configurations (Figure 20.3(d,e)) absorb the rotor torque in the water.

20.1.5 Anchoring Part

The torque and the thrust are transmitted through the substructure anchored to the seabed with tensioned wires. For final torque reaction, two or more rigid arms are necessary. For deeper water, a floating and mooring system is used. For shallow water, the anchoring part is taut or even missing, if the depth is limited.

20.2 DeepWind Concept at 5 MW Scale

The concept at 5 MW scale is illustrated in Figure 20.4 and the main specifications are provided in Table 20.1.

20.3 Marine Operations Installation, Transportation and O&M

The installation concept is to use a small ship for towing the two-bladed turbine to the site. The structure, without counterweight, can float horizontally in the water. By gradual addition of ballast such as seawater or rock material at the bottom section, the turbine is slowly righted avoiding the need for large installation vessels. Use of disruptive technologies such as a remotely operated underwater vehicles (ROVs), lifting bags and deep sea divers is intended to make these installation procedures less complicated and more reliable than existing installation of large wind turbines at sea. These procedures are expected to reduce downtime in waiting for particularly favourable weather windows for installation, commissioning, de-commissioning and maintenance operations, by enabling service activities in less favourable weather situations at sea.

20.4 Testing and Demonstration

The DeepWind technology presents extensive challenges and demands research, in many areas especially of system dynamics, blade pultrusion, validation of new airfoils, measurement of material properties, sub-sea generator, mooring and torque absorption system, and torque, lift and drag on the rotating and floating shaft foundation. The novel

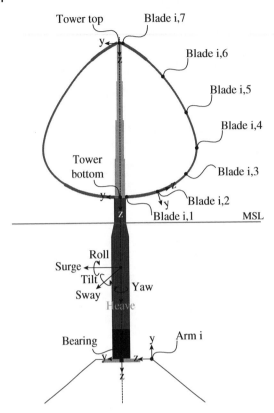

Tower top   Blade i,7

Blade i,6

Blade i,5

Blade i,4

Tower bottom   Blade i,3

Blade i,2

Blade i,1   MSL

Roll

Surge

Tilt   Yaw

Sway

Heave

Bearing   Arm i

Figure 20.4 Schematic illustration of the 5 MW concept.

Table 20.1 Baseline 5 MW rotor design.

Geometry			Performance		
Rotor radius (R)	(m)	60.48	Rated power	(kW)	5 000
Rotor height (H)	(m)	143	Rated rotational speed	(rad/s)	0.6
Chord (c)	(m)	5	Rated wind speed	(m/s)	15
Solidity ($\sigma = Nc/R$)	(–)	0.165	Cut-in wind speed	(m/s)	4
Swept area	(m²)	11 996	Cut-out wind speed	(m/s)	25

concept made fresh demands on the calculation and simulation tools needed for closing the divide between paper drawings and being able to accomplish a 5 MW conceptual design ready for industrial optimisation and evaluation of upscaling potential to a 20 MW unit.

Model testing was conducted in Roskilde fjord with a two-bladed 1 kW wind turbine (Figure 20.5), in the tank of DHI to study fluid dynamics aspects of waves and current

Figure 20.5 One kilowatt DeepWind turbine for testing (demonstrator) in different rotor configurations.

impinging on a rotating spar, to explore floater kinematics under controlled wind, wave and current conditions in an ocean laboratory of MARIN and also in the large wind tunnel of Politechnico Milano to study the performance of the inclined rotor with two and three blades.

20.5   Cost Estimations

To evaluate costs for the DeepWind design, a simplified cost analysis of a 100 MW wind farm with the 5 MW conceptual design [4] was considered. For comparison, a cost analysis derived from a utility cost estimate approach, was also performed which gave similar levelised cost of electricity (LCOE). A 5 MW unit with a steel floater was estimated to cost about 1790 €/kW, and LCOE estimates of 7–11 €/kWh. This is comparable to a cost of 1250 €/kW for onshore wind energy in 2009. The 20 MW cost was estimated at 1053 €/kW with reinforced concrete floater material, with LCOE of 0.02€/kWh in large production. It is believed that the operating expenses (OPEX) part of the 28–30 €/MWh is probably overestimated, but that reflects assumptions and uncertainties with installation and maintenance schedules. A more detailed practical set of procedures has to defined to be more precise in predicting OPEX.

Many historical VAWT designs (as discussed in Chapter 13) have had no adequate answer to the penalties of reduced rotor efficiency and increased drive-train torque (weight, cost). There is a radical conceptual difference in DeepWind from historical VAWT concepts. DeepWind offers major simplifications and cost reduction in the integrated rotor support structure system and in installation in a design that has been conceived for offshore deployment from the outset. In overall evaluation, these large advantages trade against the need for sophisticated new technology solutions to accommodate and maintain totally submerged bearings and power generating systems.

References

1 Vita, L., Paulsen, U.S., Pedersen, T.F., Madsen, H.A. and Rasmussen, F.A. (2009) *Novel Floating Offshore Wind turbine Concept*. Proceedings of the European Wind Energy Conference (EWEC), Marseille, France.

2 The 5 MW DeepWind floating offshore vertical wind turbine concept design - status and perspective EWEA Proceedings 2014, Barcelona.

3 DeepWind – from idea to 5 MW concept, EERA DeepWind'2014, 11th Deep Sea Offshore Wind R&D Conference Energy Procedia, (2014), **53**, pp. 23–33.

4 Paulsen, U.S., Borg, M., Madsen, H.A. *et al.* (2015) Outcomes of the deepwind conceptual design. *Energy Procedia*, **80**, 329–341. doi: 10.1016/j.egypro.2015.11.437

21

Gyroscopic Torque Transmission

Gyroscopes fascinate because they have a very surprising property – that a torque about one axis induces motion about a different axis. The analysis of this strange behaviour has been demanding to the extent that a comprehensive analytical explanation of the motion of a spinning top was considered to be the crowning achievement of classical rigid body dynamics [1].

Without going into the details of gyro dynamics, the effective properties in terms of exploitation in torque transmission can be summarised. Stephen Salter considered gyroscopes extensively for use in his famous wave energy device, the Nodding Duck [2], and he has developed a useful 'black box' description of gyroscopic action [3]. The black box has an input shaft A with associated input torque and speed τ_A and ω_A and an output shaft B with corresponding torque and speed, τ_B and ω_B. The box has the gyroscopic property that $\tau_A \propto \omega_B$ rather the usual property of say a gearbox, $\tau_A \propto \tau_B$. The rotational inertia, angular velocity and precession angle of the gyro in the box are respectively I, Ω and φ. It is assumed that the transmission has no losses so that the power, $P = \tau_B \, \omega_B = \tau_A \, \omega_A$.

Table 21.1 (a modified extract from Salter [3]) reveals a mapping of applied input[1] torque to output motion and vice versa that has a wide range of implications for gyro system behaviour.

In addition to the black box summary properties, it is also important to note that gyroscopic transmission is fundamentally oscillatory and, if continuous output rotation is required, the output must be rectified in some way.

In spite of such issues, it is precisely the unusual properties of gyroscopic action that make it intriguing as a possible power transmission system. One operational ideal for a wind turbine drive train is a system that converts variable, low speed, high torque input to constant high speed, low torque output that may be connected to a synchronous generator such as is routinely used in large scale power production. Can a gyroscopic variable transmission (GVT) achieve this and what would it look like?

Many inventors have patented ideas employing gyroscopic properties in torque transmission. They are generally highly complex and cumbersome to engineer. More

---

1 For convenience the terms 'input' and 'output' are employed. However, the properties are symmetrical and the concepts of 'input' and 'output' are interchangeable.

Innovation in Wind Turbine Design, Second Edition. Peter Jamieson.
© 2018 John Wiley & Sons Ltd. Published 2018 by John Wiley & Sons Ltd.

Table 21.1 Properties of a gyro 'black box'.

Input shaft A	Output shaft B
Torque	Speed
τ_A	$I\Omega \cos\varphi \, \omega_B$
ω_A	$\dfrac{\tau_B}{I\Omega \cos\varphi}$
Light damping	Heavy damping
K_A	$K_B = \dfrac{(I\Omega \cos\varphi)^2}{K_A}$
Small spring	Large inertia
S_A	$M_B = \dfrac{(I\Omega \cos\varphi)^2}{S_A}$
Small inertia	Large spring
M_A	$S_B = \dfrac{(I\Omega \cos\varphi)^2}{M_A}$

recently M. Jegatheeson (Jega) devised an ingenious system[2] of comparatively reduced complexity although still requiring a mechanical means of output rectification via clutches.

An overview of some of the issues in integrating Jega's concept [4] into a wind turbine drive train is now presented.

The essence of the concept is that via a cam or crank, the input torque of the wind turbine rotor causes the piston (Figure 21.1) to oscillate. This oscillation via the link arm rocks the axis of the gyro which is rotating in the sub frame. Oscillating the axis of rotation of the gyro, via the classic gyroscopic effect, creates a torque about an axis at right angles causing oscillation of the main frame which along with the link arm and sub frame is free to rotate (see main thrust bearing in Figure 21.1) about the piston. Thus oscillatory motion of the main frame is created. Finally a clutch pair (Jega has also patented solutions for the clutch although a magneto-rheological clutch [5] may be of interest for this application) is used to rectify the oscillatory motion of the main frame and produce a unidirectional rotational output.

The torque output of a single gyro unit as in Figure 21.1 is found to be highly irregular implying large output power variations and the gyro bearing loads are very high. In addition, if the wind turbine rotor drove a single unit as in Figure 21.1 say via a crank shaft arrangement, the stored energy in the single gyro unit would be commensurate with that stored in the wind turbine rotor! Huge torques would then be required to change gyro rotational speed or the response would be very slow thereby ruling this out as a possible power control method. All this left the challenging question of how to develop a system arrangement that may solve these problems.

A solution proposed, Jamieson *et al.* [6], was to have an axial ring cam (Figure 21.1) to create the required oscillatory input motion. Compared to the requirement for power

---

2 GL Garrad Hassan was asked to undertake an evaluation of this system. The work is ongoing and many details remain confidential. Gyro Energy Limited and the inventor Muthuvetpillai Jegatheeson (Jega) have kindly consented to publication of the discussion in this chapter.

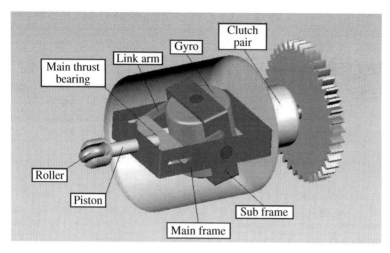

Figure 21.1 Single GVT unit.

transmission through a single directly driven unit, the angular momentum required per gyro is reduced by a factor of the number of cam lobes, n multiplied by the number gyro units, m. Thus by having 11 lobes and 6 GVTs, at a fixed design gyro speed, the rotational inertia per gyro is reduced by a factor of 66. More significantly, the torques that produce radial forces in the gyro bearings are correspondingly reduced. This is a vital advantage of the arrangement.

A detailed computer simulation model of the system in Figure 21.2 (with an 11 lobed cam and 6 gyro units having their outputs interconnected) was developed. In this model a standard synchronous generator running at 1500 rpm was connected to the GVT system output.

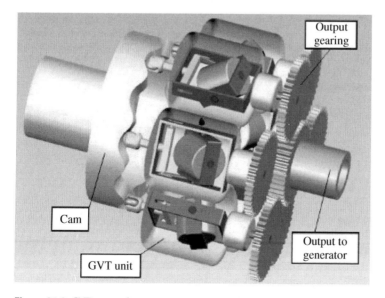

Figure 21.2 GVT system layout.

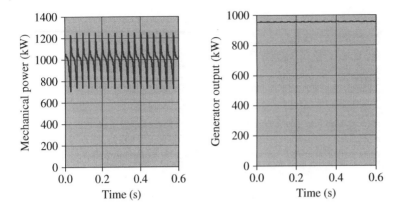

Figure 21.3 Mechanical and electrical power output.

While the summed mechanical output of the six GVT units is still highly variable (Figure 21.3), with the GVT input frequency increased in proportion to the number of cam lobes, the variations are at a significantly higher frequency than the fundamental frequency (associated with output mechanical inertia and electrical stiffness) of the synchronous generator. The generator therefore filters the mechanical torque variations which are reduced to a slight ripple and smooth output power results (Figure 21.3).

A typical history of the low speed shaft torque of a 1 MW wind turbine operating in variable speed with a conventional drive train (ambient wind speed of 18 m/s and 10% turbulence intensity) is presented in Figure 21.4. The low speed shaft torque is evidently highly variable reflecting turbulence in the wind. The pitch system acts to maintain rated power and the pitch angle changes therefore correlate approximately with the hub height longitudinal wind speed. On account of the high rotational inertia of the wind turbine rotor, the associated variation in shaft speed (Figure 21.5) is quite modest and high frequency variations in rotor speed are filtered out.

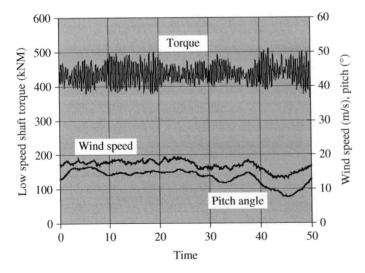

Figure 21.4 Low speed shaft torque history – typical 1 MW wind turbine.

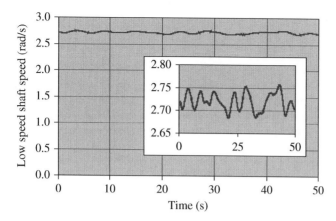

Figure 21.5 Typical low speed shaft speed history of a 1 MW wind turbine.

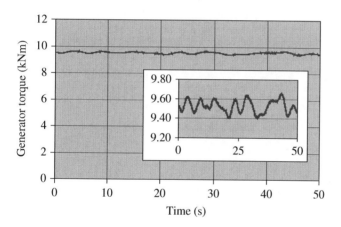

Figure 21.6 Generator torque from the GVT system at 1 MW rated output.

The same input torque and speed variations may be considered as input to a drive train with the GVT system of Figure 21.2. The outputs of the GVT system are then quite different from the conventional drive train and this is where the 'black box' properties of Table 21.1 kick in.

Due to the 'black box' effect in the mapping of torque and speed characteristics, the variation of output (generator) *torque* from the GVT system looks like the variation of the input low speed shaft *speed*. This is emphasised in the inset charts of Figures 21.5 and 21.6. Thus in the GVT system, the rotor inertia not only smoothes input speed variations but it also smoothes output torque variations. To be precise, the high frequency torque variations, as in Figure 21.3, which are associated with intrinsic features of the internal cycles of the GVT system remain but the low frequency activity that is directly due to wind turbulence (Figure 21.4) is largely eliminated.

This means that the GVT provides a remarkable transmission system with an output that is substantially insensitive to wind turbulence and with a mechanical capability of variable speed that will enable the direct connection of a standard, constant speed, synchronous generator. The price of this benefit is a system with high (but very predictable) internal cyclic loads and a demanding duty for various bearings in the system, gyro bearings especially.

In the further evaluation of the GVT system mass and cost comparisons were developed alongside a conventional drive train. The GVT system does not have substantial impacts on the rest of the wind turbine and so cost and cost of energy evaluations along the lines indicated in Chapter 9 can readily be followed with focus only on drive train comparisons. Outline evaluation was encouraging in respect of both mass and cost of the GVT system but as ever another level of more detailed design and prototype testing will be required to confirm its potential.

References

1 Crabtree, H. (1909) *An Elementary Treatment of the Theory of Spinning Tops and Gyroscopic Motion*, Longmans, Green and Company, London.
2 Salter, S.H. (1982) *The Use of Gyros as a Reference Frame in Wave Energy Converters*. 2nd International Symposium on Wave Energy Utilisation, June 1982.
3 Salter, S.H. (1980) Recent progress on ducks. *IEE Proceedings*, **127** (**5**, Part A), 308–319.
4 Jegatheeson, M. and Moore, H. (2005) GyroTorque™ Continuously Variable Transmission for Use in Electricity-generating Systems Utilising Wind and/or Wave Energy. IPENZ Engineering TreNz 2005-001. ISSN: 1177-0422.
5 Kavlicoglu, B., Gordaninejad, F., Evrensel, C.A. *et al.* (2002) *A High-Torque Magneto-Rheological Fluid Clutch*. Proceedings of SPIE Conference on Smart Materials and Structures, San Diego, CA, March 2002.
6 Jamieson, P., Jegatheeson, M. and Leithead, W. (2004) *Gyroscopic Variable Transmission for Wind Turbines*. Global Wind Power Conference and Exhibition, Chicago, March 2004.

22

The Norsetek Rotor Design

Some very early rotor designs, for example, the famous Gedser wind turbine [1], which is often regarded as the starting point of modern wind turbine technology, and the later Nibe wind turbines [1] used ties or bracing struts to interconnect the blades and share rotor loading. It is well known that a cantilevered beam of a given length that is required to support a given loading can be reduced in mass by over a factor of 10 by support systems involving struts and tension cables. This concept was taken to an extreme level in the proposed space frame wind turbine design of Watson which was evaluated by Hansen [2].

A much more recent concept (Figure 22.1), which is significantly innovative compared to the earlier designs of the braced rotor, is being developed by the Norwegian company Norsetek AS.

The aim is to achieve lighter and stiffer large rotor designs enabling further economies in the tower top structure by minimising the overhang of rotor from the tower whilst also reducing tower top mass with some benefit of tower and foundations. With more than the usual three blades, the blades are more slender and lighter but more effectively interconnected by the rotor support systems. A stiff rotor arrangement is achieved with two independent bracing systems. One is fixed in the frame of the hub, with each blade connected to a forward projecting strut but with rotary joints to each blade so that the blades may still pitch in the usual manner. Independent of that system, each blade is supported in its own right by a projecting 'A' frame and tie wires. All these elements are faired aerodynamically so as to minimise drag. Penalties from drag on the support members can be compensated for or even outweighed by the benefit in reduced tip loss from having more blades.

Preliminary evaluation involving rotor optimisations and the impact of different design tip speeds and blade numbers was undertaken using Equation 1.82 with specially derived additional terms to reflect the effect on power coefficient of drag losses associated with the rotor bracing system. Once a general parametric study had been conducted, preferred design arrangements were studied in greater detail using blade element momentum (BEM) code (GH Bladed) coupled to finite element structural models that accounted for details of the bracing and its joints. Having more than three blades enables better averaging of rotor disc loading, more options for control of fatigue loading (assuming independent pitch control) and also redistribution of loading through the interconnection of the rotor members.

In rapid blade pitching into negative incidence, as is conventional practice for emergency braking of the rotor, blade out-of-plane bending moments and rotor thrust are

Innovation in Wind Turbine Design, Second Edition. Peter Jamieson.
© 2018 John Wiley & Sons Ltd. Published 2018 by John Wiley & Sons Ltd.

Figure 22.1 The Norsetek braced rotor concept. Reproduced with permission of Norsetek AS.

often reversed. This has been discussed in the evaluation of a high-speed downwind rotor (Section 7.2). There are also other extreme load cases, some arising in fault states, giving rise to reversal of blade bending loads. Initially, a three-bladed design of 129 m diameter rated at 5 MW, was developed as a baseline for comparison with Norsetek designs. With similar operation to conventional wind turbines, the extreme reversed (negative) blade root out-of-plane bending moment of this baseline design (−14 MNm) was found to be of the order of 2/3 of the positive extreme value (23 MNm).

In the initial design configuration of Norsetek, such reversed bending loads were underestimated and a number of design options, some involving structural modifications and some exploring more effective supervisory control, were then considered. Firstly, only changes to supervisory control were investigated involving the following:

1) Maximum pitch rate,
2) Software settings, rotor overspeed level,
3) Controlled shutdown rate (of change of rotor speed),
4) Brake torque level for emergency shutdowns,
5) Time for which generator can produce torque after a grid loss event.

After careful tuning of these variables, having due regard for all impacts on machine safety, reversed bending loads were substantially reduced and a satisfactory design and operational scheme evolved. At a later stage, Norsetek re-evaluated the design, making structural modifications to best accommodate the expected loading regime and also to further optimise mass and cost reduction.

The initial evaluation has been promising in terms of potential for rotor mass reduction and cost reduction, although there are of course many challenges in the detailed design of such a system, for example, in effecting long-lasting but economical connections and in catering adequately for the reverse bending loads as have been discussed.

This is a typical example of a growing number of innovative developments in wind energy where private inventors with a strong general technological background but perhaps new to wind are seeking to bring innovative concepts to the attention of the wind industry.

References

1 http://www.windsofchange.dk/WOC-eurstat.php (accessed April 2011).
2 Hansen, A.C. (1987) Review of the Aerodynamics of the MOD-2 Emulating Space Frame Wind Turbine. Report for R Lynette and Associates, October 1987.

23

Siemens Blade Technology

Innovation as a process has been the subject of extensive academic and business-oriented examination. The focus of this book, however, has been on the technico-economic evaluation of innovate concepts without any consideration of the process – how ideas develop and how, if ever, they are adopted in an industry. The following text is almost verbatim from Henrik Stiesdal, formerly Technical Director of Bonus AS and now of Siemens Wind Power, edited a little at his request. It gives a fascinating glimpse into the requirement for innovation, the process and the realisation.

> In the early years of Bonus' wind turbine manufacturing, blades were purchased from sub-contractors, initially from Økær Vindenergi, then from Alternegy, the licencee of Økær, and after their collapse in 1986, from LM Glasfiber. During the 1990s we gradually encountered increasing problems with blade supply.
>
> From 1991 onwards I looked for alternatives, but found none – prospective suppliers were either too small, had too little competence, or had other drawbacks. Aerpac, for example, invariably landed in lawsuits with their customers – not a good sign. I had discussed the problem on several occasions with an old friend, Martin Winther-Jensen. Martin had been involved in the building of the Tvind turbine in the 1970s, we had got acquainted in 1978, and we had become friends, both working with wind power, I in the industry, he with type approvals at Risø. Martin later came to work for us in Bonus.
>
> During a summer vacation trip in 1995 in my small sailing boat, Martin and I for the nth time discussed the blade problems, saying – why couldn't one do it like this ... how about this arrangement, wouldn't that be better ... and so on. During a couple of long days' cruise in the archipelago south of the island of Fyn, we more or less hashed out the basics of an 'idealized blade', solving the issues with blades made in two pieces (as was the norm) – eliminating the glue joints, using only very few glass types, and basing it all on vacuum infusion.

The one-piece blade manufacture developed by Stiesdal and Winther-Jensen is illustrated in Figure 23.1.

> Absence from other tasks and an open mind created by the ambience surely were key elements in the establishment of the creative process. Also very important was to have somebody to ping-pong with – I had of course thought much about it, but when on your own, you often see obstacles that are in the end not real. Here

Innovation in Wind Turbine Design, Second Edition. Peter Jamieson.
© 2018 John Wiley & Sons Ltd. Published 2018 by John Wiley & Sons Ltd.

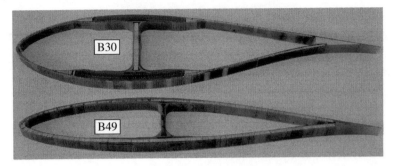

Figure 23.1 One-piece blade technology developed by Stiesdal and Winther-Jensen for Siemens. Reproduced with permission of Siemens Wind Power A/S.

it was of vast importance to spar with somebody who could see solutions. There is, however, a long way from the euphoria of being creative in a boat cockpit on a calm summer day to an industrial product, and we trundled back into the routine of supply from LM. The ideas were not dead, however. I got hold of another old acquaintance, Kaj Johansen. Kaj had for a brief period been manufacturing blades for Bonus when I joined the company in 1987, but he was too small to make it, and at one point in time he went bankrupt with his company. We later employed him as a blade repair technician, and it was on this background that I invited him to participate in the brainstorming.

Kaj's blades had also been one-piece, but based on laying the glass on a solid foam core that would remain inside the blade. This could never give a high volume content of glass and would not work for large blades. However, from his earlier work, Kaj knew of the difficulties of injecting thick layers of fiberglass. Mastering this part of the technology would be a pivotal question – either one could do it, then everything would be fine, or one could not, and we would have to forget the vacuum injection.

We set up an extension of Kaj's home workshop, and he rapidly developed various technologies which demonstrated that it would indeed be possible to inject very thick layers. Whilst not yet connected to the one-piece process, this was a key enabler and milestone. During the process of injection development Kaj, Martin and I had several discussions, and it was during these discussions that the ideas of the 1995 boating trip were brought to a realisable level, looking at each element in turn (blade root, stability, lay up technologies, vacuum bags, etc.) and determining a solution that could be integrated in one-piece manufacture. Kaj ended up making small real-life sections of how one could do the blade; they closely resembled what we ended up doing.

In 1998 we were approached by the large Danish shipyard Danyard in Aalborg. They had been manufacturing composite ships for the Danish navy, but the programme was ending and they were looking for work. Could they do something for us? This seemed like a godsend – somebody with large facilities and much competence. I invited them for a meeting, they were agreeable and clever. Kaj and I then visited them, and we more or less came to the conclusion that this would work. But then they turned around – even though the managers we spoke with were very interested, the board would not accept it. Danyard was owned by a shipping group, the shipyard would be sold off, and it could only be sold as a shipyard, not as something in a transition phase. So, a no go.

The people had seemed genuinely sorry that this would not work. So I called the shipyard CEO and asked, why not come and work for us, and bring a few key people? This is what happened, and on 1 November 1998 we formally started the development operation in Aalborg, with the aim to develop industrial level blade manufacturing technology based on the concepts developed 1995–1998.

During 1999 most of the details came into place, and even though one always suffers from the law of diminishing returns, several good contributions were made by the people brought in from Danyard or hired externally in the early phases of the project.

So, all in all, the innovative blade is the result of a long process:

- Urge for a solution arising from frustration at the given state of affairs
- The 'revolutionary part' of the 1995 boat trip which became genuine brainstorming, essentially hashing out the key elements of the 'idealised blade'
- The 'evolutionary part' of the Kaj Johansen explorations during 1996–1998
- The 'fine tuning' part of the technology project in Aalborg.

One of the key enablers was decision making capability. I was free to say – let's do this, let's hire the people. Of course I had to agree with and get the acceptance from our CEO, Palle Nørgaard, but he was as frustrated as I regarding blade supply issues, so that was no problem.

This then is a record of how a leading wind turbine manufacturer, Siemens, has used innovation in the development of in-house manufacture of wind turbine rotor blades in response to technical and commercial issues with external supply. With many innovations, the Enercon direct-drive technology may be another case in point; it is possibly critical that key technology decisions are in the hands of one or a few people who have authority to act. Much potential innovation, for which there may be quite strong technical justification, can languish between layers of management in which particular individuals are averse to any kind of perceived risk.

24

Stall-Induced Vibrations

Stall is notorious is many industries – a hazard for aircraft safety and a cause of potentially damaging vibrations in turbo machinery. In the wind industry, partly due to the more progressive nature of stall, usually from root to tip, and the generally severe fatigue environment of rotor blades in atmospheric turbulence, stall has not loomed so largely as a design concern and has been positively exploited as a means of regulating output power.

In normal operation of a wind turbine in attached flow conditions, there are very strong lift damping moderating blade vibrations. The same lift damping contributes to the stability of a ship under sail and, when the wind turbine is operating, this can also reduce the impact of wave loading on fatigue of offshore wind turbine support structures [1].

When a stall-regulated wind turbine operates in deep stall, angles of attack on the blade sections become large. This can result in operation of some of the blades at angles of incidence beyond maximum lift, where the slope of the lift curve and hence lift damping may be negative. Blade vibrations are then potentially unstable. Blades usually have generous margins on torsional rigidity and torsional vibrations have not been a significant problem. In some cases, especially with larger wind turbines, stall-induced flap vibrations have been a concern. Much attention was given to this [2] in a substantial research programme of the European Union (EU) funded under Joule III. However, there has been little sign of stall-regulated wind turbines having major problems with blade flap vibrations. Often, simulations, even with state-of-the-art unsteady aerodynamic models (including dynamic stall and stall hysteresis effects), predict more severe vibrations than are encountered in practice.

Nevertheless, an interesting problem with stall-induced edgewise vibration arose in the late 1990s. Public information on the problems encountered by wind turbine operators and the fixes produced by blade manufactures are well documented in issues of *Windpower Monthly* from May 1997 to December 1999 [3–8]. In some cases, the edgewise vibrations destroyed blades and in others significantly interfered with effective energy production. The crux of the problem can be understood from Figure 24.1.

If all three blades vibrate in phase in the edgewise direction, shaft torque variations of a magnitude three times that of each blade are produced, and torsional damping in the drive train can, in principle, limit such vibration. However, in the vibration mode of Figure 24.1, one blade is stationary, whilst the other two vibrate in anti-phase.

Innovation in Wind Turbine Design, Second Edition. Peter Jamieson.
© 2018 John Wiley & Sons Ltd. Published 2018 by John Wiley & Sons Ltd.

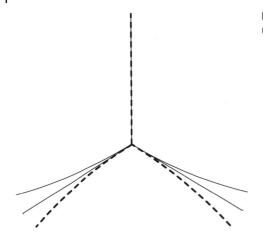

Figure 24.1 Asymmetric edgewise vibration mode.

This produces zero net torque on the rotor shaft and the vibrations cannot therefore be influenced by drive-train damping.

As usual, such problems, although simply presented as in Figure 24.1, are quite subtle in reality. While the motion illustrated produces no net rotor torque, there is a motion of the centre of mass of the rotor, along the axis of the nodally stationary blade (a vertical line for the blade position illustrated), which becomes a whirling of the rotor when the blades are rotating. Petersen *et al.* [9] show how coupling of the blade and rotor dynamics to global system dynamics has a critical influence on the level of response.

When the edgewise vibration problem first appeared in stall-regulated wind turbines operating in deep stall at wind speeds around 17 m/s, the reaction of a number of blade manufacturers was to develop blade dampers as a means of damping of the edgewise vibrations at source. The principle behind one of these systems is considered here briefly as a classic illustration of an innovation derived from basic pendulum dynamics. It should be noted that the problem at hand is not a *resonance* problem where an excitation frequency matches a specific natural frequency that is insufficiently damped. Instead, it is a *stability* problem where, whatever the natural frequency within reason, large vibrations will take place at that natural frequency through lack of damping of the relevant vibrational mode.

Dynamic vibration absorbers are engineered into many components ranging from small handheld electrical hair cutters to large buildings for earthquake protection. The concept was introduced in 1928 by J. Ormondroyd and J. P. den Hartog. An analysis of the basic two degrees of freedom system with coupled masses and springs is presented by Den Hartog [10]. The concept is to transfer the vibrational energy that would excite vibration in a major component into vibration in a minor component (the absorber).

In the context of damping edgewise vibrations of a blade, the absorber may, in principle, be a simple pendulum. Unfortunately, a simple pendulum of an appropriate natural frequency to deal with the edgewise blade instability will be found to be about as long as the blade! Anderson *et al.* [11] addressed this problem by developing a system based on a compound pendulum.

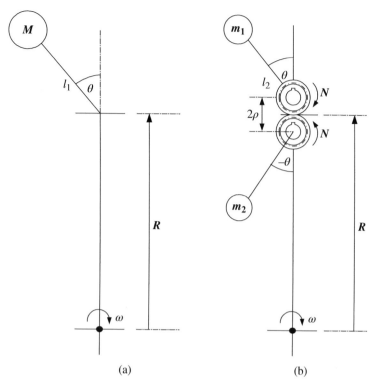

Figure 24.2 Simple and compound pendula.

In Figure 24.2, a simple pendulum is compared with a type of compound pendulum in which the masses are forced to move in contrary motion through a geared interconnection. The wind turbine rotor is rotating at frequency ω. Centrifugal acceleration, acting 'up the page' in Figure 24.2, replaces the role of gravitational acceleration in the familiar suspended pendulum. The system natural frequency is now determined in each case. For the simple pendulum (Figure 24.2(a)):

$$Ml_1^2 \frac{d^2}{dt^2}\theta = -M\omega^2(R + l_1 \cos(\theta))l_1 \sin(\theta)$$

and for l_1 and θ small

$$l_1 \frac{d^2}{dt^2} = -\omega^2 R\theta$$

$$\text{Natural frequency;} \quad \Omega_1 = \omega\sqrt{\frac{R}{l_1}} \tag{24.1}$$

For the compound pendulum (Figure 24.2(b)):

$$m_1(l_2)^2 \frac{d^2}{dt^2}\theta = -m\omega^2(R + \rho + l_2 \cos(\theta))l_2 \sin(\theta) + N \tag{24.2}$$

$$-m_2(l_2)^2 \frac{d^2}{dt^2}\theta = -m_2\omega^2(R - \rho - l_2 \cos(\theta))l_2 \sin(\theta) + N \tag{24.3}$$

Subtract Equation 24.3 from Equation 24.2 and treat ρ, l_2, θ as small

$$(m_1 + m_2)(l_2)^2 \frac{d^2}{dt^2}\theta = -(m_1 - m_2)\omega^2 R l_2 \theta$$

Natural frequency; $\quad \Omega_2 = \omega \sqrt{\left\{ \dfrac{m_1 - m_2}{m_1 + m_2} \right\} \dfrac{R}{l_2}}$ \hfill (24.4)

Since Ω_2 in Equation 24.4 depends on the mass difference, a low natural frequency can be achieved even when the pendulum arm l_2 is short. For a specified frequency Ω, if $\Omega_2 = \Omega_1 = \Omega$, then:

$$l_2 = \left(\frac{m_2 - m_1}{m_1 + m_2} \right) l_1$$

This shows that l_2 can be much shorter than l_1 for any desired Ω. Actual values of m_1 and m_2 are related to the required level of damping. Thus, this type of compound pendulum system has the property that its natural frequency depends not on pendulum length but on a mass difference. This principle was the basis of an elegant compact vibration absorber with a magnetic damping system that was used in a series of wind turbine blades.

The world's leading independent blade supplier, LM Wind Power, naturally sought to address this blade vibration problem definitively. According to an LM Newsletter [12] of 2003:

> So far, LM Glasfiber has solved these problems by introducing aerodynamic damping. This involves adding specially designed profiles to the edges or a so-called structural damper feature which consists of rubber laminations in the blades.
> Now there is a third option – a fluid damper. This new damper is made up of a U shaped tube embedded in the blade and containing a precisely measured volume of temperature resistant fluid with an operating range from −30 to +55 °C. The fluid in the damper absorbs the energy of the vibrations and so eliminates them.

A liquid system is entirely rational for a 'damping' function and should be of low/negligible maintenance compared with any mechanical system.

References

1 Jamieson, P. and Quarton, D.C. (1999) *Technology Development for Offshore.* European Wind Energy Conference, Garrad Hassan & Partners Limited, Nice, March 1999.
2 Petersen, J.T., Madsen, H.A., Björk, A. *et al.* (1998) Prediction of Dynamic Loads and Induced Vibrations in Stall. Risø-R-1045(EN), May 1998.
3 Blade Cracks Signal New Stress Problem, Preventative Investment Needed on Turbines with Large LM Blades. Windpower Monthly article (May 1997).
4 Stall Strips Cure Wobble. *Windpower Monthly* article (Feb. 1998).
5 Blade Retrofit Work on Entire Series. *Windpower Monthly* article (June 1998).

6 A Solution for Blade Wobble. *Windpower Monthly* article (Oct. 1998).

7 Fighting it Out for the Heavyweight Title. *Windpower Monthly* article (May 1999).

8 Vibrations Fix Looks Solid. *Windpower Monthly* article (Dec. 1999).

9 Petersen, J.T., Thomsen, K. and Madsen, H.A. (1998) Local Blade Whirl and Global Rotor Whirl Interaction. Risø-R-1067(EN), August 1998.

10 Den Hartog, J.P. (1985) *Mechanical Vibrations*, 4th edn, Dover Publications, pp. 87–106. ISBN: 0486647854

11 Anderson, C., Heerkes, H. and Yemm, R. (1999) *The Use of Blade-Mounted Dampers to Eliminate Edgewise Stall Vibration*. EWEC 1999, Conference Proceedings, Nice, March 1999, pp. 207–211.

12 LM Newsletter (Feb. 2003).

25

Magnetic Gearing and Pseudo-Direct Drive

25.1 Magnetic Gearing Technology

The permanent magnet generator (PMG) designs now employed in wind turbine systems usually exploit high-strength magnets, the so-called rare-earth magnets, especially the neodymium-iron-boron compound developed in parallel by General Motors, Sumitomo Special Metals and the Chinese Academy of Sciences in 1983. These magnets offer higher flux densities than the older ferritic magnets and hence can enable more compact and powerful electrical machines. Dramatic price reductions from typically €160/kg in the mid-1990s to €16/kg in 2006 have made this technology affordable, although prices have increased since to values around €50/kg (2012).

The UK company Magnomatics Limited has developed magnetic gearing technology which, according to simulation studies based on the use of such rare-earth magnets, indicates a transmitted torque density capability of the order of 50–150 kNm/m^3 comparable to that of two- and three-stage helical gearboxes.

The most obvious magnetic gear concept (Figure 25.1) is an analogue of mechanical gear designs in which the meshing teeth are replaced by adjacent permanent magnets. It will be evident both for the conventional gear and the magnetic gear of this type that the torque transmission takes place over only a small region of contact or near-contact. As the shear stress in a magnetic field is much less than the allowable stresses of metals in contacting gears, such magnetic gears (Figure 25.1) cannot transmit as high torques as conventional gears and are suited only to specialist applications.

A key advance in magnetic gear design was the invention [1] of a high-torque magnetic gear. This gear (Figure 25.2) comprises three concentric components:

1) An inner permanent-magnet array with a number of pole-pairs p_i,
2) An outer permanent magnet array with a number of pole-pairs p_o,
3) An intermediate annular component, the modulating component, with n_p ferromagnetic pole-pieces.

The entire circumference of this gear is used to transmit torque. Hence, the torque density of this magnetic gearing technology is much greater than the earlier magnetic gear design of Figure 25.1.

In order that a constant torque is transmitted between these components, the relationship between the number of pole-pairs and pole-pieces must observe

$$P_i + P_o = n_p \tag{25.1}$$

Innovation in Wind Turbine Design, Second Edition. Peter Jamieson.
© 2018 John Wiley & Sons Ltd. Published 2018 by John Wiley & Sons Ltd.

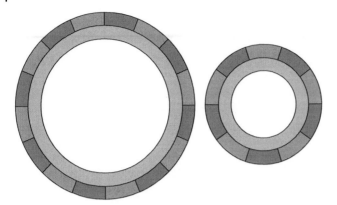

Figure 25.1 Early magnetic gear design.

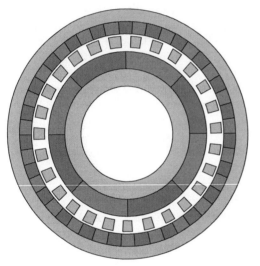

Figure 25.2 High-torque magnetic gear, with $p_i = 4$, $p_o = 23$ and $n_p = 27$.

The operation of this gear can be better understood by assessing the magnetic field in the air gaps created by both magnet arrays. For torque transmission, it is necessary that the magnetic field created by one magnet array couples with another magnetic field with an identical number of pole-pairs. Because both permanent magnet arrays have a different number of pole-pairs, there would be no torque transmission between these arrays in the absence of the modulating rotor. As an example, Figure 25.3(a) shows the magnetic field in the air gap adjacent to the outer magnets for the magnetic gear shown in Figure 25.2, where both the outer magnets and the modulating rotor are absent, that is, the field is entirely generated by the inner magnet array. The spatial harmonic spectrum of this magnetic field, which is shown in Figure 25.3(b), clearly shows that the field has p_i pole-pairs ($p_i = 4$). The waveform of Figure 25.3(a) is not exactly a pure sine wave because it reflects a realistic magnetisation of the curved magnets with only four pole-pairs. Hence, there is some minor harmonic content in Figure 25.3(b) additional to that at the principal harmonic number 4.

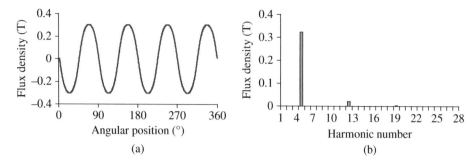

Figure 25.3 (a) Flux density, (b) spatial harmonic spectrum of (a). Field due to the inner magnets only, in air gap adjacent to outer magnets, with modulating rotor absent.

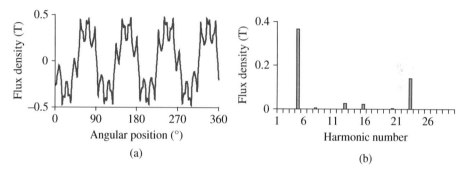

Figure 25.4 (a) Flux density, (b) spatial harmonic spectrum of (a). Field due to the inner magnets only, in air gap adjacent to outer magnets with modulating rotor present.

Due to the presence of the ferromagnetic pole-pieces, the magnetic field from each permanent magnet array is modulated such that a spatial harmonic having the same number of poles as the other array of permanent magnets is generated. This harmonic is asynchronous since it rotates at a speed different from that of the original rotor. For example, Figure 25.4(a,b) shows the magnetic field and its harmonic spectrum which are due to the inner magnet array in the air gap adjacent to the outer magnet array. This is the case shown in Figure 25.2 with the modulating rotor present but with the outer magnets removed. It can be clearly seen that in addition to the harmonic with p_i pole-pairs, an additional harmonic is created with p_o pole-pairs, which will interact with the outer magnet array to transmit torque. Because the harmonic with p_i pole-pairs moves at a speed different from that of the harmonic with p_o pole-pairs, there exists a geared torque transmission between both rotors.

The equation which relates the speeds of the different components of the gear is given by

$$P_i\omega_i + P_o\omega_o = n_p\omega_p \tag{25.2}$$

where

ω_i is the speed of the inner permanent magnet array;
ω_o is the speed of the outer permanent magnet array;
ω_p is the speed of the modulating component.

When one of the three components of the gear is held stationary, there will be a fixed gear ratio between the other two components. For example, when the modulating component is held stationary ($\omega_p = 0$), the gear ratio between the inner and outer permanent magnet array is

$$G_{r,o \to i} = -\frac{P_o}{P_i} \qquad (25.3)$$

and both shafts are rotating in opposite directions. Alternatively, when the outer magnet array is held stationary ($\omega_o = 0$), the gear ratio becomes

$$G_{r,p \to i} = \frac{n_p}{P_i} = 1 + \frac{P_o}{P_i} \qquad (25.4)$$

Both shafts are now rotating in the same direction, and a higher gear ratio is obtained. With this gearing technology, gear ratios of between $1:1$ and $\sim 30:1$ in a single stage are easily possible.

25.2 Pseudo-Direct-Drive Technology

Magnomatics has taken magnetic gear technology a stage further [2] in the integration of the high-torque magnetic gearing within a brushless PMG design. The resulting electrical machine, which is shown in Figure 25.5, has been called the Pseudo-Direct Drive (PDD®), because it has the characteristics of a direct-drive machine, although it uses a magnetic gear to achieve its very high torque densities.

The PDD® consists of three components:

1) An outer stator, comprising a lamination pack with copper windings, and stationary outer magnets with a pole-number p_o;
2) An inner permanent-magnet array with a number of pole-pairs p_i, rotating at high speed with no external mechanical connection;

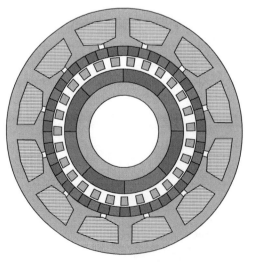

Figure 25.5 Cross section of a pseudo-direct drive.

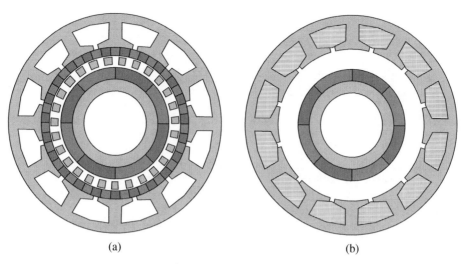

| (a) | (b) |

Figure 25.6 (a) Components in PDD® which are part of the magnetic gear element and (b) components which form the PMG.

3) An intermediate annular component, the modulating rotor, with n_p ferromagnetic pole-pieces, rotating at low speed and connected to the input shaft of the generator.

The principle of operation of the PDD® can be understood by considering the operation of the different sub-components within it, that is, the magnetic gear and the electrical machine. Figure 25.6(a) shows the elements of the device which contribute to its magnetic gear operation, that is inner magnet rotor, modulating rotor and stationary outer magnets. Figure 25.6(b) shows the elements of the device which contribute to its operation as an electrical machine, that is, the outer lamination pack with the copper windings and the inner permanent magnet rotor.

The inner rotor contributes to the operation of both the electrical machine and the magnet gear, and, as a result, the PDD® uses less magnet material than would be used by a combination of a magnetic gear and separate electrical machine. The high-speed inner rotor interconnects the magnetic gear with the electrical machine and so there is no total torque applied on this rotor, that is, the torque which is applied on this rotor by the magnetic gear is reacted by the torque applied by the electrical machine. Moreover, the electrical machine only generates the torque on the high-speed rotor, which is only a fraction of the total input torque. Thus the required magnetomotive force (mmf) in the machine, and hence the copper losses, are significantly smaller than in a conventional direct-drive generator, which would have to react to the full torque on the input shaft. Although the machine requires a relatively small mmf and current density, the designer still has full flexibility to optimise the machine for any desired terminal voltage output, typically 600 V or 3 kV. Further, the generator is controlled in a manner identical to a conventional direct-drive PMG, through the use of existing full-rated power converter technology.

The PDD® concept is certainly a very logical one with a number of advantages. The reliability issues associated with contacting gears including tooth wear and contamination of lubricants are completely avoided. With the torque transmission distributed

Figure 25.7 Magnomatics 300 kW PDD under test.

around the full circumference of each ring element, concentrated radial bearing loads that occur in systems with conventional gearing are avoided. The benefit of direct drive in providing the drive train as a single electromechanical system is achieved. In addition, the effect of gearing is achieved without full load efficiency losses of conventional gears typically of the order of 0.8% per stage. Thus, the benefit of higher speed of the generator rotor in reducing generator torque and size is realised. A case has been made (Chapter 6) and supported by the chosen development paths of a number of manufacturers in their latest drive-train technology that the hybrid drive train may be the most compact and cost-effective. The PDD system can be seen as an analogue of the hybrid type of drive train providing in effect one or two stages of gearing and multi-pole generator but in a simplified integrated system.

Progress has been made in the development of PDD systems for wind turbines (Figure 25.7). Within the European Union's Horizon 2020 research and innovation programme, Magnomatics has commenced an ERA-LEARN 2020 project (CHEG, compact high efficiency generator) in collaboration with ORE Catapult Development Services, Garrad Hassan Nederland BV, Newton Darby Ltd, EDF Energy R&D and JL Mag to develop a PDD. The PDD is also very well suited to marine propulsion. Magnomatics has a project with Innovate UK, Rolls Royce and ATB Lawrence Scott to design, manufacture and test a 2.5 MW magnetically geared propulsion motor powered by a PDD.

References

1 Atallah, K. and Howe, D. (2001) A novel high-performance magnetic gear. *IEEE Transactions on Magnetics*, **37** (4), 2844–2846.
2 Atallah, K., Rens, J., Mezani, S. and Howe, D. (2008) A Novel 'pseudo' direct-drive brushless permanent magnet machine. *IEEE Transactions on Magnetics*, **44** (11), 2195–2198.

26

Summary and Concluding Comments

Often, a technical book and invariably a newspaper report grinds to a sudden halt without any overall evaluation or conclusions when there is no more information to impart. In the context of innovation in wind turbine design, there are many diverse aspects to consider. The focus of this book is on the appraisal of concepts using top-level parametric analyses to provide insight.

It is generally apparent in design that there is still a need for further development of concept design methodology as a subject in its own right. This book does not enter the theory of conceptual design but certainly draws attention to some of the methods. 'Design' tools figure prominently in the evolution of wind technology, but most such tools are in reality analysis tools. As such, they may analyse well whatever configuration is presented but offer no guidance on what the preferred configuration should be.

Every engineer confronted with a new wind turbine concept should focus initially on power and torque – power as measure of energy capture potential and value, and torque as usually a strong indicator of weight and cost. Every engineer should also keep plenty of envelopes for back-of-the-envelope calculations. Failure to do so can sometimes lead to extended wasteful expenditure of resources on concepts that should have been challenged at a much more basic level.

Even with the very best of innovative technology, there can be a huge challenge in taking to market in all stages from funding of background research to prototype development and proving. Various 'blocks' to innovation are common. There is the manager inside a larger organisation who is doing just fine and does not want to risk reputation on something new, however promising, that may fail. Much proposed innovation inevitably does fail, and Paul Gipe in *Wind Energy for the Rest of Us* is rightly scathing about how substantial investment can sometimes be attracted to inappropriate innovations. Critical evaluation and seeking a complete rounded view of the impact of innovations (however sketchy that view may be in the early stages) is vital as opposed to driving them forward on their most conspicuous advantages blind to major obstacles. Another block is that, within the research communities in various parts of the world, with funding linked to national politics and the public view of wind energy, the wind industry understandably wants to present a unified front with key themes for research and have everyone 'singing from the same hymn sheet'. This can impede funding of more radical ideas that may have strong merits but do not quite seem to fit in at a particular point in time.

Innovation in Wind Turbine Design, Second Edition. Peter Jamieson.
© 2018 John Wiley & Sons Ltd. Published 2018 by John Wiley & Sons Ltd.

In the discussion of many wind energy debating topics presented in this book, there are no final evaluations and no moral high ground simply because one argument is supported by a little analysis when another is not. However, analysis is valuable and is self-educating – it often partly confirms and partly surprises – sometimes supporting, sometimes contradicting seemingly reasonable assumptions or 'gut feelings'. Throughout this text, the main aim is to shed light, not advocacy for one solution or another, not to substitute new received wisdom (or is it dogma?) for old but to challenge accepted wisdom and awaken new trains of thought.

Index

Innovation in Wind Turbine Design, Second Edition. Peter Jamieson.
© 2018 John Wiley & Sons Ltd. Published 2018 by John Wiley & Sons Ltd.